MINDING THE HEAVENS

THE STORY OF OUR
DISCOVERY OF
THE MILKY WAY

About the Author

Leila Belkora was born in New York City to an American mother and Moroccan father, and grew up in Geneva, Switzerland. She obtained a BA in physics and an MS in engineering physics at Cornell University, and a PhD in astrophysics, specializing in solar radio astronomy, from the University of Colorado, Boulder. She has since divided her time between science writing and teaching astronomy. She lives in Irvine, California.

MINDING THE HEAVENS

THE STORY OF OUR DISCOVERY OF THE MILKY WAY

LEILA BELKORA

Institute of Physics Publishing
Bristol and Philadelphia

British Library Cataloguing-in-Publication Data
A catalogue record for this book is available from the British Library.

ISBN 0 7503 0730 7

Library of Congress Cataloging-in-Publication Data are available

Commissioning Editor: John Navas
Production Editor: Simon Laurenson
Production Control: Sarah Plenty
Cover Design: Frédérique Swist
Marketing: Nicola Newey and Verity Cooke

Published by Institute of Physics Publishing, wholly owned by The Institute of Physics, London

Institute of Physics, Dirac House, Temple Back, Bristol BS1 6BE, UK

US Office: Institute of Physics Publishing, The Public Ledger Building, Suite 929, 150 South Independence Mall West, Philadelphia, PA 19106, USA

Typeset by Academic + Technical, Bristol
Printed in the UK by MPG Books Ltd, Bodmin, Cornwall

CONTENTS

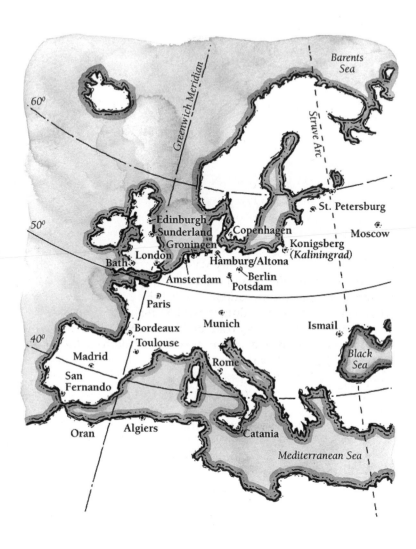

ACKNOWLEDGMENTS

My first thanks are due to my parents, Abdel Hak and Janice Belkora, and to my in-laws, Judy and Chalmer Hans, for their loving babysitting. If, as originally planned, I had finished the book before Alicia was born, things no doubt would have been easier, but the book is better for having ripened longer, and I couldn't have done it this way without their help.

In the same vein I thank my husband Randal for his encouragement, support, and patience, which he gave tirelessly even when very tired from midnight awakenings.

I might not have started on the book in the first place without the advice and encouragement of editors Meg Tuttle and Tom Quinn. It was a great pleasure to work with them. And I found that, if one has a baby in the middle of writing a book, it is very convenient if one's editors become parents at the same time.

My "readers" did almost as much as my editors to improve the book. Thanks to my parents again, Lisa Berki, Clare Topping, Richard Riley (not the Honorable Richard Riley, former U.S. Secretary of Education, but the celebrated director of Cornell's Sage Chapel Choir), and members of the Cornell Campus Club-International Women's Club. Those teachers, mentors, and editors who did the most to stamp out my bad writing habits before they became ingrained are Burton Melnick, Judy Jackson, James Glanz, and Sonya Booth. Since this is my first book, I wanted to thank them here too.

For technical assistance I thank the librarians of the Cornell Library, especially Laura Linke and Nancy Dean of the Rare and Manuscripts Division. Staff members of the British Library, especially Michael Boggan, were very helpful in advance of my visit. Peter Hingley and Mary Chibnall at the library of the Royal Astronomical Society were very kind and helpful.

For help with various bits of research, much of it by telephone and e-mail, I thank Kathryn Kjaer at the University of California at Irvine; J P Hall at the Local Studies Centre at the Sunderland City Library; Arleen Zimmerle and Lorett Treese at Bryn Mawr College; Debbie Landi at the University of Missouri; Mark Hurn at the Institute of Astronomy in Cambridge, England; Lisa Brainard and Wilma Slaight at Wellesley College; Caroline Smith at Caltech; Keith Gleason at the Sommers-Bausch Observatory of the University of Colorado; and Mahmoud Ghander and Gail Archibald at Unesco. I am also grateful to Stuart and Alexandra Rock, Alan Batten, Brian Marsden, Owen Gingerich, Michael Hoskin, Tom Gehrels, Alan Harris, Mildred Shapley Matthews, Martha Haynes, Virginia Trimble, Barbara Becker, Patrick Morris, Stephen Pappas, Todd Huffmann, Don Campbell, and Anita Watkins. Any errors in this book are mine, but these friends and contacts all contributed to the accuracy and completeness of my information.

I thank my illustrator, Layne Lundström, for his patience and enthusiasm. I consider myself lucky to have found him when the manuscript was in the early stages. Last but not least, I thank my editor at IOPP, John Navas, who solved problems and expertly guided the book to completion.

A NOTE ON SOURCES

In writing what I hope is a popular account of the discovery of the Milky Way and other galaxies, I have relied extensively, though not exclusively, on secondary sources—illuminating and often fascinating studies and biographical works by astronomers and historians of astronomy. In some cases, the work of just one or two scholars has guided my approach.

The works of Thomas Wright, the subject of my chapter 3, have been carefully analyzed—and in some cases, brought to light in the first place—by Michael Hoskin. His editions of Wright's work include *An Original Theory or New Hypothesis of the Universe* (Wright, 1750), *Clavis Coelestis* (Wright, 1742), and *Second or Singular Thoughts upon the Theory of the Universe* (Wright, n.d.).

Much has been written by and about William Herschel, and I have used a number of sources, but again, I have found the most authoritative and easily accessible work is that of Michael Hoskin, particularly *William Herschel and the Construction of the Heavens* (Hoskin, 1963).

As far as I know, there is only one biography of Wilhelm Struve besides the one I used, and it is in Russian. I gratefully acknowledge my debt to Alan Batten, author of the English-language biography *Resolute and Undertaking Characters: The Lives of Wilhelm and Otto Struve* (Batten, 1988). Otto Struve wrote a short biographical account of his father in German, and I have examined this too, but using Batten's book as a guide.

Barbara Becker's PhD dissertation on William and Margaret Huggins, *Eclecticism, Opportunism, and the Evolution of a New Research Agenda: William and Margaret Huggins and the Origins of Astrophysics* (Becker, 1993) is the most comprehensive and detailed study of Huggins' life that I am aware of, and I have also found her treatment of his research on stellar and nebular spectra and radial motion of the stars to be a valuable guide to Huggins' own publications.

A scholarly biography of Jacobus Kapteyn does not exist, in part because many of his papers, which were being assembled for a biography, disappeared during World War II. I have used his daughter's biography in E Robert Paul's translation from Dutch to English, *Life and works of J C Kapteyn* (Paul, 1993a). In 2000, a scientific conference on Kapteyn's legacy brought to light some problems with this translation. Contributors to the conference proceedings, *The Legacy of J C Kapteyn: Studies on Kapteyn and the Development of Modern Astronomy* (van der Kruit and van Berkel, 2000) helped put Kapteyn's work in perspective.

The best existing guide to Shapley's life is his own informal biography, *Through Rugged Ways to the Stars* (Shapley, 1969). Shapley's voluminous scientific output has been very helpfully discussed by Robert W. Smith in *Expanding Universe: Astronomy's "Great Debate," 1900–1931* (Smith, 1982), and by Owen Gingerich, Michael Hoskin, Richard Berendzen, and other historians of astronomy whose work is cited in chapter 8. David DeVorkin's biography, *Henry Norris Russell: Dean of American Astronomers* (DeVorkin, 2000), is a useful reference on Shapley, Russell's student.

Gale Christianson's biography of Edwin Hubble, *Edwin Hubble: Mariner of the Nebulae* (Christianson, 1995), has been my chief guide to his life and work. Helen Wright's biography of George Ellery Hale, *Explorer of the Universe* (Wright, 1966); Smith's book; and DeVorkin's biography of Russell are also, of course, useful for Hubble as well as Shapley.

In the rare instances in which I have dug up interesting tidbits on my own, I have tried to make this clear in the notes.

1

INTRODUCTION

"Astronomy, by the eminence of its subject and the flawlessness of its theories, is the most beautiful monument of the human spirit, the most distinguished decoration of its intellectual achievement."

Pierre-Simon Laplace,
Exposition du Système du Monde (1835)[1]

This book is about how we discovered that we live in a galaxy, in a universe of galaxies. The title phrase, "minding the heavens," I borrowed from one of the astronomers I write about, Caroline Herschel. She was the sister of a famous astronomer of the late 1700s and early 1800s, William Herschel. Caroline not only assisted her brother in his exploration of the Galaxy, but also was an astronomer in her own right. When her brother had to be away, she was competent to take over at the telescope and "mind the heavens" for him.[2]

The story of the discovery of our own and other galaxies unfolds through the lives of seven astronomers — and their assistants — who worked on the question of where we live in the cosmos. I was motivated to tell the story through a series of biographies in part by my own desire to know more about the astronomers who have shaped our view of the universe. Why did Wilhelm Struve, director of Russia's imperial observatory under the Czars Alexander I and Nicholas I, become an astronomer after studying philology? What kind of person was Edwin Hubble, whose portrait I saw in the astrophysics library at the California Institute of Technology while I was there doing research? I never had time to look into such questions when I was a graduate student. As a teacher of introductory college courses in astronomy, I thought the lives of the astronomers would make the scientific material more

1

interesting to my students, too. Robert L. Heilbroner's classic on the lives, times, and ideas of the great economic thinkers, *The Worldly Philosophers*, provided much inspiration in adopting a similar approach for astronomy.[3]

The only difficulty was to limit the number of astronomers profiled. I selected astronomers who, from the mid-eighteenth century to the mid-twentieth century, made key advances answering the question of our location in the galaxy and in the universe. Their insights and their blunders tell the story of our evolving understanding. I might have made the selection differently in one or two cases, and highlighted other paths to our current understanding. However, I believe that the seven I chose—Thomas Wright, William Herschel, Wilhelm Struve, William Huggins, Jacobus Kapteyn, Harlow Shapley, and Edwin Hubble—cover the territory that was most important to include.

I have found that Edwin Hubble is the only astronomer in this list that most people have heard of. Hubble was indeed an outstanding figure, most deserving of having a space telescope named after him. He established that our galaxy is only one of many galaxies scattered throughout space, and he found evidence that the universe is expanding in all directions. But it is only in the context of developments in the 200 years preceding his career that we can fully understand his accomplishments and the reason for his fame.

The story begins in the mid-1700s with the Englishman Thomas Wright, who was not so much an astronomer as a somewhat eccentric philosopher. Wright appears to be one of the first people to have thought carefully about the three-dimensional structure of our stellar system, which we call the Milky Way galaxy but which, in those days, constituted the entire known universe. He also pondered the question of whether there might be other stellar systems, or other universes. His ideas inspired other philosophers to consider the problem.

Not long after Wright promulgated his ideas, William Herschel, a Hanoverian who made England his home, began what was arguably the first scientific attempt to map the stellar system. Herschel, one of the greatest observers and telescope-makers in the history of astronomy, ventured to trace the contours of what we now call the galaxy. Unfortunately, he could not put any scale on his map, as the distances to the stars were not known.

Establishing the distances to some of the nearer stars was the work of Wilhelm Struve, the nineteenth-century astronomer in Russia's imperial observatory. Struve was among the first to measure stellar distances by the method of parallax. An admirer of William Herschel, Struve also tried to continue Herschel's research on the shape and extent of the galaxy, although with limited success.

William Huggins (there are a lot of Williams or Wilhelms in this septet) is a transitional figure who approached the study of the heavens from a completely different perspective. In the mid-1800s this self-taught amateur applied the new technique of spectroscopy, which gives information on chemical composition to the light emitted by objects in and outside our galaxy. Spectroscopy opened up new vistas in astronomy, and in fact led to the development of the so-called "new astronomy," the combination of astronomy and physics or astrophysics. The discovery of the shape and extent of our galaxy, and of our galaxy's place in the cosmos, would not have been possible without the insights that spectroscopy brought.

Jacobus Kapteyn, a Dutch astronomer of the late nineteenth and early twentieth centuries, took advantage of the new information gleaned from spectroscopy and updated Herschel's mapping technique. His representation of our stellar system, which became known as the "Kapteyn Universe," required a lifetime of patient effort to put together. The Kapteyn Universe was an important model of the distribution of stars until the 1910s, and some aspects of it survived beyond that period.

The American astronomers Harlow Shapley and Edwin Hubble, contemporaries who were notorious rivals, provide the *dénouement* in this account of our discovery of our place in the cosmos. Shapley astounded the astronomical world in the 1910s with the news that the "Kapteyn Universe" was only a small part of a vast galaxy. Hubble, a successor of Shapley at the Mount Wilson Observatory in California, established that our galaxy was beyond doubt only one of many similar systems of stars, or galaxies, scattered throughout space.

Viewing the galaxy from within

The word "galaxy" is a familiar one. Today even elementary-school children know that we live in a galaxy—a system of billions or even trillions of stars, bound by gravity and orbiting a massive center—and that our Sun is one of the lesser lights in the Milky Way galaxy. One can even buy T-shirts showing stars in the classic whirlpool pattern, with the words "YOU ARE HERE" and an arrow pointing to the Sun's location in one of the spiral arms.

How do we know we live in a galaxy? Many of my students seem to think we know because we have seen pictures of it. This is not an unreasonable assumption in light of the stunning photographs collected by the Hubble Space Telescope and other ground-based and satellite-based telescopes. In the so-called "Hubble deep field" photograph, for example, space looks positively crowded with galaxies (see figure 1.1). There are a few bright points in this image that represent foreground stars, but the rest, yellow, blue or reddish in color, are galaxies. Some are wide, flat spirals that we see nearly face-on, presenting a disk-like appearance. Some look spherical. Some appear as thin lines—these are the disk-shaped galaxies seen edge-on. The variety of colors stems from the different chemical compositions and ages of the stars making up the galaxies, and the presence of dust and gas clouds among the stars, which lend a reddish hue to the galaxy.

Such photographs make it seem eminently reasonable that we live in one such galaxy, in our own group or cluster of galaxies. In fact, although we have a good idea of what our galaxy must look like from a distance, and we know quite a bit about neighboring galaxies in our group, no one has ever seen a photograph of the Milky Way galaxy in its entirety. We cannot get far enough away to put our stellar system in perspective. Our most far-flung robotic eye, the Voyager 1 spacecraft, was launched in 1977. Traveling through space at hundreds of millions of kilometers (or hundreds of millions of miles) per year, Voyager 1 is scheduled to reach the outer edge of the solar system—not even as far away as the nearest star—in the first quarter of the twenty-first century. To pass through the disk, rise above the plane of the Galaxy, and look back with

Hubble Deep Field
Hubble Space Telescope · WFPC2

PRC96-01a · ST ScI OPO · January 15, 1995 · R. Williams (ST ScI), NASA

Figure 1.1 The "Hubble Deep Field" — a view taken with the Wide Field and Planetary Camera 2 on board the Hubble Space Telescope. As described in the text, most of the objects seen here are distant galaxies. A foreground star, within our own galaxy, has "rays" extending from it — an artifact of the imaging system. The view is actually a synthesis of separate images in red, green, and blue light. (See color section.) (Credit: Jeff Hester and Paul Scowen (Arizona State University), and NASA.)

a cosmic bird's-eye view across the entire span of its spiral arms would require billions of years more travel time.[4]

What we know about the shape and size of our galaxy emerged from the efforts of many astronomers, beginning in the late eighteenth century and culminating in the early part of the twentieth century. Detective work of an astronomical sort was required to make sense of the available information. The problem of studying our galaxy from within it is like trying to learn about a crowd of people from a vantage point inside the throng. Consider, for example, that you are part of a graduation procession at a large school. Looking to your left and right, you might see only one or two neighbors, while the head and tail of the line may be out of sight. Clearly you are in a line of people, but your perspective gives you only limited information about the size and shape of the procession crowd. Similarly, from our vantage point in a spiral arm of the Galaxy, we have some information about the nearby disk, while some parts of the galaxy are obscured from view. And to complicate matters, astronomers have had to devise methods of estimating distances that allow them to gauge the extent of the starry congregations without leaving the surface of the Earth.

The most important clue to the distribution of stars is the phenomenon we call the Milky Way. The term "Milky Way" has two possible, related meanings: it refers to our home galaxy, and it also means the misty band of milky-white light we see arching across the sky (figure 1.2). Residents of countries in the northern hemisphere see the Milky Way band of light most prominently in the late summer, fall, and winter. Southern hemisphere observers see it best in spring and summer.

The Greeks gave us the term "Milky Way," a translation of "kiklos Galaxias" or milky circle. The story behind this name is that the infant Heracles (Hercules in the Roman version) tried to suckle at the breast of the goddess Hera (Juno, to the Romans). In what nursing mothers everywhere recognize as a sign of a powerful let-down reflex, some of the milk sprayed out, missing Heracles' mouth. By failing to latch on to this divine stream, Heracles missed out on his chance for immortality. The milk that spurted up into the sky formed the Milky Way.[5]

When Galileo first turned a telescope to the Milky Way in 1609, a tapestry of close-packed stars sprang into view. He

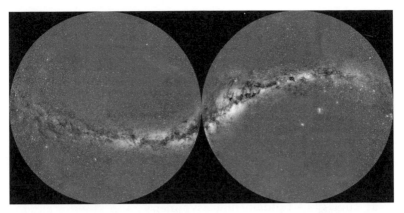

Figure 1.2 The Milky Way in the northern and southern hemispheres (left and right panels, respectively). Mosaic assembled by Axel Mellinger from 51 wide-angle photographs taken over the course of three years. (See color section.) (Credit: Axel Mellinger. Reprinted with permission.)

correctly inferred that the misty glow of the Milky Way is nothing other than the combined light of these stars, much more tightly condensed in this region than in other parts of the sky. For him, the question of the Milky Way was nicely settled by this telescopic view and left no more to wonder about. "All the disputes which have vexed philosophers through so many ages have been resolved, and we are at last free from wordy debates about it," Galileo wrote in his popular booklet, the *Starry Messenger*. "The Galaxy is in fact nothing but congeries of innumerable stars grouped together in clusters. Upon whatever part of it the telescope is directed, a vast crowd of stars is immediately presented to view, many of them rather large and quite bright, while the number of smaller ones is quite beyond calculation."[6] Galileo also noted that several other "nebulous" or cloudy patches of light could be seen scattered about the night sky, and that the telescope revealed these, too, to be groups of stars.

For more than a century after Galileo's pronouncement, few astronomers or philosophers seem to have been interested enough in the Milky Way to suggest that more could be learned about it. What intrigued Thomas Wright, the first person profiled in this book, was what the crowding of stars revealed about the system in which our Sun is embedded. As we shall see, Wright imagined various configurations that our stellar system might have—for example, the stars might be arranged in a spherical shell—that would result in the view we have of the Milky Way. By Herschel's time already, astronomers understood that the stellar system or galaxy has the shape of a watch, wide and flat. Not until the twentieth century, however, did astronomers have the means to map our own galaxy reliably from within.

Early theories of the universe

The crowding of stars in the narrow band of sky we call the Milky Way suggests a fundamental asymmetry on a grand scale: the stars do not lie scattered equally in all directions. For reasons that may never be completely clear to us, for we have the advantage of hindsight, observers and philosophers alike overlooked this clue to the structure of the stellar system until the middle of the eighteenth century. Galileo complained about the philosophers' "wordy debates," but these disputes referred to the nature of the diffuse light of the Milky Way, not to the structure of the system of stars.

Early astronomers pictured the stars distributed on the surface of a solid sphere. In this scheme, which originated in the fourth century BCE (before the Christian era) and which Aristotle and Ptolemy developed through the second century BCE, the Earth occupied the center. The Moon moved in a sphere encompassing the Earth, and the Sun and planets orbited in their own successively distant spheres. The last planetary sphere was that of Saturn. The whole system came to an abrupt end at the sphere of fixed stars (see figure 1.3). In the third century BCE, the Stoic school of philosophers imagined a modified system in which the spherical realm of stars lay embedded in an infinite void. The Stoics essentially stripped away the Aristotelian outer boundary to avoid the problem of defining an edge to space.

For centuries, philosophers preserved the essential symmetry of the Aristotelian system even as they modified the details. Both Arab and Western Christian scholars elaborated on the moral correlate to the physical system, associating the outermost sphere of stars with a divine mover and the terrestrial center with all that is mortal and impure.

8

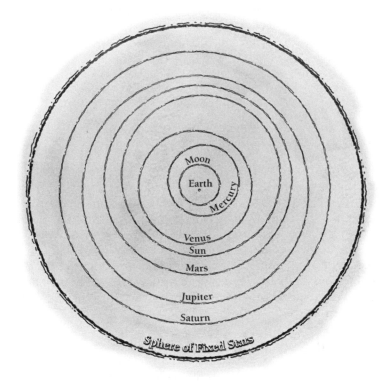

Figure 1.3 An Earth-centered system with the order of the planets as given by Ptolemy. The system ends with the sphere of fixed stars. (Credit: Layne Lundström.)

When in 1543 Copernicus put the Sun at the center of the system and moved the Earth to one of the encompassing spheres, man's conception of his place in the universe changed radically. Furthermore, with this Copernican revolution, the sphere of fixed stars lost some of its significance as the moral antithesis to the mundane realm, and philosophers began to consider alternatives to the sphere of fixed stars. Not long after Copernicus, the Englishman Thomas Digges published his own Copernican or heliocentric system, with the stars completely dispersed throughout an infinite void (see figure 1.4). In Digges' conception, the

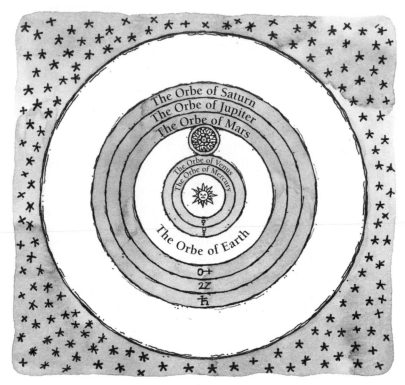

Figure 1.4 System imagined by Thomas Digges (c. 1546–95) and drawn to accompany his *Perfit Description of the Celestiall Orbes*. The label for his outermost sphere says that "the orbe of starres fixed infinitely up extendeth hit self in altitude spherically." This space is also the court of celestial angels and a site of endless joy. (Adapted by Layne Lundström.)

distribution of stars in space was more or less uniform and symmetrical.

Whether they imagined the stars as fixed to an outermost sphere or dispersed in an endless space, astronomers found little reason to dwell on them as anything other than a fixed and rather uninteresting backdrop to the more changeable elements of the night sky. Until the nineteenth century and even beyond, astronomers devoted far more attention to the wandering of the planets, the transient appearance of comets, the occasional bursting forth of a "new" star, and to small irregularities in the Earth's orbital motion around the Sun than they did to the stars *per se* and to the possibility that they might be distributed in some structured way.

The theory of island universes

The theory of island universes — which has been around in some form for a long time, but which philosophers and astronomers debated with renewed vigor in the eighteenth and nineteenth centuries — draws attention to the difficulty with the word "universe." To most of us, it means "everything." The universe consists of hundreds of billions of galaxies, and a lot of dark, mostly empty space in between. The universe in this sense is the cosmos, including both what is known and observed and what is unobserved. But in earlier times, it often meant the system of stars we see around us; before the term "galaxy" became current in the twentieth century, astronomers referred to our system as the universe. Thus in 1914, when the great English astronomer Sir Arthur Eddington wrote a book on *Stellar Movements and the Structure of the Universe*, he meant the structure of what we would now call our galaxy. But in 1933, he used the term in its modern sense when he wrote *The Expanding Universe*.

The theory of island universes states that systems of stars, or galaxies, are scattered at great distances from us, like islands in an ocean of space. Some philosophers, like Thomas Wright, saw the existence of other worlds as a natural consequence of an infinite cosmos. In his *Original Theory or New Hypothesis of the Universe*, printed in 1750, he wrote that "we may conclude in Consequence of an Infinity, and an infinite all-active Power; that as the visible Creation is supposed to be full of sidereal [starry] Systems and planetary Worlds, so on, in like similar Manner, the endless Immensity is an unlimited Plenum of Creations not unlike the known Universe."[7] Wright attempted to draw some of these creations, or "a finite view of infinity" as he called it (see figure 1.5).[8] His creations or universes are not all alike, but each has a supernatural or divine center, represented by an eye. In some of these island universes, one can discern a spherical shell of stars, Wright's preferred conception of the Milky Way system.

11

Figure 1.5 Wright's "Plenum of Creations." Wright attempted to show, in cross-section view, a number of "creations" filling the immensity of space. The eye symbols at the centers of the spheres represent the "divine Presence." In some cases, the stars are grouped in nested spheres or shells around their respective centers. (Adapted, with permission, from Hoskin (1971).)

Immanuel Kant, the German philosopher who gave the idea much currency (although not the appellation "island universes," which came later), suggested that some of these creations or starry systems might be visible to us as cloudy patches in the night sky. Galileo had shown that the diffuse light of the Milky Way arose from innumerable close-packed stars, and in a similar way, Kant and others supposed, distant island universes composed of stars like our Sun might appear as milky disks or circles. Kant was not far off the mark: as we shall see in chapter 2, a few of the nearest galaxies do appear as cloudy spots, even to the unaided eye.

The story of our discovery of the Milky Way and other galaxies is in many ways the chronicle of the popularity, demise, and renewal of the theory of island universes. None of the astronomers whose work is described here could study our own galaxy without thinking about the ramifications of his conclusions for the island universe theory. The theory of island universes itself is a minor character in the drama of our understanding of the Milky Way and other galaxies, always in the background, sometimes moving into the spotlight. Only very recently, in the middle of the twentieth century, did we come to appreciate the size, structure, and even the history of our spiral disk galaxy, the Milky Way; and we learned at the same time that our magnificent system has its counterparts both near and far, in the billions of galaxies stretching into remote space and distant time.

2

THE NAKED-EYE
VIEW OF THE SKY

"No more of the universe is visible to our unaided eyes than to the eyes of our Neanderthal ancestors. But science, the product of our imagination, has immensely extended the range of our imagination. Our inward eye can range beyond the dome of visible stars to the unseen realm of the nebulae and galaxies."

Chet Raymo (1982)[1]

On late fall or winter nights I like to go out and look for the Andromeda nebula. This fuzzy patch of light among the glittering stars gets its name from its appearance and location: nebula is Latin for cloud, and this small cloud-like object appears in the constellation Andromeda. When the winter night is moonless and there is no haze of light pollution from cities, the Andromeda nebula is so prominent that you don't need a telescope to see it. It has been known to skywatchers since the tenth century at least, when the Persian astronomer Al-Sufi included it as a "little cloud" in his catalog of the heavens.

Al-Sufi, who worked in a great medieval observatory in Baghdad, didn't have any optical instruments, but didn't have to worry about electric light pollution, either. Today the Andromeda nebula can be tough to see without binoculars or a telescope. The night must be very dark, as it is in rural areas or at sea. To the unaided eye—the "naked" eye, as astronomers like to call it—the nebula looks like a faint fuzzy star. With the small binoculars I use mainly for birdwatching, its oval shape just barely becomes apparent. But it is worth searching for because it is the most distant thing one can see with the naked eye. It's an entire galaxy, an "island universe," a system of billions of stars held together by gravity. When I look at the

14

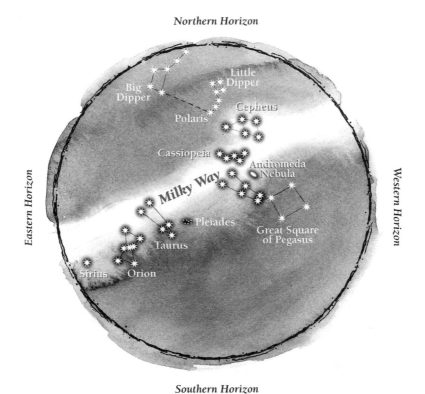

Figure 2.1 Locating the Andromeda nebula. The Andromeda nebula (oval symbol) can be found near Andromeda and Cassiopeia. (Credit: Layne Lundström.)

Andromeda nebula, I am looking past the stars of our own galaxy, and across a great gulf of apparently empty space.

To locate the Andromeda nebula, start with the constellation Cassiopeia, which many people know (see figure 2.1). The stars of Cassiopeia form a giant W or M, depending on your perspective. To the east of Cassiopeia — under the W, if you see the stars that way — one can pick out a long, narrow, slightly crooked "V" of stars: Andromeda. The stars along the side of the "V" nearest to Cassiopeia are dimmer than those of the opposite branch. The "V" terminates in the Great Square of Pegasus, another well-known constellation. The trick to finding the nebula is to locate the second pair of stars defining the "V" (down from the open end of the letter) and to look to the right about the same distance away as the separation of stars in the "V." Once your eyes have had time to adjust to the dark, you may find the nebula there.

The Andromeda nebula is so far away that to quote its distance in miles — about 12 trillion million — seems a bit silly. Numbers that great are better expressed in terms of the light-year. We're not ordinarily aware of light flitting through space. When we flick on a light switch, the room floods with light almost instantaneously. But light, which is a form of electro-magnetic radiation, travels at 186 000 miles per second. It covers 6 trillion miles a year, and that is a useful yardstick for distances in space. So the Andromeda nebula is as far away as light can travel in 2 million years, or 2 million light-years distant.

From the northern hemisphere, the Andromeda nebula is the only galaxy most people can see with the unaided eye. Southern hemisphere observers can also see two small irregular galaxies that are companions to our own, the Large and Small Magellanic Clouds. They keep us company a mere 200 000 light-years away. Of course, there are many more galaxies in all parts of the sky, visible with telescopes. Their distances are staggering, even expressed in millions of light-years.

The Andromeda nebula, an island universe, plays an impor-tant role in this book. Our tour of the night sky will help put its distinguishing characteristics in context.

The stars

Our eyes need the contrast with very dark, moonless nights to appreciate the brilliance of stars. Electric lights surrounding city dwellers effectively blind them to all but the brightest objects in the sky. Indeed, many people in North America have never seen the Milky Way, or mistake it for a plume of smoke when they are out in the country. This is a sad consequence of urban sky glow, upward-shining electric light reflecting off water molecules or smog. Sky glow washes out one's view of the fainter stars and the Milky Way, just as room lights wash out the image from a slide projector.

From the city, one might see the Moon and the planets Venus, Mars, and Jupiter, all close neighbors by astronomical standards, shining by reflected light from the Sun. Venus can be so bright, in fact, that one urban legend has it that an air traffic controller mistook it for an approaching airplane and wanted to give it "permission to land."

Stars bright enough to see from light-polluted cities include the red star Betelgeuse and the bluish star Rigel in the familiar constellation of Orion. ("Beetlejuice" and "RYE-gel" are acceptable pronounciations in English. See figure 2.2 for a chart of the stars in Orion.) The name Rigel appropriately comes from the Arabic for "foot," as it appears in the foot of Orion. Betelgeuse comes from the Arabic *bayt-al-jauzaa*, which translates to "house (or room) of the twins." The fact that this name makes no reference to the giant Orion may mean that this star was once seen as pertaining to another constellation, or that there has been some shifting around of star names in translation. Other stars visible despite city light pollution include Aldebaran (al-DEB-a-run), the brightest star in the horns of Taurus the Bull, and Canopus (ca-NO-pus), which dominates the spring sky for urban observers in the southern hemisphere. Many people are surprised to learn that Polaris, the famous North Star that one can find using two stars of the Big Dipper, is not particularly bright and might be washed out by city lights. Its importance derives from its position marking north, and not from its brightness.

Away from city lights, the unaided eye can see thousands of stars—about 3000 on any given night. The richness of the sky, when viewed from the desert, the rural plains, or the sea, is as breathtaking to astronomers as to anyone else. After about 20 minutes, the time it takes the eye to adapt to low light levels, details emerge: the sky is brighter in some directions that in others, stars come in different colors, and one might see little fuzzy patches of light that don't resemble stars.

A natural classification of the stars, developed by the Greek astronomer Hipparchus in the second century BCE, is to call the brightest stars "first magnitude" stars, those that appear noticeably less bright "second magnitude," and so on. According to this system, stars fainter than the third magnitude are difficult to see from inhabited areas, and stars fainter than the sixth

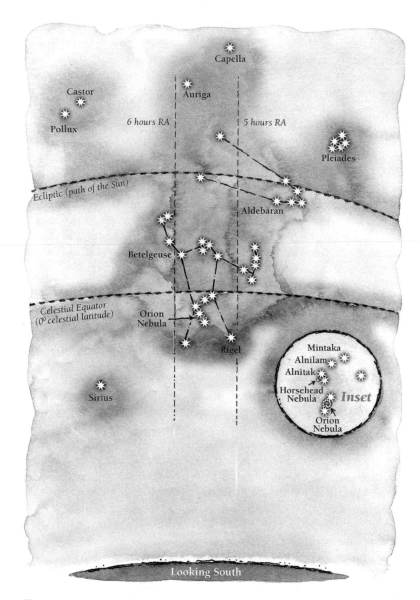

Figure 2.2 Orion. The famous Orion nebula is found in the ''sword'' hanging down from Orion's belt. The figure of Orion serves as a useful guide to the location of several reference points on the celestial sphere. (Credit: Layne Lundström.)

magnitude are not visible to the unaided eye. Hipparchus compiled a catalog of some 850 bright stars, giving data on their positions and estimates of their magnitudes.

Because it is based on how the human eye perceives brightness, Hipparchus's classification is awkward to use in an era of photo-electric devices. The difference between two magnitudes is an inconvenient factor of 2.512 in brightness. Awkward, too, is the fact that the Sun and many of the planets are much brighter than the average first magnitude star, and so require negative numbers to be represented on the same scale.

The magnitude system we inherited from Hipparchus, and revised in modern times, runs from −27 for the Sun, through 0, and beyond the sixth magnitude for the faint stars. Binoculars with lenses 40 to 50 mm wide, or a telescope with equivalent opening diameter (about 2.4 inches) will show objects at eighth or ninth magnitude, depending on the level of urban sky glow. Research telescopes on Earth see galaxies fainter than 23rd magnitude, and the Hubble Space Telescope's Wide Field and Planetary Camera 2, in orbit above the Earth's atmosphere, can discern objects as faint as 28th magnitude.

Many astronomers quote brightnesses using the modern version of Hipparchus's system, and it is useful to recognize a few reference points of the magnitude scale. Sirius, the brightest star in the sky, has a magnitude of −1.5. Vega, a brilliant star that rises high overhead in the summer for viewers in the northern hemisphere, has a magnitude almost exactly 0. The brightest stars in the Big Dipper and in the constellation Andromeda are all about the second or third magnitude. To see stars just at the naked-eye limit of sixth magnitude, find the bright star Vega with a star chart or planetarium software and look northeast. Keen eyes will distinguish two faint objects at about fifth magnitude; through binoculars, each of those turns out to be a double star. This is the famous "double double" star in the constellation Lyra.

The reason many of the brighter stars have Arabic names such as Betelgeuse and Rigel, even in European star lore, can be traced to the fate of Hipparchus's catalog. Three historical catalogs, those of Hipparchus, Ptolemy, and Al-Sufi, form links in the chain of inheritance. Ptolemy of Alexandria, the greatest astronomer of late antiquity, flourished around the year 150, some 300 years after Hipparchus. Ptolemy drew heavily on Hipparchus's work in creating his own encyclopedia of the stars, which he called the *Mathematical Compilation*, but which is more commonly referred to as the *Almagest*. About 800 years later, Al-Sufi translated Ptolemy's great work and synthesized Ptolemy's star catalog with Arabic names and traditions in his *Book of the Fixed Stars*. Medieval Europeans read the works of Al-Sufi and other Arabic astronomers in translations from the Arabic to Latin. They simply Latinized the Arabic star names. Thus the star Acrab in Scorpius comes from the Arabic Al-'Aqrab for Scorpion, and Alnitak in Orion comes from An-Nitaq, for belt.

Constellations

Some patterns of stars are so distinctive that people from many parts of the world and many historical eras have independently named them as a group or constellation. The pattern that many North Americans see as a long-handled ladle and call the Big Dipper — technically an asterism, or collection of stars, within the constellation Ursa Major — is called the drinking-gourd by some African-Americans, and the plow or plough in England. The seven bright stars of the constellation are known as the Sapt Rishi, or seven wise men, in India. The Basques spin a complicated tale around the constellation: the four stars of the dipper cup are two stolen oxen and two thieves, and those of the handle include the owner of the oxen, his servant, his housemaid, and his dog, all in pursuit of the thieves.

In some cases, the patterns remind viewers of familiar sights. Stars grouped like a backward question mark in Leo do look a bit like a lion's mane, and many cultures have seen a warrior or hunter in Orion. On the other hand, some constellations don't bear any resemblance to the figure or object they are named for. Ancient people chose to honor a god or mythical figure with a piece of celestial territory, and the name stuck.

Many of our modern Western constellations were first labeled by the people of Mesopotamia in the third millenium BCE. Capricornus, Sagittarius, Scorpius, and Leo are among constellations depicted on stone tablets many thousands of years old. But it is to the ancient Greeks that we owe some of our favorite stories about the constellations.

Greek astronomers of antiquity handed down one legend that connects at least six of the constellations in the northern sky: Cassiopeia, Andromeda, Cepheus, Cetus, Perseus, and Pegasus. Cassiopeia was a queen of Ethiopia, married to king Cepheus. According to this legend, Cassiopeia boasted that she was more beautiful than the Nereid sea-nymphs. The gods, displeased by this vanity, chained Cassiopeia's daughter, the princess Andromeda, to a sea-side cliff, where she would surely be devoured by the monstrous Cetus. The hero Perseus came to her rescue, carrying the severed head of Medusa. Medusa's blood dripping into the sea gave rise to the winged steed Pegasus.

While many constellations are associated with legends thousands of years old, they have all been redefined in the modern age. For centuries it was up to the creators of star-atlases to set the exact boundaries between the constellations and to name star patterns too far south for the Greeks or Mesopotamians to have seen. Some parts of the sky don't have bright stars or a distinct pattern, and these were devoid of constellations. The International Astronomical Union brought some order to the situation in 1930. This non-governmental organization for promoting the study of astronomy established precise boundaries for 88 constellations, covering the entire celestial sphere so that every star or object is within a constellation. This is the same international organization, headquartered in Brussels, that assigns names for newly discovered astronomical objects such as asteroids.

Whether or not we mentally "connect the dots" in a constellation, the apparent grouping of stars gives us the impression that all the stars in a single constellation are at about the same distance from us. In some cases, it is true: the stars of the Pleiades—the "seven sisters" asterism—are in fact close together in space. More often the stars of a constellation lie at a variety of distances. In the Big Dipper asterism, Alkaid, the last star of the handle away from the dipper cup, lies about twice as far away as Dubhe, the uppermost of the two stars in the outer edge of the cup. In the mind's eye, we see the stars projected on a two-dimensional plane.

The Milky Way

The band of light we call the Milky Way arches across the sky like a river of light, narrow in some places, wide and irregular in others. Its hazy but unmistakable appearance has led observers around the world to create legends around it. As we noted in chapter 1, the Greek story says the Milky Way flowed from the goddess Hera's milk. The Chinese think of it as the celestial counterpart to the great Yellow River. Siberians call it the seam in the tent of the sky. In some East African legends, it is the smoke of ancient campfires. To some Australian aborigines, the Milky Way is a river, while the dark rifts in and around it are riverside lagoons. A number of tales from western Asia and south-central Europe portray the Milky Way as a path of scattered straw.

Northern hemisphere dwellers see the Milky Way in summer, fall, and winter. A late summer or early fall view takes in the brightest and richest span of this celestial river. At that time of year, the Milky Way stretches from the constellations Cassiopeia and Cepheus in the north, across the eastern half of the sky and through the set of stars we know as the summer triangle, and plunges toward the horizon through the constellations Sagittarius and Scorpius. Between the summer triangle and Sagittarius, dark clouds obscure the central swath of the Milky Way, making it appear to be split into two streams. Near Sagittarius and Scorpius, the Milky Way is particularly dense and bright, for this is the direction toward the center of the galaxy. Dark clouds of dust are prominent in this direction too, giving this area of the Milky Way a heavily mottled appearance.

In spring—specifically, in late April or May for northern hemisphere viewers, late November and early December for southern hemisphere viewers—the Milky Way lies low around the horizon in the evening hours and can't be seen. When we look up in the night sky at this time, we are looking out of the disk of the Galaxy and into "deep space."

Southern hemisphere observers have a better view of the Milky Way overall, and see a particularly dazzling show in the southern hemisphere winter. In July, when Scorpius is nearly overhead, the Milky Way stretches from southwest to northeast. The irregular dark clouds cleaving the Milky Way into two uneven streams are so prominent that some southern hemisphere peoples have named them "black constellations." For example, a roundish dark spot near the constellation of the southern cross looks like a partridge-type bird to Quechua peoples in the Peruvian highlands.

Southern hemisphere observers are also privileged to see the Large and Small Magellanic Clouds, high in the sky in November and December. Both are small, irregular galaxies, close to our own galaxy. They form faint but extended patches, a bit like tatters from the main ribbon of the Milky Way. The Large Magellanic Cloud is as wide as Orion's waist.

The celestial sphere

For thousands of years at least, observers of the night sky have contemplated it as a great solid dome, or the convex surface of a spherical shell centered on the Earth. The stars, in this conceit, stud the inner lining of the "celestial sphere." If you lie on your back and gaze upward at the stars in some quiet, dark location, you might even feel that you can *sense* the slow rotation of the sphere, carrying the stars across your field of view from east to west, and slinging them underneath you on the other side of the Earth.

The apparent daily movement of the sky is, of course, due to the rotation of the Earth, not the sky, and we know that the stars are not all at the same distance, as they would be if they were fixed to the inner surface of a shell. However, for many practical purposes, the model of the stationary Earth and the celestial sphere rotating around it works well. Navigators, for example, do not need to know the true distances of the stars, but only where they are located on the two-dimensional surface of the sphere.

The axis around which the imaginary celestial sphere rotates is simply an extension of the Earth's real axis of rotation. The Earth's axis, projected into space from the north pole, strikes the celestial sphere at the north celestial pole, and similarly strikes the south celestial sphere when extended from the Earth's south pole. Thus anyone who is familiar with using two stars of the Big Dipper to find the "north star" Polaris already knows how to find the north celestial pole.

The fact that a relatively bright star marks the north celestial pole is due to happy circumstance. People who live in the southern hemisphere have no equivalent south pole star, although they can use the constellation Crux, the Southern Cross, to find the south pole. The four primary stars of the Southern Cross are visible from sites south of 27° north latitude, and are up all night for viewers as far south as Australia, South Africa, or Argentina. The constellation is so cherished that five countries—Australia, New Zealand, Western Samoa, Brazil, and New Guinea—feature it on their national flags. A line through the long arm of the cross, extended about $4\frac{1}{2}$ times the length of the cross, passes very close to the south celestial pole.

Celestial latitude and longitude

The constellations are a practical way for stargazers to orient themselves, and are convenient for locating conspicuous objects such as planets, stars, and star clusters. An almanac might state that the planet Mars will be "in" the constellation Capricornus during the month of November, for example. But for greater precision in specifying locations, astronomers use celestial latitude and longitude, called, respectively, declination and right ascension. Lines of declination and right ascension circle the celestial sphere. "Declination" comes from the Latin for "bending away" or inclination along a line from the equator to the pole, while the term "right ascension" refers to the system of lines at right angles to the plane of the equator, ascending or increasing to the east around the celestial sphere.

Lines of declination correspond directly to latitude on Earth. The celestial equator has declination 0°, just as the equator is at latitude 0°. The celestial north pole is at declination +90°, and the south pole at −90°. Observers in Boulder, Colorado, at latitude +40°, will see the summer constellation Cygnus (declination about +40°) pass directly overhead. A degree of declination is divided into 60 arcminutes or minutes of arc, and each arcminute into 60 arcseconds.

Lines of longitude, reaching from pole to pole, are called meridians. By international agreement in 1884, the origin or zero-point for longitude on Earth is the meridian that passes through an historic telescope at the Royal Greenwich Observatory, England. In fact, the so-called prime meridian was originally defined very precisely by the cross-hairs in the eyepiece of this telescope, which was built by Astronomer Royal Sir George Airy in 1850. From the prime meridian it is 360° (or twice 180°, east and west) around the globe.

Using the Greenwich meridian as the zero-point for right ascension would be impractical, because a star or other fixed point on the celestial sphere would have a constantly changing celestial longitude during the course of a day. Instead, the origin of right ascension is fixed in the sky, at a point in the constellation Pisces. From there, it is 360° around the celestial sphere, or 24 hours of right ascension, with each hour divided into minutes and seconds.

The familiar winter constellation of Orion can serve as a guide to the celestial sphere and the declination and right ascension system (see figure 2.2). The celestial equator (0° declination) runs through Orion's belt. The three belt stars form a distinctive group because they appear about equally bright and regularly spaced. From east to west they are Alnitak, Alnilam, and Mintaka. One can trace the celestial equator by sweeping one's arm from due east on the horizon, through Orion's belt, and down to the western horizon. For stargazers in Ecuador, Kenya, or Singapore, near Earth's equator, this arc traced out in the sky will pass overhead. For all other viewers, the arc will tilt toward the south or north horizon, depending on whether the viewer is in the northern or southern latitudes.

The bright red star Betelgeuse in Orion's shoulder and the bright blue-white star Rigel in Orion's foot are a little less than one hour of right ascension apart. The meridian running north–south near Betelgeuse is that of six hours right ascension, and that running between Rigel and Orion's bow corresponds to five hours.

The zodiac and the ecliptic

Of the 88 modern constellations, the 12 constellations of the classical zodiac—Sagittarius, Capricornus, Aquarius, Pisces, Aries, Taurus, Gemini, Cancer, Leo, Virgo, Libra, and Scorpius—are

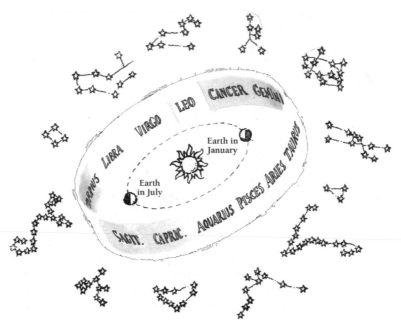

Figure 2.3 The Zodiac. The constellation an object appears "in" depends on the line of sight from the earth. (Credit: Layne Lundström.)

undoubtedly the best known. The Mesopotamians of antiquity used these to mark the passage of the Sun and planets around the celestial sphere.

Figure 2.3 illustrates how the Sun appears to "travel" through constellations on the celestial sphere. The figure shows the position of the Earth in its orbit around the Sun at two times of the year, January and July. In January, the constellation Gemini is overhead at night, when the observer is on the shaded side of the Earth. The constellation Sagittarius is in the line of sight to the Sun, behind the Sun on the celestial sphere. The stars of Sagittarius would not be visible, hidden in the glare of daylight. However, ancient stargazers kept track of the order of the constellations and the seasonal changes in the sky, and knew which constellation rose with the Sun, even if it was not visible. Thus careful observers would have known which constellation the Sun was "in" during the daytime.

26

As the Earth orbits the Sun and the seasons progress, the line of sight to the Sun changes, and the Sun appears against a different backdrop of constellations. In July, Scorpius and Sagittarius are up at night, while the Sun has reached Gemini. The Sun completes one turn through each of the constellations of the zodiac in one year. The line it follows is called the ecliptic. The only difference between the system the Mesopotamians envisaged and the current one is that we have delineated the constellation boundaries somewhat differently, and the Sun's path or ecliptic now takes it through a corner of the constellation Cetus, between Pisces and Aries, and through Ophiucus, between Scorpius and Sagittarius.

A similar diagram could be drawn to show the changing line of sight from the Earth to any of the planets. Because of the relative motion between the Earth and the planets in their own orbits, and because the planets orbit more slowly with increasing distance from the Sun, the planets do not appear to move smoothly through the constellations, nor do they take one year to complete a turn. Mars, for example, is in the constellation Aquarius in January 2002, in Pisces in February, in Aries in March and April, in Taurus in May, in Gemini in June and July, in Cancer in August, in Leo in September and October, and in Virgo in November and December. It does not return to its starting point in Aquarius until July 2003. Neptune, moving very slowly in its distant orbit, appears in the constellation Capricornus from 1999 to 2010.

Figure 2.4 shows how the declination and right ascension coordinate system, ecliptic, and Milky Way relate on the celestial sphere. The celestial poles lie above the corresponding terrestrial poles, and the celestial equator mirrors that circling the Earth. The Milky Way girdles the celestial sphere at a steep angle to the celestial equator. The figure shows the Milky Way denser and broader in the direction of the center of the galaxy (in the constellation Sagittarius, not shown on this figure), and shows the rift caused by obscuring clouds of dust, which make the Milky Way appear to divide into two streams. The Large and Small Magellanic Clouds are shown near the south celestial pole.

The ecliptic — the path of the Sun — is tilted at an angle of $23\frac{1}{2}°$ to the celestial equator. It intersects the celestial equator at two points, the vernal and autumnal equinoxes. The origin of right

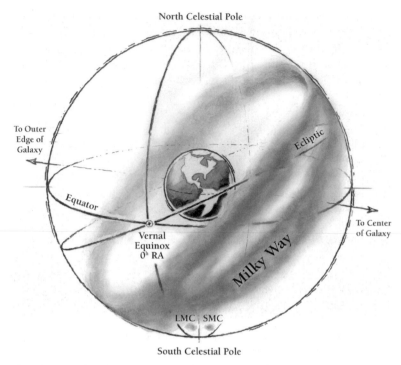

Figure 2.4 The Celestial Sphere. The diagram shows the relationship of the Milky Way to the lines of 0° declination and 0 hours Right Ascension on the celestial sphere. (Credit: Layne Lundström.)

ascension, analogous to the Greenwich meridian on Earth, is at the vernal equinox point of intersection.

Star clusters and nebulae

The night sky abounds with objects besides individual stars and planets: comets, making their occasional but much-noticed appearances; various kinds of nebulae or cloudy patches of light; and stars in clusters of two or three to hundreds of thousands. These objects aroused the curiosity of early stargazers and provided important clues to modern astronomers seeking to understand the structure of the Milky Way.

28

Figure 2.5 The Pleiades. Only about half a dozen stars in the Pleiades open cluster are visible to the naked eye, but many more appear through telescopes or in long-exposure photographs. In telescopic views, one can see a blue veil near some of the brighter stars—a reflection nebula caused by dust. (See color section.) (Copyright Anglo–Australian Observatory/ Royal Observatory, Edinburgh. Photograph from UK Schmidt plates by David Malin. Reproduced with permission of David Malin Images.)

The Pleiades or "seven sisters" group of stars east of Orion is one of the best known star clusters (see figure 2.5). The naked eye discerns six members of the cluster. It is something of a mystery why legends from different eras and widely separated parts of the world, from ancient Mesopotamia to modern indigenous Australia, refer to seven stars, or six stars and one "missing" member. A telescopic view reveals several hundred stars, along with a blue-tinged veil of dust trailing through the cluster.

The Pleiades group is an example of an *open* or *galactic cluster*—an asymmetric, loosely bound association of stars, located in the disk of our galaxy. The Pleiades cluster lies only about 380 light-years away. Some of the naked-eye nebulae

Ptolemy mentioned in his *Almagest* turned out, when seen with optical aids, to be open clusters. One he described as "following the sting of Scorpius" — a brilliant open cluster, seen against the backdrop of the Milky Way. It is sometimes known as Ptolemy's cluster, but the seventeenth-century comet-hunter Charles Messier gave it the catalog number by which it is best known today, M7. Ptolemy also described the so-called Beehive cluster, M44, in the constellation Cancer. Galileo was the first to turn a telescope to this nebulous object and note the presence of dozens of bright stars. Modern telescopes have revealed at least 200. The Beehive cluster is easily seen with binoculars.

More spectacular even than these jewels are the *globular clusters*, consisting of tens of thousands or even millions of stars. This number is far less than the billions of stars composing our galaxy, but greater than the number of stars in a typical galactic cluster. The stars in a globular cluster gravitate strongly toward the center of the cluster, forming a ball that is usually noticeably brighter in the middle. The compact structure of globular clusters contrasts with the loose structure of open or galactic clusters.

Globular clusters (see figure 2.6) do not concentrate in the plane of the galaxy, as galactic clusters do. They can be found in all directions in the sky, and lie at distances of 10 000 to 60 000 or more light-years. In the northern hemisphere, two of the better known globular clusters that can be seen with the naked eye under good conditions, and that are particularly rewarding with binoculars and small telescopes, are M4 and M13. M4 sits about a degree west of the red star Antares in Scorpius, while M13, also known as "the great globular in Hercules," lies about halfway between the bright stars Arcturus and Vega. But these globular clusters, impressive as they are, offer no match to the two biggest globular clusters visible from the southern hemisphere, Omega Centauri (in the constellation Centaurus) and 47 Tucanae, in the constellation Tucana, near the Small Magellanic Cloud. Omega Centauri has a magnitude of 3.8, and so does not even require particularly good conditions to be seen; 47 Tucanae is very prominent, but lies so far south that astronomers did not observe it telescopically until the late 1700s.

A number of cloudy-looking objects are truly nebulous in character, and not just very distant or close-packed groups of

Figure 2.6   The globular cluster known as M5 (the fifth item on the list of nebulae compiled by the French astronomer Messier in the late eighteenth century). A globular cluster is an assembly of hundreds of thousands or millions of stars, orbiting the center of our galaxy. (Copyright Anglo-Australian Observatory. Photograph by David Malin. Reproduced with permission.)

stars. The "Great Nebula" in Orion's sword (see figure 2.7) is no doubt the best known. South of the belt star Alnitak, in a line angled away from Orion's body, are three stars forming Orion's sword or dagger. The central "star" is the Great Nebula, bright but hazy. Binoculars or a small telescope reveal a cluster of stars embedded in the nebula. (If you don't have binoculars or a telescope, try looking through your hand cupped like a tube. Some say this technique does improve the view a bit, although you won't see the stars embedded in the nebula this way.) This nebula is of the *emission* or star-forming kind. Hundreds of stars are in various stages of condensing out of a thick, turbulent cloud of hydrogen and helium gas. The Orion nebula is really a stellar nursery.

31

Figure 2.7 The Orion nebula, the most famous example of a diffuse emission nebula. The nebula shrouds from view a stellar nursery, where stars are condensing out of hydrogen and helium gas. (See color section.) (Copyright Anglo-Australian Observatory/Royal Observatory, Edinburgh. Photograph from UK Schmidt plates by David Malin. Reproduced with permission of David Malin Images.)

Reflection nebulae also consist of clouds of gas and dust, but they shine by reflected light of the stars nearby, rather than glowing from being heated up by those stars. The blue veil surrounding the Pleiades, visible through a moderate-sized telescope, is one such reflection nebula.

William Herschel, one of the astronomers profiled here, named a third kind of nebula, quite different in character from the star-forming region in Orion or the dusty environment of the Pleiades. Herschel saw smooth round nebulae, actually formed from shells or cocoons of gas blown off dying stars. The Cat's Eye nebula in Draco (see figure 2.8) is a fine example, visible

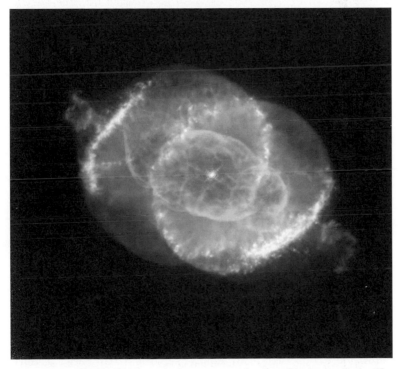

Figure 2.8 The Cat's Eye nebula, an example of a planetary nebula. The name comes from the fact that the gas and dust surrounding the star looks, through a small telescope at least, like a disk or planet. (See color section.) (Credit: J.P. Harrington and K.J. Borkowski (University of Maryland), and NASA.)

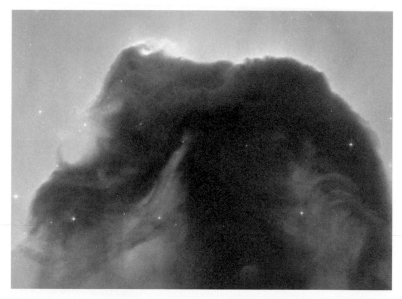

Figure 2.9   The Horsehead nebula in Orion. The Horsehead nebula is a dark nebula, silhouetted against the brighter light from the emission nebula in Orion. (See color section.) (Credit: NASA, NOAO, ESA and The Hubble Heritage Team. (STScI/AURA.)

with small telescopes. Ignorant of their true nature, Herschel named them *planetary nebulae* because they looked superficially like the disks of distant planets. Through telescopes, planetary nebulae reveal a diversity of shapes and colors. The Dumbbell Nebula in the constellation Vulpecula and the Ring Nebula in Lyra are accessible to viewers with small or moderate telescopes.

Dark nebulae are clouds of interstellar dust and gas so thick that no stars can be seen through them. Northern hemisphere observers with access to a telescope can see a fine example in the Horsehead Nebula in Orion—a dark nebula superimposed on the glow of an emission nebula (see figure 2.9). Southern hemisphere observers can easily see the Coal Sack, a dark nebula that looks like a hole in the Milky Way. It abuts two of the stars of the Southern Cross.

These various types of nebulae and star clusters recur throughout our story of the discovery of the Milky Way galaxy, as clues and signposts to the seven astronomers profiled.

3

THOMAS WRIGHT: VISIONARY OF STELLAR SYSTEMS

"Who in England is so peculiar as to be bothered by the apparent irregularity of the Milky Way!"

Abraham Gotthelf Kaestner,
unsympathetic reviewer of Wright's book,
An Original Theory or New Hypothesis of the Universe, 1752[1]

On a damp and windy night in September 1729, around the time of his eighteenth birthday, Thomas Wright ran away from his first job as an apprentice. Scandalous rumors about his involvement with a young woman buzzed about the northern English town of Bishop Auckland. Wright had pleaded his case, but could not convince his master, clock- and watchmaker Bryan Stobart, of his innocence. So, clutching the astronomy books he had spent most of his pocket money on, Wright picked his way across the fields west of town, "intending for Ireland" as he wrote in his journal, and trying to avoid stumbling into treacherous pit-holes left by coal miners.[2]

Wright was in a pickle, and not for the first or last time. His great talents and originality would secure him a place in the history of astronomy; his drawings would adorn astronomy textbooks more than 200 years after his death. But his intellectual and emotional intensity made it difficult for him to settle down and find his niche in society. On a recent trip home, when he was recovering from a broken collar-bone and reading all the astronomy books he could buy or borrow, his father had grown impatient with his zealous study and burned all the books he could find. His mother subsequently supplied him with money for books,

35

but Wright, in his impetuous flight across the fells and away from the scandal behind him, didn't feel like going back to take refuge at home in Byer's Green, only a few miles away.

Fortunately for the runaway, he soon encountered a kindly miller with a comfortable guest bed in his mill. By sunrise the next morning, when the miller filled his pockets with bread and cheese, Wright was reconsidering his original plan to make for the west coast and sail for Ireland. The next day he presented himself at his father's friend's house in Sunderland, a port on the east coast, "inexpressibly weary and fatigued" after walking more than 60 miles. "Next day writ to my father," he recorded in his journal, "and to my great joy was sent for Home."[3]

Thomas Wright (figure 3.1) was born 22 September 1711 in the village of Byer's Green not far from the town of Durham.

Figure 3.1   Thomas Wright (1711–1786). (Reproduced by permission of the British Library.)

His family included two older brothers and at least one sister, also older than he. His father John owned a small parcel of land and thus enjoyed the status of yeoman, or independent farmer, but earned a living primarily as a carpenter. Wright's mother Margaret we know little about. She probably had some education as a child, for her sister Mary grew up to become headmistress of a boarding school in Yorkshire. Wright described Mary as "a very great Scholer but not Rich."[4]

To the delight of historians, Wright confided details of his life and work in a journal, which passed to a friend after his death and eventually found its way, along with other Wright manuscripts, to the British Library. Wright's brief but frank descriptions of his circumstances and motivations help shed light on the development of his ideas about the structure of the universe. They also reveal the character of a man who was not a professional astronomer either by our standards or by those of his day, but who had a significant insight into the three-dimensional arrangement of the stars of the Milky Way system.

The friend who acquired Wright's journal, George Allen, did modify Wright's text in one way, which makes some of the quotations from the journal in this chapter confusing: in preparing a biographical note about Wright for the *Gentleman's Magazine*, Allen went through the text and generally, though not always, over-wrote the first-person pronouns to change them to the third person. Thus Wright's description of the scandal that drove him from Bishop Aukland reads, in part, "Soon she came to Bed to him upon which fearing the consequence of Forfeitin[g] his Indentures ec [etc.] left her and complain'd, upon which the Man swore to be the death of him." The original text said "me" and "my" in place of "him" and "his."[5]

We learn from Wright that as a child he was "Very wild & much adicted [to] Sport." He attended a community-supported or "private" school from the age of 4 or 5, then a tuition fee-charging or "public" school, where he studied Latin. However, he notes under the heading 1719, when he was 7 or 8, that he was obliged to leave the Latin school, "being interrupted by a very great Impediment of Speach."[6]

Wright's sister taught him arithmetic and writing for a while. She evidently did not make much headway with his writing style, which is not noteworthy, but Wright's talents in mathematics must have appeared at this time, for his family enlisted the help of a tutor. Wright notes that he studied with a local man by the name of Thomas Munday, "a wrgt [right] good Accomptant and an Astronomer." An entry opposite the date 1723 says in his typically brief style, "Much in love with Mathematicks."[7]

When Wright was 13, his father bound him as an indentured apprentice to the watchmaker Bryan Stobart. Normally such a contract would have lasted seven years, until the apprentice was about 21 years old, and ready to become a journeyman. During Wright's abbreviated 4-year residence with Stobart, before the scandal, he would have begun to learn all aspects of watch- or clockmaking, including the use and maybe even the manufacture of a watchmaker's tools, the construction and fitting together of watch movements, and the technical illustration of the movement, showing the proportional sizes of the toothed wheels. This last element of the trade was particularly well suited to his artistic bent. He wrote in his journal that in his spare time he was "Very much given to ye Amusement of Drawing, Planning of Maps and Buildings."[8] This was the germ of a lifelong passion for illustration that he would later apply to explaining astronomical concepts.[9]

Background: the age of Newton, Halley, and Cassini

We do not know what Munday taught Wright about astronomy and its affiliated field of mathematics, but he seems to have been well-supplied with books, which Wright eagerly borrowed. "Mr. Munday reports I have stole all his Mathematicks from him," Wright observed in 1729.[10]

Munday introduced Wright to astronomy at a time of great renewal for all scientists or "natural philosophers" as they were then known. Some 40 years earlier, Isaac Newton had published his revolutionary treatise, *Philosophiae naturalis principia mathematica*, the "mathematical principles of natural philosophy," better known by its abbreviated Latin title *Principia*. In this work he formulated the theory of gravitation as the force tending to attract massive objects toward one another, and he articulated the mathematical laws of motion that govern everything from the fall of an apple from a tree to the motion of a planet around the Sun. His theory explained such important phenomena as the Moon's orbit of the Earth and the establishment of ocean tides.

Mathematicians and astronomers, including Newton himself, immediately began applying the laws of motion and the theory of gravitation to elucidate the previously cryptic motions of solar system bodies such as comets, Jupiter's satellites, Earth's Moon, and the Earth itself, which shows small irregularities in its motion. Edmond Halley identified his famous recurring comet, for example, by applying Newton's laws. Halley gathered his predecessors' observations of the positions of comets and calculated the implied orbits, or paths, of these comets around the Sun, under the influence of gravity. Among the many curved paths he drew up, the orbits of comets seen in 1531, 1607, and 1683 struck him as very similar, and in 1705 he correctly inferred that a single comet had reappeared at intervals of about 76 years. His prediction of the next return of the comet would not be borne out until years after his death in 1742. When the comet appeared in December 1758, as he had foretold, the public acclaimed him anew as a great astronomer, but mathematicians had already recognized the success of his application of Newton's laws.

Wright began reading extensively in astronomy and mathematics just around the time of Newton's death in 1727 and the publication, in 1729, of an English translation of the Latin *Principia*. These events may have prompted Wright to read Newton's original works at that time, but whether he read Newton then or later, he almost certainly came across simplified explanations of Newton's concepts by authors such as William Whiston, a Cambridge professor and associate of Newton, and John Keill, a promoter of Newton's work at Oxford. Whiston was, to judge from Wright's quotations and borrowings, one of his favorite authors.

Wright would also have read of telescopic discoveries in the previous half-century, notably those of the Italian-French astronomer Gian-Domenico Cassini at the Paris Observatory. In the late 1600s Cassini captivated the public with large-scale maps of the Moon, showing its heavily cratered surface in unprecedented detail. He uncovered a new feature in Saturn's ring—a ring that Galileo had first seen only as a set of ear-like appendages on either side of the planet. Cassini saw and described a flat ring with a dark gap separating it into two parts, a gap still known as the Cassini division. Most importantly for the understanding of man's place in the cosmos, Cassini fixed the scale of the solar system itself, using the method of parallax (described in chapter 4). Although his results were subject to uncertainty, he established that the Sun lies on the order of 150 million kilometers (93 million miles) from the Earth.

Wright certainly was aware, too, of the circumstances that had led to the establishment of France's Observatoire de Paris in 1667 and England's Royal Observatory in 1675: the longitude problem. All too often, European ships carrying home the bounty from overseas colonies lost their way at sea or ran aground at night, for want of knowledge of their east–west position. (India's great non-telescopic observatories at Jaipur and elsewhere were completed in Wright's lifetime, but they were not dedicated to solving the longitude problem.)

Mariners, and mapmakers charting new territories in the Americas, desperately needed a way to keep track of time at sea. To calculate their longitude, they must compare their local time of day, as determined by the Sun, to Greenwich or Paris time, as determined by some kind of clock; the difference in time translated to a difference in longitude. Astronomers had vowed to decipher the complex motions of Earth's Moon and the motions of Jupiter's satellites so that these systems could be read as celestial "clocks." Analysing these celestial timekeepers required new, more precise charts of the background stars, however, and so they petitioned their governments for telescopes and facilities.

John Harrison solved the longitude problem in the mid-1700s with a mechanical timepiece built to survive long sea voyages, and clocks eventually supplanted celestial means of marking Greenwich time aboard ship. But when Wright took up his mathematical and astronomical studies, the longitude problem still loomed in the astronomical community as the most pressing practical problem, and focused astronomers' attention on the accurate charting of the locations of the stars and the study of the planetary motions. Although it was surmised that the stars were distant suns, quite probably endowed with their own planetary systems, the stars themselves, the obvious differences among them in color and apparent size, and their tendency to form clusters did not elicit much interest.

Explorations

At 18, Wright was still far from making his mark on the world. Home again at Byer's Green, while his father and three justices of the peace negotiated a legal and financial termination to his apprenticeship, Wright dove back into his books.

A note in his journal from this period, when he was living with his parents and applying for work in nearby towns, shows that his interest in astronomy at this time had less to do with an ambition to observe or chart the stars and planets for himself than a desire to understand the Creator behind them. He wrote, "Reflecting upon almost every object, conseive may find Ideas of ye Deaty and Creation." In this endeavor he was not alone, but following a certain vogue among churchmen of philosophical bent.[11]

When Newton formulated his laws of motion, he not only provided new tools for astronomers, but also spawned a new line of theological work. The great mathematician himself doubted that the universe's apparent stability could have arisen without God's design or intervention. The same gravitational attraction that gave rise to planetary orbits would cause the stars — indeed all matter — eventually to clump together. In the *Principia*, Newton suggested (rather unconvincingly, to the modern reader) that God prevented this catastrophe by removing the stars from each others' gravitational sphere of influence: "[L]est the systems of the fixed stars should, by their gravity, fall on each other mutually, he hath placed those systems at immense distances one from another," he wrote. Elsewhere, he similarly suggested that the Creator would form and repeatedly re-form a system such as our solar system, to remove the irregularities of motion that would inevitably creep in as the planets and comets exerted gravitational effects on one another.[12]

Even those philosophers who would argue that God had designed a perfect, stable universe at the outset — and so had no need to intervene — liked to look for evidence of divine harmonies in the workings of nature. A number of books appeared on this astro-theological theme of searching for evidence of the Deity: in 1669, for example, John Craig published *Mathematical Principles of Christian Theology*, self-consciously modeled on Newton's *Mathematical Principles of Natural Philosophy*. Whiston, one of

41

Wright's favorite sources of quotations and inspiration, first brought out his influential work *Astronomical Principles of Religion* in 1717. William Derham enjoyed the success of his *Astro-theology* during Wright's youth.[13] Wright shared the same impulse as these authors; a starting point for one of his proposed later tracts was, "If you can not Believe in god and his infinite power which you do not see; you may still believe in his Nature and Wisdom which you do see."[14] The note in his journal about looking for signs of the Deity and Creation shows that his interest in astronomy was from the outset intertwined with his search for a moral principle in the universe.[15]

Wright's reading filled his time as he waited for responses to his queries about employment. Some two months after leaving Stobart, having failed to find any post despite widening his search to Newcastle and towns in Yorkshire, he gave up on business and took up the study of navigation. He resolved to be a "saylor," and persuaded his father to support a trial voyage. In January 1730, he set sail for Amsterdam, full of hope for his new career.

Wright delighted in his view of the horizon at sea. With no mountains to clutter the view, and only monotonous gray swells in all directions, he must have found it easy to picture the Earth as a perfect globe. The trip broadened his horizons in a metaphorical sense, too. He took "great Notice" of Amsterdam's town hall or Stadhuis on Dam Square, and what he called its "Geographical Pavement Figure."[16] It's not hard to imagine that this magnificent edifice, which locals nicknamed the eighth wonder of the world, inspired some of Wright's later efforts to design astronomically-themed buildings for the English aristocracy; the Stadhuis was everything Wright loved, and on a grand scale.

The citizens of Amsterdam completed their new town hall, then Europe's largest government building and a model of the new Classicist style, in 1665. The outer facade glorifies Amsterdam's trade with the four continents of Europe, Asia, Africa, and America, and features a large sculpture of Atlas carrying the world on his shoulders. Inside, the great Citizen's Hall is perfectly proportioned according to the classical Greek ideals revived and promoted by the architect Andrea Palladio. There, Wright admired the "Pavement Figure" — three circular maps in the marble floor, representing the eastern and western hemispheres and the stars of the northern hemisphere. (Paintings of the southern skies were added later. Halley had already charted some of the southern stars.) The corners of the hall bear depictions of Aristotle's four essential elements: earth, water, air, and fire. A guide to the building, now called the Royal Palace, sums up a description of the Citizen's Hall this way: "The decorative scheme of the entire room is based on the universe, with Amsterdam at its center."[17]

Wright probably would have enjoyed more travels abroad, but the life of a sailor didn't suit him. The sea disagreed with him, and he felt his duties were too dangerous. "A very bad, tedious voage [voyage]," he noted upon his return. "Very near being cast away (by Canting ye Ballast) in a very great storm."[18] He returned to Sunderland and settled down—to the extent he was capable of settling—as a teacher of mathematics and navigation.

In winter there were plenty of seamen around to enroll for his classes because coal ships were laid up in port for the season, and Wright prospered. He suffered only a temporary set-back in the summer of 1730; he had fallen in love with a "Miss E. Ireland," but, although she agreed to marry him, he could not win her father's approbation in the face of "two Rich Rivales." Wright bolted to London, "[t]he disappointment not siting easy" as he noted candidly.[19] Just as he was about to board a ship bound for Barbados, a friend of his father somehow found him out and prevented his departure. Wright tarried in London for a while, working for makers of mathematical instruments, then accepted money from his father for the trip home and returned to teaching mathematics and navigation.

After the loss of "E.," Wright appears to have made an effort not only to supplement his teaching income but to make a name for himself, at least locally. His efforts to create and distribute an almanac illustrate his tenacity and the supportive role of his family in this regard.

An annual almanac generally included the times of sunset and sunrise, a calendar of lunar phases, data on solar and lunar eclipses, and other information of interest to gentlemen-farmers or mariners. Some also included astrological prognostications. A well-known astronomical almanac, similar to the one Wright proposed, was available from Oxford, but Wright hit upon the idea of calculating his astronomical calendar to the longitude of

Durham, the nearest town to Byer's Green, and so tailoring it to the uses of his fellow north-countrymen.

Wright's calculations for 1732 were ready to be engraved and printed in the fall of 1731, but he had not allowed enough time for the printing and distribution. The Company of Stationers in London, the chartered organization that would have published the almanac, promised to do business with him the next year, if he could complete his manuscript earlier and come up with 500 subscribers. In 1732, he was ready in the spring with his calculations for 1733, and had found 900 subscribers. This time, however, the Stationers in London unexpectedly balked when he showed up in their offices, declining to compete with the venerable Oxford almanac. Wright demanded that they explain the situation in the Durham newspapers, to satisfy his subscribers there.

Wright was almost out of money when the Stationers turned him down, and he set out to cover the 200 miles home to Byer's Green on foot. Friends and relatives gave him shelter, food, and money along the way. "Meet with uncommon Sevilities [civilities] upon ye Road," Wright noted in his journal.[20]

Wright was determined to revise his almanac and have it printed in Scotland. His long-suffering father, who had bailed him out so many times before, did not think this plan worth pursuing, but Wright set out on foot again "with a small assistance from his [i.e., my] Mother and syster." Again his extraordinary luck on the road held out; he was accosted by two "Highway Men," but they turned out to be of the most sympathetic sort. As Wright tells the story, they "oblige him to sit down by them upon a green Hill, ask him many Question Relating to ye various states of Life, is satisfied with his Reasoning and answers, and makes no attempt to Rob him but directs him the Best Way."[21]

In Edinburgh, Wright encountered yet another delay in publishing his almanac, and yet another promise to publish the next year's. In the meantime, he published "with great Sucess" a calculation of the upcoming total eclipse of the Moon, on November 20, 1732, and was "very Fortunate in citing ye time." Unfortunately, the engraver and printer in Edinburgh proved to be "a Rogue" and took Wright's money without producing the second almanac, which Wright had re-calculated to the longitude of Edinburgh.[22]

By February 1733, a slightly older and much wiser Wright, back in England, had found employment as a companion to the Rector of Sunderland, and the tide of his affairs was beginning to turn. The commissioners of the river Wear, the port authorities, sponsored his invention of a "composition" or set of various types of sundials, and erected them on a pier in Sunderland. The local authorities paid for a printed description and explanation of them. And in the fall, at the age of 22, Wright finally achieved a kind of social and financial breakthrough: his employer Daniel Newcome, the Rector of Sunderland, introduced him to the Earl of Scarborough, who in turn invited him to London and became one of his patrons.

Early London years: astro-theological musings

Wright moved to London in the fall of 1733, and lived there for about 30 years before returning to the family estate at Byer's Green. It was in 1750 in London that he produced the innovative astro-theological and geometrical treatise for which he is now best known, *An Original Theory or New Hypothesis of the Universe*. According to the introductory pages to this work, he began constructing an astro-theological theory much earlier—in 1734, soon after moving to London, in fact—but for some reason he did not venture to publish his thoughts, eventually much revised in *Original Theory*, until he had seen several other, more purely astronomical, works in print.

During Wright's first year in London, 1733–1734, he appears to have been busy finding people who would pay in advance for, or subscribe to, a mathematical instrument he called the "*Pannauticon*." No copy of the *Pannauticon* survives, so its function is unclear, but a notice on one of Wright's later works informs us that Wright had "lately publish'd by Subscription the *Perpetual Pannauticon or Universal Mariner's Magazine*, being a Mathematical Instrument," and a different notice adds that the instrument explains "the Lunar Theory and motion of the Tides." A printed key to the instrument mentions that it consists of a number of drawings or "schemes" showing "divers Circles."[23]

In promoting his *Pannauticon*, Wright had help from one of the first people he met in London, Roger Gale, who was to become a good friend. Gale was an antiquary who served as Treasurer of the Royal Society, an organization of natural philosophers chartered in 1662 to further the study of nature and to foster experimentation and mechanical or mathematical invention. Gale introduced Wright to the Royal Society, which allowed Wright to communicate his *Pannauticon* to that august body. From then on, Wright garnered subscribers left and right; the Earl of Scarborough even obtained permission for Wright to dedicate his instrument to the king, George II, and procured for him a subscription from the Prince of Wales.

Wright's sparse journal entries for the winter of 1733–34 indicate that his efforts to find subscribers kept him busy throughout the season. In the spring of 1734, he was occupied with preparing the copper plates of his *Pannauticon* schemes for the press, engraving himself those he could not afford to have engraved for him, and seeing the work through publication. Thus it was probably in the late summer of 1734 that he had time to think about his astro-theological view of the universe. After delivering copies of the *Pannauticon* to subscribers in London, he traveled north to the Sunderland and Durham area to visit his family and the Reverend Newcome, and to deliver copies of the *Pannauticon* to subscribers there. He returned to London in the fall.

Whether in London or in the north of England at the time inspiration struck, Wright found himself mulling over the possibility of a multitude of worlds. The theoretical existence of other inhabited planets, or even of solar systems around other stars, had been debated for centuries. Aristotle had scoffed at the idea of an infinite universe replete with other Earth-like planets, but his authority on the subject had been contravened by the Bishop of Paris in 1277, who ruled that to deny the possibility of an infinite universe, or of other creations, was to limit God's power. The idea of other worlds became even more attractive in the sixteenth century after Copernicus argued for a Sun-centered system of planets; indeed it was difficult to see the stars as other suns without contemplating also the likelihood of other planetary systems.

The problem for Wright was to locate Heaven and Hell in such a populated space. Some of the writers he admired proposed that Hell lay in the infernal center of the Sun, and the "Throne of

God" beyond the stars — in which case one might infer that each solar system had its own Hell, but shared the region of Heaven with other solar systems. Wright thought otherwise. The hypothesis he developed at this time presented a highly structured, symmetric universe centered on a "Sacred Throne of Omnipotence" and with room for "myriads" of planetary systems distributed around their respective suns.[24]

As was by now his habit, Wright created a drawing to explain the details of his hypothesis: for him, to think was to draw. Indeed, many of his publications were primarily artistic compositions, with the accompanying text, if any, playing the subordinate role of an explanation or key to the drawing. In the 1730s particularly, his aim seems to have been to provide a synoptic view of an entire field on very large pieces of paper. In 1731, for example, before he left the Sunderland area for London, he conceived "a General Representation of Euclid's Elements in one Large Sheet: and the Doctrine of Plain and Spherical Trigonometry all at one View, on an other." In 1737, he was to create a work titled *The Universal Vicissitude of Seasons*, which he described as "exhibiting by inspection at one view, the various rising and setting of the Sun to all parts of the World, with the hour and minute of day-break, length of day, night and twilight etc every day in the year."[25]

To illustrate his hypothesis about the plurality of worlds, Wright created nothing less than a cross-sectional view of the entire universe. Unfortunately, the illustration is no longer extant. However, we can glimpse what he had in mind from two documents relating to the hypothesis, the latter possibly dating to 1738, and some undated sketches found among his papers. These two documents provide an important framework for understanding not only the 1730s hypothesis, but also the theory of the universe that was to follow in 1750. In both the earlier and later versions of his hypothesis, the Sun is embedded in a spherical shell of stars.

A sketch (figure 3.2), probably made later for the *Original Theory*, depicts two views of a universe similar to the one Wright conceived of in the 1730s. In the bottom view we see a small sphere at the very center, emblazoned with a triangular symbol similar to one Wright described in one document as a "Hyroglyphic" or "Emblematic Trigon" representing the divine

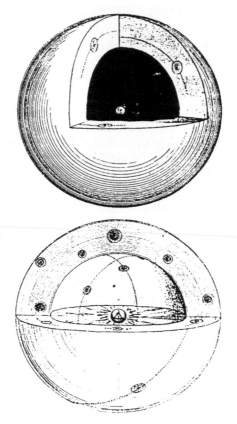

Figure 3.2 Two views of a "universe" of stars arranged in a spherical shell about a divine center. Undated sketch found among Wright's papers, probably made in preparation for his *Original Theory* of 1750. (Reproduced with permission from Hoskin (1971).)

presence.[26] At some distance from this seat of God is a shell filled with stars, each apparently with its own planetary system shown by a set of circles. Dotted lines show the orbits of the stars, which are not confined to a plane as the planetary orbits are in his scheme, but circle the inner sphere in all directions, like strands of yarn in a knitting ball. The top view is a variant on the bottom one.

The first document describing the idea is a single paragraph, a brief description of his solution to the problem. The solution is "represented in a section of ye Universe twelve feet radius, extending from the Imperial Seat or *Sedes Beatorum* to ye verge of chaos bordering upon ye infinite abiss." Surrounding the *Sedes Beatorum*, also called the Sacred Throne of Omnipotence, is a "Region of Mortality" in which planetary bodies such as the Earth, together with their suns, "circumvolve" around the divine center. Enclosing both these concentric spheres is dark space, "supposd to be the Desolate Regions of ye Damnd."[27] As Wright makes clear, his "section" of the universe shows but a slice of the total, like the spoke of a wheel, with the central *Sedes Beatorum* on one side of the broadside and the dark and desolate regions at the other.

The second document, which reads like a lecture interpreting his mathematical and emblematical figures, seems to refer to at least two posters, or perhaps the 12-foot section drawing and an accompanying scheme.[28] This text specifies that the Sun, along with the other stars, is in orbit about the *Sedes Beatorum*, and is to be found "near ye center of ye middle region," i.e. near the middle of the spherical shell of Mortality enclosing the inner divine sphere. The stars filling the region of Mortality include those visible from Earth; Wright notes that the brightest, those of the first magnitude, are closest to our system and the rest "proportionable removed" according to their appearance. Those stars visible to the naked eye all lie within a circle of a certain radius around the Sun. Beyond them, Wright says, are the telescopic stars and beyond them more stars, "by no means perceptible to ye human eye."[29] Thus the inhabitants of Earth are aware only of the nearest stars in the Region of Mortality, and cannot directly see even the bright glow of the divine center, or the dark empty spaces of the abyss.

This hypothesis made it possible for him to preserve a structure and spherical symmetry to the universe, and to account for a plurality of worlds—although not, strictly speaking, an infinite number, since the volume of space in the Region of Mortality is limited. In envisioning the stars as being in orbit around a divine center, and not static, his hypothesis also incorporated the recently discovered phenomenon of moving stars.

The stars had for centuries been viewed as fixed in place, forming an unchanging backdrop to the comings and goings of the Moon and planets. However, in 1717, Halley had discovered that some stars had moved since their positions were recorded by Hipparchus. Reviewing Hipparchus's data in Ptolemy's catalog, Halley found that the stars Aldebaran (then known as Palilicium), Sirius, and Arcturus had shifted their positions in ways that no other stars had, and the shifts could not be explained by any known instrumental or atmospheric factor.[30] Perhaps, the feeling was, all the stars move, but long intervals of time were required to detect the movement, and only those stars nearest to us have a great enough displacement to be discerned over a span of several hundred years. For some astronomers and philosophers, including Wright, these motions explained the stability of the universe; the stars were not compelled to drift into one large agglomeration over time, under the influence of gravity, since they were constantly in motion and experiencing a continuously changing gravitational environment from the other stars.

However pleasing Wright found his vision of the grand plan of the universe, incorporating spherical symmetry, a plurality of worlds, and moving stars, he apparently kept it to himself or shared it only with friends for the time being. Perhaps his admiration for the Royal Society, and his desire to be taken seriously by its members, made him reticent on the subject of astro-theology. A draft of the Royal Society's statutes written in 1663 stated explicitly that its business was "not meddling with Divinity, Metaphysics, Morals," or other unscientific fields such as politics.[31] Perhaps he simply lacked the nerve to assert himself as an authority in metaphysics. Whatever the reason, Wright's early drawings representing the moral and physical world together never circulated as a book. When he finally published a revised version of his theory in 1750, it was in part because he believed he had solved a very long-standing, purely astronomical problem: how to account for the appearance of the Milky Way.

Man of many talents

Life in London was good. Wright settled into lodgings in Piccadilly, the fashionable area near St. James' Palace, at that time the official royal residence. Having brought his *Pannauticon* project to a successful conclusion, he delved into the vibrant social and intellectual life of the city, giving a course of lectures in astronomy at Brett's coffeehouse in Charles Street near St. James's Square. Thinkers and businessmen liked to congregate in these types of establishments, which had been popular since the 1600s, to hold meetings, read a newspaper, or listen to a lecture. (Indeed, in later times, coffeehouses were sometimes known, somewhat pejoratively, as penny universities, because of the low cost of admittance.)

Wright readily made friends and found patrons, besides the Earl of Scarborough, in this new environment. The Earl of Pembroke became one of Wright's chief supporters, granting Wright the use of his library and nurturing his interest in architecture. John Senex, a well-known mapmaker who had published Halley's celestial and terrestrial maps and Whiston's astronomical diagrams, commissioned work from Wright and sent private students his way. The Duchess of Kent took Wright under her wing, ensuring that Wright moved in fashionable circles.

Gradually Wright evolved a routine of teaching private students mathematics or astronomy in winter and visiting aristocratic families, often at their country estates, for weeks at a time the rest of the year. He found himself in demand as a private tutor, mainly to ladies, for whom it was fashionable to study mathematical sciences and the use of globes. Thus he taught geometry to a Mrs. Townshend, the daughters of the Duke of Kent, and the daughters of Lord Cornwallis, and under his tutelage, the Duchess of Kent surveyed her garden and made a plan of it.

Wright evidently owned or had access to a number of scientific instruments, for he made astronomical observations and conducted several of his own surveys during his London years. His journal mentions no telescopes, but some of his works and magazine articles mention lens-based and mirror-based telescopes—a "tube of two convex glasses" in one instance, a "five Foot Focus Reflector" in another—most likely less than about 2.5 inches diameter aperture, affording a view of ninth or tenth magnitude stars.[32] He used these telescopes to observe the stars, and also to study the appearance of comets. For surveying, he would have used a theodolite, a telescope equipped with a scale to measure precisely the horizontal or vertical angle between two observed objects.

Even before he left northern England, Wright had discovered that surveying could be a profitable sideline for someone with his knowledge of astronomy and mathematics. The owners of country mansions were just beginning to be swept up in the eighteenth-century English craze for "naturalistic" gardens enlivened with serpentine ponds, grottos, and artfully placed groves of trees; this was the era of gardening as landscape painting, with allusions to classical themes. On an occasional basis, Wright surveyed and made maps of the country estates of his friends and patrons, laying the basis for their landscaping plans. This type of mapmaking exercise no doubt contributed to Wright's unique combination of interests and spatial skills that meant he was one of the first to consider the effect of geometrical perspective on our perception of the universe of stars.

The technique of triangulation, obtaining the distance of a marker by observing it from two different positions, is at the heart of surveying. A surveyor first lays out a baseline, a straight line of known length (see figure 3.3). He views the object whose distance is to be determined from the two ends of the baseline in turn. A distant object forms the apex of a long triangle, while a nearby object forms the apex of a short, squat triangle. The known length of the baseline, and the measured angles between the baseline and the lines of sight to the marker, pinpoint the location of the marker. As Wright roamed the gardens he surveyed, mapping the locations of trees and ponds and considering the views from different angles, he developed the art of mentally relating three-dimensional landscapes to two-dimensional maps—an art that would later inform his efforts to understand the structure of the Galaxy.

Teaching, visiting friends and patrons, and surveying estates kept Wright pleasantly occupied and intellectually stimulated. Particularly during his first few years in London, he seems to have experienced a surge in creativity. He devoted his spare time to his old passion of astronomical and architectural drawing. Some of his projects existed only in design, on copper plates and prints, and some he executed. One of his largest sculptural works was undoubtedly his model of the solar system, a "system of ye Planetary Bodies in true Proportion Equal to a Radius of 190 feet. (all Brass)," which he presented to the Earl of Pembroke. He designed a "Hemesphereum," probably a domed ceiling with astronomical embellishments, and drew plans for different kinds of sundials. He devised something he called an

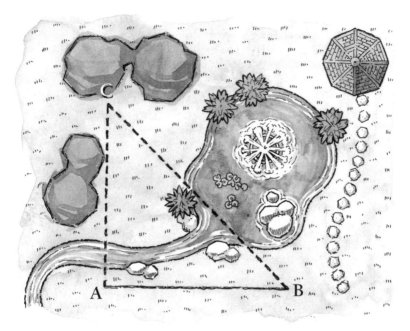

Figure 3.3 Technique of triangulation. A surveyor lays out a baseline AB of known length. From each end of the baseline, the angles to the object at C are measured. (If the length of the baseline AB is known, and the angles CAB and CBA are known, the distance AC or BC can be found from geometry, without actually pacing it out.) (Credit: Layne Lundström.)

"Astronomical Fan." In 1737 his journal records, "His invention at this time run to(o) fast for Execution."[33]

Wright's mother died in 1741 and his father in 1742, so he could no longer turn to them for support. But by this time he was well established in London, and had found his niche in society.

Astronomy and astro-theology

A number of Wright's "schemes," as he called his illustrations, served as the basis for didactic works on astronomy, including *The Universal Vicissitude of Seasons*, articles about comets in the

Gentleman's Magazine, and diagrams explaining the appearance of lunar and solar eclipses. These broadsides would have made impressive visual aids for a lecturer, and are works of art in their own right.

In 1742, he finished a set of four plates, each 2 feet by 3 feet, illustrating a large number of basic astronomical phenomena. One of them shows, in the center foreground, the solar system—including not only the Sun and planets, but also many comets, and moons of Jupiter and other planets. In the corners of the plate, as background to the current conception of the solar system, he represented the Pythagorean, Ptolemaic, and other historical systems. The top of the plate bears a depiction of the Sun, and the bottom a map of the Moon.[34]

The set of four plates formed the basis of an elementary astronomy textbook. The full title of the work indicates that the text was merely the key (Latin "clavis") to the diagram: *Clavis Coelestis, Being the Explication of a Diagram entituled a Synopsis of the Universe: or, the Visible World Epitomized*. Although Wright's illustrations are striking, the content of the work is mostly conventional. The *Clavis Coelestis* treats "plainly and simply," as its author says, of the planets, laws of motion, properties of light, the cause of seasons, calendars, tides, and the phenomena of eclipses and occultations. On the subject of a plurality of worlds, Wright is succinct. He says simply that all modern astronomers consider the stars to be "great Globes of Fire like the sun," and that they "may very possibly be the Centers of other Systems of Planets like ours, since we have no Reason that can contradict it, and many that may induce us to believe it."[35]

One unusual aspect of Wright's overview of the solar system in *Clavis Coelestis* reminds us that he always had effects of geometrical perspective in mind. He not only gives the basic astronomical data on each of the planets—such as orbital period and distance from the Sun—but also conjures up for his readers what the view of the night sky would be from another location. He notes for example that "To Mercury, the Sun and Venus are the only two great Bodies of the Universe. He views Venus and all the rest of the Planets, as we do Saturn, Jupiter and Mars; but Venus shines upon him with great Lustre, and 'tis probable, her great Light in opposition to the Sun, serves him instead of a Moon."[36]

After the publication in 1742 of his *Clavis Coelestis* and accompanying diagram, and an account of a comet which he furnished to the *Gentleman's Magazine*, Wright's interest in astronomy *per se* appears to have waned. His situation was secure enough that he refused an offer that same year to become "Chief Professor of Navigation" at the Imperial Academy of St. Petersburg. He continued to teach, draw, and survey, but he also found time for touristic travel to sites such as Stonehenge and the garden at Moor Park.

In 1746, a trip to Ireland shifted the main focus of his enthusiasms to antiquities, the relics and monuments from the Stone Age to medieval times. He stayed in Ireland for several months as a guest of Lord Limerick, Governor of County Louth, and Bishop O'Gallagher of Raphoe, a town in County Donegal just a few miles from the prehistoric Beltony stone circle. Wright had a gift for sketching "on the spot" both the perspective views and cross-section plans of the castles and remains that he saw (see figure 3.4 for an example of such a combination drawing, made on a different tour, in England). Upon his return to London in the summer of 1747, he began preparing what was to be his most popular work in his lifetime: *Louthiana*, published in 1748, a description of the antiquities of County Louth. A second edition of this successful work was to follow in 1758.

Only after the success of *Louthiana* did Wright publish his astro-theological theory — or rather, a revised version of it — in the *Original Theory* of 1750. His journal entries end with the Irish tour, so we have no information on his activities immediately leading up to the publication of the *Original Theory*. What is clear is that something prompted Wright to ponder the Milky Way, and to try to reconcile this apparent ring of light circling the celestial sphere with his conviction that the stars of all creation orbit a divine center. Thinking like a clockmaker, a surveyor, and an architect, he kept turning the problem over in his mind: what does the appearance of the night sky, and in particular the swath of densely clustered stars in the Milky Way, say about the structure of the universe? In the *Original Theory*, he finally arrived at a world view that included the important elements of his astro-theological theory of the 1730s, but that also accounted for the appearance of the Milky Way based on geometrical arguments.

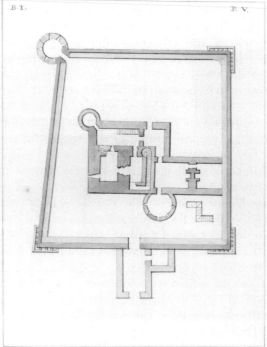

Figure 3.4 Wright perspective and plan drawings. Wright's artistic talent is evident in sketches and plans he made "on the spot" on tours of historic sites in England and Wales. (Reproduced with permission from the British Library.)

He wrote of his motivation: "This luminous Circle has often engrossed my Thoughts, and of late has taken up all my idle Hours; and I am now in great Hopes I have not only at last found out the real Cause of it, but also by the same Hypothesis, which solves this Appearance, shall be able to demonstrate a much more rational Theory of the Creation than hitherto has been any where advanced."[37]

The *Original Theory* takes the form of nine letters addressed to an unnamed friend, and a preface addressed to the public in which Wright begs "every kind of Indulgence" for offering a work "entirely upon a new Plan."[38] Wright is conscious of breaking new ground merely by considering the problem of the Milky Way. He notes with justification that "this amazing Phaenomenon which have been the Occasion of so many *Fables*, idle Romances, and ridiculous Opinions amongst the Antients, still continues to be unaccounted for, and even in an Age vain enough to boast Astronomy in its utmost Perfection."[39] Indeed, one of the mysteries of the history of astronomy is how astronomers from Ptolemy to Aristotle to Galileo could have given the Milky Way what one historian has called the "silent treatment." Ptolemy contented himself with a description of the Milky Way and its location in the sky; Aristotle gave a patently unsatisfactory explanation of it as a kind of fog. Even Galileo, who saw it through a telescope and found it to be nothing but a congeries of stars, had nothing more to say about this unique clustering phenomenon.

Wright approaches the solution to this long-dormant problem slowly, guiding his reader through a number of concepts vital to an understanding of his hypothesis. His solution — his "original theory" — is based on the spherically symmetric universe that he conceived of in the 1730s, but this of course was his private rumination, not generally known. Thus the first letter seeks merely to convince his friend that the idea of a multitude of suns and planetary systems is not far-fetched or controversial. "[T]hat the Stars are all Suns, and surrounded with planetary Bodies...is not a Thing merely taken for granted, but has ever been the concurrent Notion of the Learned of all Nations," Wright says.[40] He quotes Newton and Derham, the author of *Astro-Theology*, among other astronomer-philosophers, to emphasize that a Creator whose power and wisdom are without bounds may very likely create multiple systems. His interlocutor appears to have had some difficulty accepting this premise, for Wright was obliged to revisit the problem. Letter the Fourth begins, "Sir, You tell me you begin to be a tolerable good *Copernican*, and would now be glad to have my Opinion further upon the Nature of the Sun and Stars, with regard to the Suggestion of their being like Bodies of Fire. This you say will go a great Way towards confirming you in the Notion you have begun to embrace of a Plurality of Systems, and a much greater Multiplicity of Worlds than our little solar System can admit of."[41] If nothing else, this repetition of themes reminds us that the opinions of "professional" astronomers in Wright's day did not necessarily carry much weight with the general public, and that Wright was more mindful of this public than of astronomers who might be reading his work. Wright's style, too, shows that he was tailoring his arguments to the poets and philosophers among his friends. In contrast to the *Clavis Coelestis*, which is written very straightforwardly and concisely, the *Original Theory* is rife with literary quotations.

In Letter the Second, Wright addresses his friend's concern for method. The ideas of the *Original Theory* will not stand up to purely mathematical arguments or "infallible Demonstration," Wright admits, but he intends to make use of analogy as an alternative, if weaker, way of reasoning. In particular, Wright attempts to convince his friend and reader that when spherical symmetry prevails in a system, it is possible to infer facts about the whole system from only a partial view. "[F]rom a very small part of *orbicular Things*, we are able to determine the Form and Direction of the Whole," Wright notes.[42]

By way of introduction to the science, Letter the Third lays out the "now-established Astronomy of Copernicus."[43] Wright reviews the principal facts about the planets—their orbital parameters, and their relative sizes—in the Sun-centered system and describes the trajectories of comets. This information is not critical to an understanding of his explanation of the Milky Way, except perhaps in the fact that the comets, with their orbits taking them far out of the plane of the solar system, provide a kind of precendent for assuming that the stars may have similar orbits about the divine center, as Wright imagined in his 1730s hypothesis. Letter the Fourth, as we have noted, returns to the arguments for the stars as suns with their own planetary systems, all at very great distances.

The fifth and sixth letters discuss the Milky Way, laying the groundwork for Wright's main hypothesis, which is contained in Letter the Seventh. Wright reminds his reader that the Milky Way circles the celestial sphere, nearly bisecting it, and that it is "very irregular in Breadth and Brightness, and in many Places divided into double Streams"[44] (see chapter 2). He recalls the fanciful explanations of this zone of light given by ancient peoples: it represents the soldering of the two hemispheres of the sky, or "*Juno's* Milk, spilt whilst giving Suck to *Hercules*," or the track of Phaeton's wild ride with the chariot of the Sun.[45]

That the Milky Way actually consists of a large number of "small" or distant stars, Wright can argue from his own observations with "a very good Reflector" telescope.[46] The light from these stars crowding together in the distance, Wright says, is combined; the rays of light coalesce so that the overall effect is like a river of milk. This phenomenon is repeated on a smaller scale in a number of nebulae, which, Wright asserts, are known to consist of stars. He mentions for example the nebula we now call the Beehive Cluster in the constellation Cancer, an open cluster which he believed contained 36 stars, and which we now suspect contains more than 200. He also quite properly lists the southern hemisphere's Magellanic Clouds, which he himself had never observed, as examples of cloudy regions consisting of many stars close together.

To give some idea of the distances of the stars, Wright invokes an experiment by the seventeenth-century astronomer Christiaan Huygens. Huygens had compared the brightness of the Sun and Sirius by covering his telescope, while it was pointed at the Sun, and allowing the light to enter only through a small hole. Judging from the amount by which he had to reduce the light from the Sun for it to appear comparable to the light from Sirius, and knowing how light diminishes with distance from the source, he inferred that Sirius must be at least 2 trillion miles away. Wright admits that these figures boggle the mind; he notes in connection with this that "few People can range their Ideas with such Perspicuity, as to arrive at any adequate Notion of any Number above a thousand."[47]

In Letter the Sixth, Wright asks his reader to grant him one postulate: "That all the stars are, or may be, in Motion."[48] In fact, no less an authority than Halley provides the evidence that this is so. Wright quotes from Halley's paper in which the astronomer announces the discovery of what we now call the proper motion of stars. As mentioned earlier, Halley had been studying the star catalogs of Hipparchus as transcribed by Ptolemy, and had noticed that three stars—Aldebaran, Sirius, and Arcturus—had shifted their positions by more than half a degree in the intervening centuries. These shifts could not be accounted for by inaccuracies in observation, errors in transcription, or systematic changes in the positions of all stars due to a wobble in the Earth's spin axis. For Wright, as for many astronomers, the news suggested that all stars move. Only the nearest could be seen to move appreciably over the time span of human observation. Wright incorporated this astronomical fact in envisioning the stars as in orbit around a divine center.

Throughout these preliminaries Wright has hinted that the apparent lack of order and symmetry in the distribution of stars is the effect of our geometric perspective, and that from another location, we might view the disposition of the stars as orderly. In Letter the Seventh he states this explicitly. As viewed from the Earth's off-center or eccentric position in the solar system, he reminds his reader, the orbits of the planets are complex and their motions appear irregular. When viewed from the perspective of the Sun, at the center of the system, the planets are seen to move in relatively simple, orderly paths. Similarly, he says, "nothing but a like eccentric Position of the Stars could any way produce such an apparently promiscuous [i.e., disorderly or random] Difference in such otherwise regular Bodies." There may be "one Place in the Universe," Wright suggests, from which the order and motions of the stars appears "most regular and most beautiful."[49] The favored perspective belongs, of course, to the divine center of a spherical universe. The illustrations to Letter the Seventh show a universe much like that he had described in the 1730s, with a divine presence at the center, surrounded by a spherical shell of stars orbiting in all directions.

The explanation for the Milky Way, the very heart of Wright's original theory, emerges from a consideration of the view from within the shell of stars. To drive home his point, Wright draws a close-up view of the shell of stars, showing a star such as the Sun near the middle of its thickness (see figure 3.5). Because the shell's thickness is small compared to the

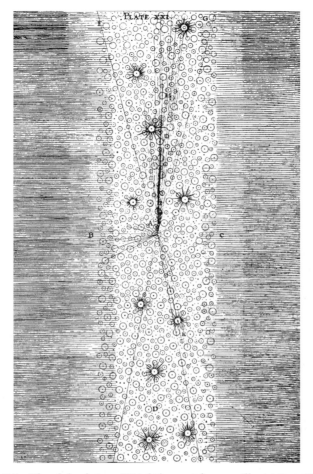

Figure 3.5 The slab of stars, Wright's most famous illustration. The slab represents a close-up view of a small section of the entire system shown in figure 3.6. The slab, in other words, is a section of a thin shell which lies at a great distance from the divine center of the "creation." (Reproduced with permission from Hoskin (1971).)

radius of the sphere it defines, the curvature of the shell is barely perceptible, and the close-up view is a flat segment, with two parallel lines delineating the inner and outer surfaces of the shell. This close-up view looks very much like a slab of stars and, based on this diagram, later generations of astronomers

61

sometimes mis-attributed to Wright the first modern conception of the Milky Way galaxy as a flat disk. However, it is clear from the text that Wright intended the slab to represent a small section of the "orbicular" or spherical system he had in mind.

Imagine the stars scattered about and filling the space in this segment, Wright asks his reader: "consider what the Consequence would be to an Eye situated near the Center Point, or any where about the middle Plane, as at the Point A" (in our figure 3.5). An observer would see the stars "promiscuously dispersed on each Side, and more and more inclining to Disorder, as the Observer would advance his Station towards either Surface, and nearer to B or C."[50] That is, if the Sun and Earth were in the shell, but near the surface at B, the distribution of stars would be very uneven, with most lying in one hemisphere. If the Sun were near the middle of the starry region, all parts of the sky would contain stars.

Looking into the shell, on a tangent to the great sphere surrounding the divine center or in the direction of H or D, the observer would see the effect of innumerable stars extending to a great distance. Another of Wright's illustrations (see figure 3.6) shows the entire system from which the cross section was taken, and lines AD and AE, showing the direction in which the Milky Way would be seen. (In this figure, Wright has chosen to show two concentric shells of stars about the divine center, or what he calls "a Creation of a double Construction."[51] Since an observer near the Sun sees only a limited region of his own shell of stars, indicated by the circle at G in the upper panel of our figure 3.6, there is no observational constraint to having more than one shell of stars.) In one long sentence, Wright summarizes the effect of the crowding of stars as seen from an observer at A: "Thus, all their Rays at last so near uniting, must meeting in the Eye appear, as almost, in Contact, and form a perfect Zone of Light; this I take to be the real Case, and the true Nature of our Milky Way, and all the Irregularity we observe in it at the Earth, I judge to be intirely owing to our Sun's Position in this great Firmament, and may easily be solved by his Excentricity, and the Diversity of Motion that may naturally be conceived amongst the Stars themselves, which may here and there, in different Parts of the Heavens, occasion a cloudy Knot of Stars, as perhaps at E."[52] In other words, the irregularity of the distribution

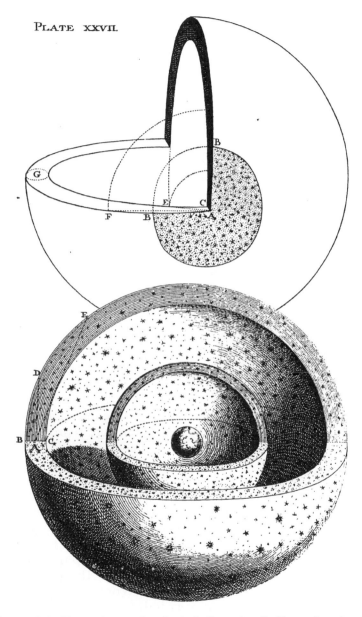

Figure 3.6 Two views of spherical "creation." (Reproduced with permission from Hoskin (1971).)

of stars in the sky, with a large number aggregated in the zone we call the Milky Way, is the effect of our immersion in the shell, which puts us in an eccentric position; viewed from the divine center of the system, the distribution of stars is actually regular and symmetric. Furthermore, as viewed from the Earth, the chance alignment of a number of stars as they pursue their separate orbits in the shell, around the divine center, may give rise to the appearance of a nebula or star cluster.

Wright is evidently highly pleased with this elegant and genuinely original solution to the problem. But he has to admit that the same logic would apply to an alternative distribution of stars about the divine center, one with the symmetry of a ring rather than a thin spherical shell. The phenomenon of the Milky Way may equally be accounted for, he notes, if the stars are arranged "in the Manner of *Saturn's* Rings."[53] Figure 3.7 shows a top view of such a system, with the extent of the *visible* creation from an Earth-like planet limited to a shaded area. Indeed, if more than one creation is allowed, Wright is inclined to believe that there may be various systems of stars, some arranged spherically and some in rings about their respective divine centers. But it is clearly the spherical system that he prefers, and dwells on.

In Letter the Eighth, Wright tries to give his friend, to whom the letters were originally written, some idea of the scale of the solar system; apparently the friend had wondered why the Earth was not shown explicitly in Wright's schemes of the entire creation. Wright answers that if the Sun were represented by the dome of St. Paul's church, a sphere 18 inches in diameter in the West End neighborhood of Marylebone could stand in for the Earth, and a globe of about 12 feet diameter in the town of Chelmsford, some 50 miles from London, for Saturn, the last known planet. But, Wright notes, "if you will take into your Idea one of the nearest Stars; instead of the Dome of St. Paul's, you must suppose the Sun to be represented by the gilt Ball on top of it, and then will another such upon the top of St. Peter's at Rome represent one of the nearest Stars."[54] Thus in drawing up his general scheme of the universe, Wright says, he judged the seat of the Earth to be "of very little Consequence."[55]

In the concluding Letter the Ninth, Wright returns to the idea of a multitude of worlds. From our perspective in the twenty-first

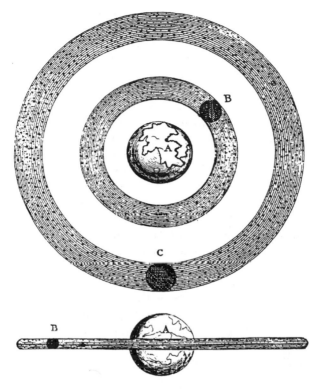

Figure 3.7 An alternative "creation" imagined by Wright; the stars lie not in a spherical shell, but in one or more rings around the divine center. (Reproduced with permission from Hoskin (1971).)

century, one of his more interesting speculations is that "the many cloudy Spots" in which "no one Star or particular constituent Body can possibly be distinguished" are "external Creations, bordering upon the known Creation," or, if we can be permitted to equate his creations with stellar systems, other galaxies.[56]

The rest of Letter the Ninth is given to a consideration of the moral aspects of his hypothesis. The contrast with the more straightforward account of the spherical structure of creation is a bit jarring, but it serves to remind us that Wright's purpose was to integrate a physical and moral view of the universe. In the focus or center of Creation he sees a primitive fountain, overflowing with divine Grace, and it is here that "the virtues of the meritorious are at last rewarded and received into the full Possession of every Happiness, and to perfect joy." The ideal worlds that Wright imagines have been created for the reward of the meritorious bear a striking resemblance to the magnificent "naturalistic" gardens he surveyed and helped plan among his friends in the aristocracy; he sees worlds "fill'd with Grottoes and romantick Caves," and others "with vast extensive Lawns and Vistoes, bounded with perpetual Greens, and interspersed with Groves and Wildernesses, full of all Varieties of Fruits and Flowers."[57]

Kant reads Wright

Far from London in the East Prussian capital of Königsberg, the philosopher Immanuel Kant, 13 years younger than Wright, read a review of Wright's *Original Theory* of 1750. The review appeared in a periodical, and summarized Wright's models without reproducing his engraved illustrations. Therein lies the root of a deep misunderstanding, because without Wright's illustrations to guide him, Kant formed an erroneous impression of the details of Wright's models.

Kant's imagination was fired by a statement in the summary comparing the system of stars to our solar system of planets orbiting the Sun. The summary did not properly emphasize that in Wright's view, the stars orbit their divine center—a spiritual space, devoid of matter—and do so at some distance from it. Kant immediately conceived of a disk system, similar in fact to the modern conception of our galaxy, in which the stars are spread throughout the system. In his cosmological dissertation *Universal Natural History and Theory of the Heavens*, printed anonymously in 1755, he wrote:

"Mr. Wright of Durham, whose treatise I have come to know from the Hamburg publication entitled the *Freie Urteile*, of 1751, first suggested ideas that led me to regard the fixed stars not as a swarm scattered without visible order, but as a system which has the greatest resemblance with that of the planets; so that just as the planets in their system are found very nearly in a common plane, the fixed stars are also related in their positions, as nearly as possible, to a certain plane which must be conceived as drawn through the whole heavens, and by their being very closely massed in it they present that streak of light which is called the Milky Way."[58]

Because the Milky Way appears to encircle the sky, Kant added that "our sun must be situated very near this great plane," or be part of it.

Having reasoned to his satisfaction that the stars form a disk-shaped system, and that a stellar disk surrounding us accounts perfectly for the appearance of the Milky Way, it was a short step for Kant to propose that similar stellar systems, which we would now call galaxies, dot the infinite space of creation. He had read descriptions of "nebulous stars," and believed these were best explained as distant stellar agglomerations. These stellar systems might also be disk-shaped; indeed, some references in the astronomical literature seemed to provide support for his idea.

In particular, Kant had read of observations by the astronomer Pierre de Maupertuis. De Maupertuis had represented nebulous objects, Kant noted, as "small luminous patches, only a little more brilliant than the dark background of the heavens; they are presented in all quarters [of the sky]; they present the figure of ellipses more or less open; and their light is much feebler than that of any object we can perceive in the heavens."[59]

De Maupertuis's description of the shape of nebular stars, ranging from roughly circular to elongated or elliptical, seemed to fit what one would expect for distant stellar disks — not, as Wright would have it, spheres or rings — observed from a variety of viewpoints. If viewed edge-on, the disk of stars would appear highly elongated; if the disk happened to face the observer, it would appear round (see figure 3.8). Disks at intermediate angles, neither face-on nor edge-on, would appear elliptical. The feebleness of the light simply meant, to Kant, that the systems were at "inconceivable" distances.

"I easily persuaded myself that these ['nebulous'] stars can be nothing else than a mass of many fixed stars," Kant wrote in his preface.[60] Without bothering to check them out for himself, the philosopher admitted them as evidence and laid out his case.

"I come now to that part of my theory which gives it its greatest charm, by the sublime ideas which it presents of the plan of creation," Kant declared in the first chapter. He asked his

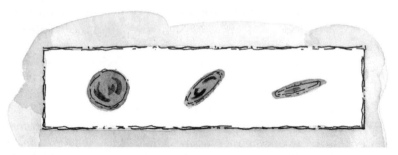

Figure 3.8 A disk of stars, or galaxy, viewed face-on (left view) and edge-on (right view). The middle view is for an intermediate viewing angle. (Credit: Layne Lundström.)

reader to imagine the consequences if "a system of fixed stars which are related in their positions to a common plane, as we have delineated the Milky Way to be, be so far removed from us that the individual stars of which it consists are no longer sensibly distinguishable even by the telescope." He concluded, "if such a world of fixed stars is beheld at such an immense distance from the eye of the spectator situated outside of it, then this world will appear under a small angle as a patch of space whose figure will be circular if its plane is presented directly to the eye, and elliptical if it is seen from the side or obliquely. The feebleness of its light, its figure, and the apparent size of its diameter will clearly distinguish such a phenomenon when it is presented, from all the stars that are seen single."[61]

So an early version of the concept of "island universes" hatched, a product of Wright's pioneering effort to explain the appearance of the Milky Way and Kant's bold synthesis of observation and theory. Kant's argument is not as strong as it might at first seem, in adducing de Maupertuis's data to support the concept of a multitude of stellar systems similar to our own. It turns out that only one of the "nebulous stars" described by the astronomer—the Andromeda Nebula—is, in fact, a galaxy, a stellar system comparable to our own; the other nebulous objects included a small number of globular clusters and the Orion nebula, which hardly anyone could describe as round or elliptical. Further examples of true "island universes" or galaxies besides our own weren't published until 1755. Nevertheless, both Wright's and Kant's insights, riddled as they were with misunderstandings, represented real progress in scrutinizing what lies beyond the solar system.

Kant, like Wright, had no doubt that his effort to explain the appearance of the Milky Way was new. Kant was extremely pleased with his ideas, but so mindful of their simplicity that he wondered why no one else had come up with them. He remarked that Wright did not sufficiently observe the potential applications and ramifications of his stellar system model, and took all astronomers to task for not even considering the problem posed by the Milky Way. In his characteristically long-winded style Kant wrote:

"Whoever turns his eye to the starry heavens on a clear night, will perceive that streak or band of light which on account of the multitude of stars that are accumulated there more than elsewhere, and by their getting perceptibly lost in the great distance, presents a uniform light which has been designated by the name Milky Way. It is astonishing that the observers of the heavens have not long since been moved by the character of this perceptibly distinctive zone in the heavens, to deduce from it special determinations regarding the position and distribution of the fixed stars."[62]

Second thoughts

In 1762, unaware of the extended lease on life Kant was giving his *Original Theory*, Wright moved back to Byer's Green. His parents had both died about 20 years earlier, and the house where Wright was born had passed to his elder brother John. Wright bought the property in 1755 and made room for his own stylish gardens and plantations by purchasing land from his neighbor. He pulled the existing house down and built one of his own design, a small Roman-style villa with a front terrace from which he could watch the Sun set. A friend later described how his unusual house reflected its inhabitant's unusual personality: "There was something flighty and eccentric in his notions, and a wildness of fancy followed even his ordinary projects; so that his house was not built or fitted up, upon the model, or in the manner, of other men's buildings."[63] Wright's own description of the house mentions "a laboratory for the purposes of mechanical and other experimental philosophy" and paintings and prints or mottos decorating virtually every wall, ceiling and staircase.[64]

Wright lived there comfortably almost 30 years, until his death in 1786 at age 74. He designed buildings and gardens, made occasional astronomical or meteorological observations, and assembled his papers for publication. He raised money among local gentlemen to restore a nearby Roman "circus" or race-track. About 1764, a natural daughter of his, Elizabeth, was born, but he appears to have remained unmarried and rather isolated. Some of his former students who were in the area made a habit of inviting him to dinner, but he enjoyed his "perfect tranquillity" among his books and prints.[65]

In the last two or three years before Wright's death, articles by the astronomer William Herschel, the subject of the next chapter, appeared in the *Philosophical Transactions* of the Royal Society. Wright does not appear to have read these articles on the shape and extent of the stellar system, or to have commented on them. Perhaps he was no longer following developments in astronomy as closely as he once had. In any event, in a curious twist, Wright had by then changed his mind about the structure of the moral and physical universe he had described in the *Original Theory*.

Among Wright's papers from his later years at Byer's Green is a set on *Second or Singular Thoughts Upon the Theory of the Universe*, written sometime after 1771. It is difficult not to be disappointed at his second thoughts, because they involve a return to the medieval notion of a solid framework for the celestial sphere. In this later model, which seems clumsy compared to the ethereal spheres of the *Original Theory* model, the stars are the mouths of volcanoes embedded in the solid sky. Variable stars, that is, stars whose brightness changes over months or years, arise from active volcanoes, and comets are volcanic ejecta. The Milky Way, in this model, is "a vast chain of burning mountains forming a flood of fire surrounding the whole starry regions."[66]

As he acknowledges in his manuscript, Wright formed his new ideas not as the result of "labour, or intense study, nor indeed any other intirely new discoveries," but because his attention was drawn to an "accident" or "great operation of

Nature which before was either insufficiently attended to or perhaps overlooked."[67] The accident he refers to is the great earthquake of 1755, centered off the coast of Portugal. Modern reconstructions of the event put the magnitude of this earthquake at 8.5 or 9, and the death toll in Lisbon at 30 000–60 000. For those who witnessed it directly, or saw its strange effects on inland and shoreline water levels as far north as the Netherlands, the earthquake marked a major milestone in European history.

The earthquake focused Wright's attention on theories of the interior of the earth, and these theories in turn reminded him of Whiston's concept, in his *Astronomical Principles of Religion*. Whiston had remarked that if the Sun, planets, comets, and stars all existed within a great cavity inside a larger body — itself a giant star or comet — then the inhabitants of the planets could not have "any Philosophical Evidence that there is such an External World at all."[68] Wright's new model built on just this idea; he nested the Sun and planets inside a greater sun, which itself might be the central body of a greater planetary system, and so on. Viewed from outside our celestial sphere, in other words, our sky is the inner surface of another sun, volcanically fueled. Most bizarrely, Wright proposed that "the Heaven of one state or creation" — the sun at the center of a given sphere — "may prove little more than the Hades of another, and so on ad infinitem both ascending to infinity and descending to negation: magnitude and miniature having no proportion or distinction in ye ideas of God."[69]

The difficulties with Wright's *Second Thoughts* serve as a reminder that his primary purpose was to illuminate God's creation with reference to the harmonies of nature. He was not primarily an astronomer — even by the standards of his day, when a self-educated person could make advances in fields such as optics or mechanics, and a good telescope could propel an observer to the forefront of the field. Wright made no systematic studies of natural phenomena, and his telescopic observations were not significant. He was rather a kind of visionary, and his fertile imagination in combination with his strong spatial sense led him deeper into the question of our position among the stars than even the most eminent astronomers of his day. But the same fertile imagination could as easily distract him from his greatest insight and most original contribution to astronomy:

71

the explanation of the appearance of the Milky Way as due to our immersion in a thin, flat layer of stars.

At his death, Wright left his house and its contents to his daughter Elizabeth. She survived him only 18 months. Wright's friend George Allen acquired many of his prints and copper plates, and these apparently were lost in a fire after Allen's estate passed to the Newcastle Literary and Philosophical Society.

In his own lifetime Wright was probably better known for his landscape and architectural designs and for his book on Irish antiquities, *Louthiana*, than for his astro-theological books and lectures. Despite the fact that the *Original Theory* was reviewed in a German newspaper, the work was not translated into German; plans to do so evidently fell through. Kant's acknowledgment of Wright may have elicited some interest in the Englishman, but Kant's *Universal Natural History and Theory of the Heavens* was, due to the bankruptcy of its publisher, not widely read until after copies of the books became available in about 1766.

In the nineteenth century there were scattered admiring references to Wright and his drawing of the plane section of stars giving rise to the appearance of the Milky Way. For the most part, these authors failed to grasp Wright's ideas or the context in which he presented them. Constantine Samuel Rafinesque, a naturalist educated in France and Italy but working as a professor of botany and natural history in the United States, came across a copy of the *Original Theory* and was so impressed with it that he issued an American edition in 1837. However, his edition did not include Wright's all-important plates, and, like many of Rafinesque's own writings, did not circulate widely. Rafinesque clearly overstated the case when he described Wright as an astronomer of the stature of Plato, Copernicus, and Newton. On the other hand, the explorer and naturalist Alexander von Humboldt mentioned Wright in his encyclopedic work *Cosmos* (part 1 of which was published in 1845) only as a philosopher or a kind of armchair astronomer.

A resurgence of interest in Wright in the twentieth century is due in part to the efforts of Friedrich Paneth (1887–1958). Though a chemist, Paneth had wide-ranging interests, and he happened to live first in Kant's hometown of Königsberg and later, in the 1940s and 1950s, in Wright's native Durham. Surprised to find that Wright was virtually unknown, even in his country of origin, Paneth set about tracking down Wright manuscripts and inviting scholars to reconsider his *Original Theory* and other work.

In 1966, the Cambridge historian of science Michael Hoskin learned that some 800 Wright papers were to be auctioned by Sotheby's in London. He persuaded a dealer, Dawsons of Pall Mall, to buy them, and to let him sort them and publish them before they were sold again to Durham University.[70] In the course of sorting these papers, Hoskin came across the "wholly-unsuspected" fragmentary manuscript of Wright's *Second Thoughts*, which provided a fuller context for Wright's work and underscored the importance he placed on integrating both the moral and the physical worldview.[71]

4

WILLIAM HERSCHEL: NATURAL HISTORIAN OF THE UNIVERSE

"It is a far cry from the facile imaginings of the philosopher to the rigorous demonstrations of exact science, and the true structure of the universe is not yet known."

George Ellery Hale, 1926[1]

Figure 4.1 Friedrich Wilhelm Herschel (1738–1822). (Credit: National Portrait Gallery, UK.)

The streets of Bath, a fashionable spa town in southwestern England, lay in darkness as Dr. William Watson made his way home one evening in December 1779. As he turned a corner, he caught sight of a curious figure in the gloom: a man standing to the side of a long rectangular wooden tube pointed at the sky (see figure 4.2). The tube rested on a stand in the street, in front of a modest house. It was pivoted near the bottom, and the man appeared to have his head pressed to the side of the tube at the upper end.

Intrigued, Watson drew to a stop. He maintained a respectful silence until the man took his eye off the instrument, a telescope

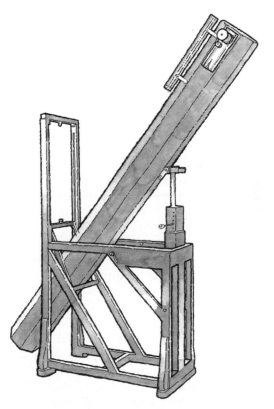

Figure 4.2 Herschel's 7-foot focal length telescope, of aperture about 6 inches. This was the instrument Herschel was using in 1779 when he met William Watson in the street in front of his house. He was using it in 1781, too, when he discovered the planet Uranus. (Credit: Layne Lundström.)

75

7 feet long and about 6 inches in diameter. Then he begged the favor of a glimpse through the instrument, and in response to a courteous invitation to see the Moon, the observer's object of study, he stepped up to the telescope and approached his eye to the eyepiece, a small cylinder protruding near the top of the tube.

Watson's first reaction was a sense of being dazzled. An intense beam of moonlight — reflected off the curved mirror at the bottom of the wooden tube, intercepted and reflected sideways by a small flat mirror near the top, and funneled through the eyepiece lenses — shone into the pupil of his eye. He blinked in the unexpected glare. At the same time, he exclaimed in admiration as the Moon's cratered surface sprang into focus.

Watson thanked his new acquaintance for sharing his view of this stunning lunar panorama. He returned to the house the next morning to express his gratitude again and to introduce himself properly. The telescope afficionado, he learned, was William Herschel, a professional musician in his early forties. Herschel had made the instrument himself, in his spare time, with help from his brother Alexander and his sister Caroline.[2]

Watson, the son of an eminent medical doctor and experimenter in electricity, recognized in Herschel a natural philosopher in need of an intellectual circle. Watson himself was a member of the Royal Society in London, and involved with the establishment of a similarly oriented philosophical society in Bath. He invited Herschel to join.

As Watson might have predicted, Herschel's contacts with the society helped him expand the scope of his research and taught him how to present his findings to an educated audience. More importantly, Watson's support helped Herschel make connections with members of the scientific establishment in London. Herschel soon needed their backing. Although neither Herschel nor Watson knew it at the time of their first meeting, the view through Herschel's home-made telescope was superior even to that afforded by telescopes at the Royal Greenwich Observatory. Herschel's skill as an observer and his familiarity with the night sky put him in a class by himself. His curiosity eventually led him to explore the universe as no one had done before. And his first discovery with this telescope — a new planet — challenged the world's leading astronomers to give credit to a relatively unknown amateur.

Herschel (figure 4.1) was born 15 November 1738 in what was then the independent port city of Hanover, Germany. The third of six children, William appears to have been warmly attached to his siblings—his older sister Sophia and brother Jacob, and his younger siblings Alexander, Caroline, and Dietrich. All of them except Sophia remained close to him in adulthood, and lived with him at different times after he settled in England.

William's father Isaac, an oboist in the Hanoverian Guards military band, had little formal schooling, but encouraged all his children's learning. Despite the family's limited finances, he provided tutors so that his sons could study mathematics, languages, and music beyond the level taught at the garrison school they and their sisters attended. Isaac must have read or studied on his own, too, for he relished intellectual discussions with his children. Caroline later recalled impassioned exchanges between him and her brothers: "Generally their conversation would branch out on philosophical subjects," she wrote in a family history. "[M]y brother William and my father often argued with such warmth that my mother's interference became necessary when the names Leibnitz, Newton, and Euler sounded rather too loud for the repose of her little ones, who ought to be in school by seven in the morning."[3]

Each of the four Herschel boys started training for a musical career as soon as he could hold a child-size violin. William's best instruments were the oboe and violin. At 14, he joined his father and Jacob as a bandsman in the Hanoverian Guards. He continued his private studies for a few years, however, having demonstrated exceptional talents in languages and mathematics and a desire to cultivate them. In his own memoir, he recalled that his tutor, a "man of Science," inspired him to study logic, ethics, and metaphysics, and inculcated a love of learning. "[A]ltho' I loved Music to an excess & made a considerable progress in it," he wrote, "I yet determined with a sort of enthusiasm to devote every moment I could spare from business to the pursuit of knowledge which I regarded as the sov[er]eign Good, & in which I resolved to place all my future views of happiness in life."[4]

The tumultuous 1750s did not allow Herschel much time for metaphysics. Europe's so-called "wars for empire" pitted the forces of the northern German states and England against those of France, Spain, Austria, Bavaria, Swabia, and Russia. In 1756,

Herschel, his brother Jacob, and his father Isaac were sent to England in anticipation of a French invasion of that country. Yet even under these rough conditions, Herschel seized opportunities to widen his horizons. While encamped in England for the better part of a year, he made the acquaintance of local musical families who were to provide valuable assistance when he and Jacob later returned there to live. And he used what little money he had to buy a copy of the philosopher John Locke's book *An Essay Concerning Human Understanding*, which articulated the empiricist view so essential to his own scientific outlook, that ideas are not innate but come from experience and reflection.

Early in 1757, the Hanoverian Guards took up new positions on their home territory, and even the bandsmen found their situation precarious as their regiment faced the French in the brutal battle of Hastenbeck. Herschel noted that "nobody had time to look after the musicians—they did not seem to be wanted."[5] He went home and made plans to leave the country, following his father's admonition to get out of harm's way. His sister Caroline later recalled, "I can now comprehend the reason why we little ones were continually sent out of the way, and why I had only a chance glimpse of my brother as I was sitting at the entrance of our street-door, when he glided like a shadow along, wrapped in a great coat, followed by my mother with a parcel containing his accoutrements."[6] Because of his youth, Herschel had never formally enlisted in the army, but he had to leave carefully. As added security, he later received formal discharge papers from the Hanoverian King of England, George III.

A career in music

Both Jacob and William fled to England. Jacob found employment as a music teacher, while William began by copying musical manuscripts for a living. Both struggled to get by, and Jacob, always more interested than William in maintaining a comfortable lifestyle, eventually gave up and returned to Hanover, where he landed a position in the court orchestra. But William soldiered on, accepting temporary posts all over the country, and by the time he was 28 years old he had substantially improved his social standing. He was known as a composer of symphonies and choral works and as a performer on the oboe, organ, and violin; he had dined with the Scottish Enlightenment philosopher David Hume, who had attended one of his solo performances in Edinburgh; and in Leeds, where he lived for about four years, he was solidly booked with private lessons.

In 1766, Herschel accepted a permanent post as organist and choral director of the Octagon Chapel in the city of Bath, which could fairly be called England's most important artistic and cultural center outside of London. He found additional outlets for his talents in that city, where well-to-do visitors liked to attend concerts and plays after "taking the waters" in the Roman baths. In the Pump Room by the Roman baths, and in the town's Assembly Rooms, concerts might be heard during the day as well as in the evening. Herschel recalled later, "My Situation proved a very profitable one, as I soon fell into all the public business of the Concerts, the Rooms, the Theatre & the Oratories, besides [having] many Schollars and private Concerts."[7]

Herschel's greatest musical gift seems to have been his performance on the oboe, although he also pleased audiences with his style on the violin, harpsichord, and organ. Contemporary listeners were favorably impressed by his "chaste" interpretation of the works of Corelli and Haydn, contrasting with the florid embellishments most performers allowed themselves to add to the printed music. Two of Herschel's compositions for oboe have been called "superb" by a modern chamber music conductor, although other critics have commented that his talent for music was not as great as his talent for exploring the celestial realm.[8]

Settling into a long-term position for the first time, Herschel had the leisure to develop a hobby: stargazing. He had learned something of the constellations from his father, according to his sister's family history. As a boy, he apparently familiarized himself with the coordinate system of the celestial sphere, for Caroline also remembered that their father had helped the young William create a "neatly turned 4-inch globe, upon which the equator and ecliptic were engraved" by her brother.[9] Now he included in his diary brief notes on his astronomical observations, such as the time and place he observed Venus or a lunar eclipse. His interest in astronomy was emerging, but had not yet developed into an all-consuming passion.

Figure 4.3 Caroline Lucretia Herschel (1750–1848). (Credit: National Portrait Gallery, UK.)

William's father Isaac Herschel died in 1767, and a few years later, in 1771, William brought his younger brother Alexander to Bath, securing a position for him as a cellist in one of the city's popular orchestras. In the same year, he wrote to his sister Caroline (figure 4.3) in Hanover to suggest that she also join him in Bath, where she might serve both as a singer to accompany his oratorio concerts and as his housekeeper. Whether he planned it consciously or not, Herschel was setting the stage for an efficient division of labor in his home and work that would allow him to expand the scope of his non-musical interests.

Caroline

A greater ray of hope than William's invitation Caroline could not have dreamed of. Since her father's death, she had fallen, as she put it in her memoir, in "a kind of stupefaction." Her father had supported her attendance at the garrison school — despite the fact that women of her station and era did not commonly learn to read and write — and had taught her to play the violin. After his death, although she longed for a job as a governess, her activities were constrained to sewing and knitting, and helping her mother keep house. Her mother curtailed her efforts at self-improvement. As Caroline saw it, "she had cause for wishing me not to know more than was necessary for being useful in the family; for it was her certain belief that my brother William would have returned to his country, and my eldest brother not have looked so high, if they had had a little less learning."[10]

At the age of 22, the unusually short-statured but vigorous and capable youngest daughter of the family felt utterly hopeless and frustrated at her prospects. William's extraordinary suggestion roused her to begin practising for a musical career — but in secret, for the prickly Jacob, her older brother, "began to turn the whole scheme into ridicule." When family members were away from home, she stuffed a gag between her teeth, and imitated "the solo parts of concertos, shake and all," as she had heard them played on the violin. "[I]n consequence I had gained a tolerable execution before I knew how to sing," she explained later.[11]

Caroline's mother acquiesced to the plan with difficulty, but Herschel made it more palatable when he arrived in Hanover to pick up his sister. "My mother had consented to my going with him, and the anguish at my leaving her was somewhat alleviated by my brother settling a small annuity on her, by which she would be enabled to keep an attendant to supply my place," Caroline wrote later. Herschel would support members of his family in a similar way throughout his life.[12]

Brother and sister arrived at Herschel's house in Bath in late August 1772. Bath's high season, which ran from fall to Easter, was just beginning. William and Alexander, who lodged with William, were busy with rehearsals, and William also with his many private students, young ladies and men. Caroline, who did not yet speak English, found she must manage without much help. At the breakfast table, William explained bookkeeping and other aspects of managing the household. Caroline was sent out alone to do the grocery shopping — although she discovered later that Alexander trailed behind her to make sure she returned home safely.

All through the winter Caroline looked forward to the end of the busy season, when William would have more time for her own daily music lessons. She struggled with homesickness and loneliness, and complained in her memoir about her brother's "hot-headed" servant, who made her life difficult.[13] But as spring approached, she observed that her brother's attention was tugging him in another direction. Her memoir records the dawning of a new passion for astronomy in William:

"The time when I could hope to receive a little more of my brother's instruction and attention was now drawing near," she wrote. "But I was greatly disappointed; for, in consequence of the harassing and fatiguing life he had led during the winter months, he used to retire to bed with a bason of milk or glass of water, and Smith's 'Harmonics and Optics,' Ferguson's 'Astronomy,' &c, and so went to sleep buried under his favorite authors; and his first thoughts on rising were how to obtain instruments for viewing those objects himself of which he had been reading."[14]

Herschel himself explained that his bed-time reading had led him from the study of music theory to astronomy. He wrote, "The theory of Music being connected with mathematics had induced me very early to read in Germany all what had been wrote upon the subject of Harmony; & when, not long after my arrival in England the valuable Book of Dr. Smith's Harmonics came int[o] my hands I perceived my ignorance & had recourse to other authors for information by which means I was drawn on from one branch of the Mathematics to another. [...] Among other mathematical Subjects optics and Astronomy came in turn & when I read of the many charming discoveries that had been made by means of the telescope I was so delighted with the subject that I wished to see the heavens & Planets with my own eyes thro' one of those instruments."[15]

Herschel's first telescopes were crude spyglasses that he and Caroline assembled in 1773 with lenses that he was able to order from manufacturers. Caroline, who had no inkling yet that her own lifework lay in astronomy rather than music, later wrote of her exasperation with the changing demands made of her: "I was much hindered in my musical practice by my help being continually wanted in the execution of the various contrivances, and I had to amuse myself with making the tube of pasteboard for the glasses [glass lenses] which were to arrive from London, for at that time no optician had settled at Bath."[16]

Herschel's preoccupation with seeing the stars and planets for himself dovetailed with the talent he and his brother Alexander obviously both had for mechanical invention, and led to his experimentation with new forms of telescopes. At first he worked with lens-based or "refracting" telescopes. He soon became dissatisfied with the lenses, however. As white light passes through a lens and is brought to a focus, the different colors composing the light disperse at different angles, so that a colored "halo" surrounds the image. In the late eighteenth century, lensmakers had not yet discovered a way to reduce this unwanted effect, called chromatic aberration.

"Reflecting" telescopes, based on mirrors or on a combination of mirror and lens, were known to give a less distorted view. In the fall of 1773 Herschel rented a reflecting telescope of a type known as a Gregorian, partly so he could use it and partly so he could take it apart and study its construction. In the Gregorian design, already about a hundred years old, a curved mirror at the rear of the telescope tube collects the light from the star or planet and reflects it back up the barrel of the instrument. This is the primary mirror, so called because it is the first to intercept the light from the sky. A smaller, curved secondary mirror is suspended in the barrel in front of the primary, blocking a small (and negligible) part of the incoming light. This secondary redirects the light from the primary back down the telescope and through a small hole in the primary, to a lens-based eyepiece at the back of the telescope. The user puts his eye to the back end of this telescope, as he or she would with a spyglass. (See figure 4.4 for an illustration of various telescope types mentioned in this chapter, including the Gregorian.)

Herschel liked the rented Gregorian telescope and wanted one of his own made—even bigger in diameter than the one he had rented, if possible. He understood that the true power of a telescope is determined by its aperture, the area of the primary mirror or lens collecting light from a distant object. A telescope with an opening of 4 inches diameter, for example, collects four

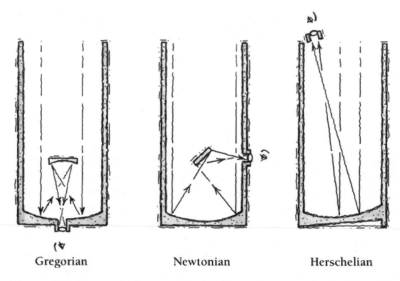

Gregorian   Newtonian   Herschelian

Figure 4.4 Types of telescope. In the Gregorian type of reflecting tele-scope, shown at left, light from the primary mirror is reflected to a curved secondary mirror, and from there to the eyepiece at the bottom of the telescope structure. The secondary mirror blocks a small part of the incoming light, but does not distort the image formed by the tele-scope (because the blocked light does not represent a specific geometric area of the source of light, but comes from the entire source). In the Newtonian design, shown in the middle panel, a flat secondary mirror directs the light to an eyepiece on the side of the telescope; aside from the difficulty of creating a perfectly flat secondary mirror, the design is similar to the Gregorian. In the Herschelian design, shown at right, the observer is rather awkwardly situated at the front end of the telescope. The advantage of the Herschelian design is that the light is only reflected once; upon reflection from a secondary mirror, some light inevitably would be lost. (Credit: Layne Lundström.)

times as much light as one with an opening 2 inches in diameter, because its geometric area is four times greater. The bigger the primary, Herschel knew, the fainter the objects one could see with the telescope.

Herschel also knew that the magnifying power — which many people mistakenly assume is the measure of the quality of a telescope—is an indication only of the apparent size of the image formed. The magnification of the image depends on the eyepiece used, and telescopes usually come with a variety of interchangeable eyepieces to provide a range of magnifying power. (A higher power eyepiece will not improve a fuzzy image—it just creates a bigger fuzzy image. But a larger diameter telescope, even with a low-power eyepiece, improves the detail and crispness of the image.)

The price a local merchant quoted to make a larger aperture reflecting telescope seemed "extravagant" to Herschel, but he soon thought of a way around the problem. "I formed the resolution to make myself one," he wrote, "as, not aware of the difficulty, it appeared to me from some former mechanical attempts that with the assistance of Dr. Smith's optics I should be able, in time, to accomplish such a work." Herschel indeed underestimated the difficulties of obtaining high-quality mirrors and mounting them in good alignment, but he did not underestimate his own skill or perseverance. In a period of a few years in the late 1770s, he made more than 200 primary or objective mirrors before declaring himself satisfied with the results.[17]

Whether or not Herschel already dreamed of making his own discoveries, he knew at least from the history of astronomy in Dr. Smith's *Compleat System of Opticks* that improvements in telescopes often led to new insights, even in the study of familiar solar system objects such as Saturn. In the early 1600s, shortly after the first telescopes were developed, Galileo had noticed mysterious "ears" on either side of Saturn, which he thought might be stationary moons of that planet. In 1659, the Dutch astronomer and mathematician Christiaan Huygens, with an improved telescope, clarified that the "ears" were rings; then in the 1670s, Cassini, with his improved optics, found the gap in Saturn's rings and discovered new moons circling the planet, too. It was worth building better telescopes even to look at known objects.

Throughout the winter of 1773 and into the spring and summer of 1774, when he was not occupied with concerts and rehearsals, Herschel devoted every spare moment to making his own telescopes. Alex, when he was home, assisted him; Caroline refers to his particular talents at least twice in her memoir. She wrote, for example, that "he took much pleasure to execute some turning or clockmaker's work for his brother."[18]

To her dismay, Caroline found her housekeeping duties must evolve to accommodate the fact that virtually every room in the house was transformed into a workshop. Alex worked a lathe in one of the bedrooms, making eyepiece lenses. In the basement, William toiled over mirrors — "speculum" mirrors made of alloys of tin, copper, antimony, and silver. A local manufacturer cast the mirror blanks for him, but William ground them into precisely curved shapes before polishing them to a high gloss with silk soaked in a tar-like concoction. "Every leisure moment was eagerly snatched at for resuming some work which was in progress, without taking time for changing dress, and many a lace ruffle was torn or bespattered by molten pitch," Caroline complained. At mealtimes William was absent-minded, "contriving" or sketching new apparatus. If he had many hours of manual polishing to do, Caroline skipped her own music practice and read to him from *Don Quixote*, *Arabian Nights*, or some novel.[19]

Herschel's reward for this unceasing effort and experimentation took the form of a Newtonian design telescope (see figure 4.4). The Newtonian design differs from the Gregorian in requiring only one curved or figured mirror instead of two: a flat secondary mirror near the middle or top of the telescope tube shunts the light from the primary mirror to an eyepiece at the side of the telescope. The aperture diameter of his first satisfactory telescope was about 4 ½ inches, which today would be typical of an amateur's first telescope.

On 1 March 1774, he inaugurated his first review of the heavens through this telescope, turning it first to the beautiful Orion nebula (see chapter 2, figure 2.7). His view of Saturn and its ring like "two slender arms" encircling the planet, he wrote in his memoir, gave him "infinite satisfaction."[20] The seriousness of his purpose is evident from the fact that he no longer used his diary to record his observations, but began a separate astronomical journal.

Herschel found it increasingly difficult to keep his mind on his private music tutoring, concerts, and accompaniments. He wrote of this period in his memoir, "Nothing seemed now wanting to compleat my felicity than sufficient time to enjoy my telescopes to which I was so much attached that I used frequently to run from the Harpsichord at the Theatre to look at the stars during the time of an act & return to the next Music."[21]

As he sprinted to his telescope during borrowed intervals of time, Herschel's imagination was likely to dwell on the mountainous landscapes of the moon or the mottled, reddish surface of Mars. In the early years of his forays into astronomy, Herschel made mostly solar, lunar, and planetary observations, following the lead of his favorite textbook authors. He initiated a long-term series of observations of the planet Venus aimed at determining its diameter, its period of rotation, the nature of its surface, and whether or not the planet has an atmosphere. Bright white spots on Mars, that planet's icy polar caps, aroused his curiosity in 1777; repeated observations allowed him to observe their seasonal changes in size. Saturn, which he called "one of the most engaging objects that astronomy offers to our view," he returned to again and again.[22] Initially Herschel, like his contemporaries in astronomy, viewed solar system objects as the natural concern of the astronomer; he paid little attention to the more distant stars unless they called attention to themselves by varying in brightness or by moving with respect to the fixed pattern of stars. (Recall that Halley had found some examples of stars with "proper motion," and that Wright had incorporated this finding in his models of the stellar system.) Later, Herschel himself was to open up the realm of the stars to exploration.

Herschel's self-education in astronomy and telescope construction is all the more remarkable considering that during the late 1770s his musical career was still gaining momentum. In 1776, he took on the responsibilities of Director of the Public Concerts at Bath. Meanwhile, Caroline's duties at home became more complex as William and Alexander turned their rented house into a telescope factory. In fact, her domestic burden increased in the late 1770s when her younger brother Dietrich, a high-spirited but irresponsible young man, came to live with his siblings following an abortive attempt to sail for a life of adventure in India. She was not a little relieved when he returned to Hanover in 1779. But musical demands on Caroline's time did not slacken. She became an accomplished singer and even received an invitation to perform at a festival in the city of Birmingham—although she declined it because she had decided not to sing in public unless her brother conducted.

From amateur to professional

The three-year period from 1779 to 1782, culminating with the discovery of the seventh planet of the solar system and events in the wake of that discovery, marked Herschel's transformation from amateur to professional astronomer. His friend William Watson played an important role in helping him make this transformation by introducing him to the philosophical societies in Bath and London. But even before he met Watson on the street outside his house in Bath, Herschel had begun to formulate lofty goals for his telescopic observing program — goals that would have seemed ambitious even to the Astronomer Royal at Greenwich Observatory. Herschel's discovery of Uranus not only made him unexpectedly famous, but set in motion a change in his circumstances that made it possible for him to devote more of his time to the pursuit of his astronomical ambitions.

Bath's Literary and Philosophical Society was just forming and had not yet held its first meeting when Watson encountered Herschel examining the lunar surface through his telescope. Such societies, which became popular in England and the United States in the 1700s, modeled themselves after the Royal Society in London, chartered in 1662. In this pre-Industrial Revolution era the Royal Society welcomed men as disparate in education as the navigator James Cook, who joined the navy without attending university, and the Oxford-educated philosopher and economist Adam Smith. Until 1840, all that was required for membership in the Royal Society was an interest in natural philosophy, however "amateur" or self-taught. (The French Académie des Sciences, established in 1666, limited membership to the more mathematically trained scientists from its inception.) In England, the local societies such as Bath's provided an opportunity for scholars, skilled merchants, and craftsmen alike to present their ideas and technological experiments to each other.

Herschel, of course, became one of the Bath society's most active members. He contributed a number of papers on metaphysics and epistemology — the philosophy of "how we know what we know" — but his reports on astronomical observations elicited the most approbation. The Bath society forwarded papers of particular interest to the Royal Society — which, as a venue for the most learned scientific dialogue and a central repository of information, often served as an advisory body to the monarch. Two of Herschel's first papers to receive the distinction of being forwarded to the Royal Society concerned the variable star Mira Ceti, which brightens and dims with an irregular period of about 11 months, and the lunar mountains.

Watson continued to mentor Herschel after introducing him to the Bath society. He read and commented on Herschel's manuscripts before Herschel submitted them, curbed Herschel's enthusiasm for unfounded speculation and provided feedback on how the papers were received at Royal Society meetings in London. Some of the claims Herschel made about his telescopes in his early papers seemed, indeed, rather fantastic to members of the Royal Society.

In this period Herschel's musical duties began to appear as frankly onerous. He had, by about 1780, set his sights on three main problems, each of which might very well have occupied him full time: the measurement of the distances to the stars through the phenomenon known as parallax; the construction of even bigger and better telescopes; and the discovery of new nebulae like the well-known "Great Nebula" in Orion, and other interesting, faint sources. He knew already that his home-made telescopes allowed him to see the stars, planets, and nebulae as few had ever seen them. No wonder Caroline recorded in her journal that the endless round of music lessons, concerts, and rehearsals seemed "an intolerable waste of time."[23]

Parallax is the apparent shift of position of nearby objects when viewed by an observer who is in motion or who observes from different locations. The principle is easily demonstrated with one's own hand and eyes: one holds one's thumb still at arm's length and views it alternately with the right eye open and left eye closed, and vice versa. The thumb appears to "jump" back and forth with respect to background objects. The "jump" occurs because each eye observes from a slightly different vantage point. The nearer the thumb to one's face, the bigger the jump. Indeed, we unconsciously gauge distances and acquire depth perception, at least qualitatively, through this effect of parallax.

Parallactic shifts or jumps can be used to measure distances. The technique is similar to the surveyor's method Thomas Wright used to measure distances or heights by triangulation. Observations of a target object are made from two stations, separated by a baseline. In the example of the thumb at arm's length, the thumb is the target, about 40 centimeters away, and the eyes are the two stations, separated by a baseline of about 5 centimeters. Mathematically, one can relate the distance to the target to the size of the parallactic shift and the length of the baseline. In the astronomical case, a target star would be observed from two different locations and its position with respect to some "fixed" background stars carefully recorded.

If the stars chosen as fixed reference points were not, in fact, more distant than the target star, the technique would not work because these stars would also shift their positions slightly. The search for stellar parallax, which astronomers had sought for more than a century already, thus required a selection of suitable target stars. In the absence of any clues to stellar distances, astronomers generally assumed that the brighter stars were the nearer ones, and focused their efforts on these as targets.

The parallactic shift of nearby objects is such a well-established phenomenon that for hundreds of years the *absence* of detectable stellar parallax gave some skeptics reason to doubt the Copernican or Sun-centered model of the solar system. As the Earth orbits the Sun, it describes a circle in space. (Actually, it describes an ellipse, but the ellipse is very nearly circular.) Observations of a target star made six months apart—say, in January and June—are separated by some 300 million kilometers [186 million miles], the diameter of Earth's orbit. With such a long baseline, the skeptics asked, shouldn't some of the stars show yearly parallactic shifts? For some time, in fact, astronomers looked for evidence of parallax among the stars as proof that the Earth circles the Sun. By Herschel's time, the Sun-centered model was no longer in doubt, but the search for parallactic shifts continued, as a quest in its own right. In 1760, Nevil Maskelyne, the Astronomer Royal, specifically urged his fellow professional astronomers to intensify their efforts to find parallax.

The problem is that the stars are very far away, and parallax is a very small effect to be sought among thousands of cataloged bright stars. The farther away the star, the smaller the "jump," so the success of the technique depends on being able to record very slight apparent displacements. Pinpointing the locations of many stars in the stellar equivalent of latitude and longitude and re-measuring these coordinates at a later time in the year is time-consuming. Furthermore, the measurements of stars in different parts of the sky might be affected differently by atmospheric conditions, to name just one problem.

In the seventeenth century, Galileo thought of a way to make the search for parallax more efficient. Instead of measuring the positions of bright (and thus presumably nearby) stars, and checking them periodically to see if any had moved back and forth on a yearly timescale, he suggested one should find double stars — that is, two stars next to each other in the sky — and monitor their separation. If one member of the pair is actually near and the other far, the near one will appear to narrow the gap, then increase it again (see figure 4.5).

Herschel appreciated the practicality of Galileo's method. The task of searching for parallax was reduced to a single measurement, the angular separation of two stars, rather than a set of right ascension and declination measurements for each star observed. Galileo's method had never been implemented, even by its originator, however, for it requires a catalog of double stars to use as targets, and a method of measuring the separation of two stars in the telescope field of view.

In 1778, shortly before he met Watson and became involved in the philosophical society, Herschel decided to try his hand at searching for parallax by Galileo's untried method. He would measure the separations of the stars with a micrometer, an instrument for measuring small distances. Micrometers took many different forms in the eighteenth and nineteenth centuries. (A simple one is illustrated in figure 4.6.) In the common wire micrometer, a vertical wire, visible in the eyepiece of the telescope, is fixed; the telescope can be pointed so that one of the two stars in a double is seen to lie on this wire. The position of a second, movable wire can then be adjusted until it matches that of the second star. A scale to the side of the micrometer allows the astronomer to record the linear separation of the wires, which corresponds to the angular separation of the two stars. Herschel also used a different kind of micrometer of his own invention, a lamp micrometer. This was an artificial double star made from two lamps shining through two pinholes in a wooden board. He would observe the real double star with his right eye at the

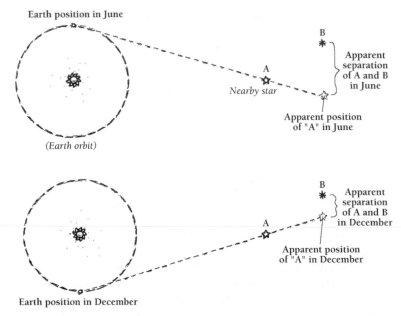

Earth position in June

(Earth orbit)

B
Apparent
separation
of A and B
in June

A
Nearby star

Apparent position
of "A" in June

B
Apparent
separation
of A and B
in December

A

Apparent position
of "A" in December

Earth position in December

Figure 4.5 Parallax. Top panel: From the Earth's orbital position around the Sun in June, an observer sees, on the plane of the sky, the nearby star A to the right of the more distant star B. Bottom panel: Six months later, in December, the Earth is at the opposite end of a baseline formed by the diameter of Earth's orbit. (The diameter is twice the Sun–Earth distance, or two Astronomical Units (AU), or 1.5×10^8 km, or 93 million miles.) From this new vantage point, the perspective on the stars is different, and star A appears to have moved closer to star B. Six months later, in June of the following year, the apparent separation of A and B will have increased again. (Credit: Layne Lundström.)

telescope, and the lamp micrometer's artificial double with his left eye. He adjusted the separations of the lamps until the real and artificial doubles matched, and from this set-up he would calculate the angular separation of the real stars.

Herschel did not find parallax immediately among his favorite targets, such as the double star Castor in the constellation Gemini. In fact, he searched for parallax in vain over the course of a 40-year career in astronomy. But the search prompted him to compile catalogs of double stars that might make good target

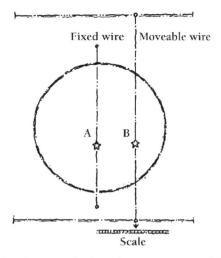

Figure 4.6 A simple type of wire micrometer as used in the late 1700s and early 1800s. The observer positions the telescope so that a fixed wire, which is visible through the eyepiece of the telescope, appears at the same location as a star or other object of interest. The position of a second, moveable wire, can be adjusted so that it appears to rest on the second star. A scale attached to the micrometer indicates the separation of the wires, and hence the angular separation of the stars in the sky. (Credit: Layne Lundström.)

objects for future generations of astronomers, who might be able to achieve better precision in their measurements and isolate smaller shifts. As the years went on his double star catalogs became ever longer and more comprehensive, and were recognized as valuable contributions to the field.

The second of Herschel's three main goals, formulated during the period when he was making the transformation from amateur to professional astronomer, brought him more gratification than the search for parallax. He set out to make bigger telescopes, pushing the limits of what could be accomplished with reflectors in the pursuit of faint or nebulous sources of light. Alexander and Caroline provided essential support in this extension of his already complex telescope-making operation.

Previously, Herschel had obtained mirror blanks about 6 inches in diameter from a local craftsman. What was left for

Herschel to do was to grind them to the required figure and polish them. But when his supplier of blank mirror disks could no longer keep up with his requirements, Herschel learned how to cast the blanks himself. First he experimented with trials of different mixtures of metals to see which yielded a mirror surface that was highly reflective but retained its shape over the wide range of temperatures prevailing during the course of a night's observing. Some of the best mirrors were so brittle and sensitive to temperature changes that touching them with a warm hand on a very cold night might shatter them. Herschel found an alloy that pleased him in these respects, although the hardest and most reflective metals tended to tarnish quickly. He found it necessary to keep for each of his telescopes a spare mirror polished and ready to be exchanged with a tarnished one at short notice.

Casting the mirrors proved to be a peculiarly tedious and sometimes dangerous undertaking. Following the advice in Dr. Smith's four-part volume on optics, Herschel melted the copper, tin, and other metals together and poured the molten mixture carefully into a heat-resistant mold to cool. Herschel's recipe for the mold itself called for loam, a soil rich in clay and sand, mixed with dried horse manure. Caroline noted in her memoir that "an immense quantity" of the horse dung had to be "pounded in a mortar and sifted through a fine sieve." Alexander and Dr. Watson took their turns at the mortar and pestle.[24]

The furnace for casting the mirrors shared space in Herschel's flagstone-floor basement with the house's kitchen. It was early in 1781 in this basement room, opening directly onto the garden, that Herschel, his assistants, and Alexander discovered the perils of working with molten metal. They were attempting to cast an exceptionally large mirror, 36 inches in diameter. Caroline wrote, "[T]he mould, &c., in readiness, a day was set apart for casting, and the metal was in the furnace, but unfortunately it began to leak at the moment when ready for pouring, and both my brothers and the caster with his men were obliged to run out at opposite doors, for the stone flooring (which ought to have been taken up) flew about in all directions, as high as the ceiling. My poor brother fell, exhausted with heat and exertion, on a heap of brickbats."[25]

Herschel survived the exploding flagstones episode and eventually succeeded in casting larger mirrors. During this

94

transition period of the late 1770s, however, his largest telescope had an aperture of 12 inches. He referred to this instrument by its length as the "20 foot telescope," and later as his "former" or "small 20 foot" to distinguish it from a subsequent 20-foot long telescope of 19 inches aperture. His 6.2 inches aperture telescope, which he called his "7 foot" telescope, also served him well, for it was more portable. This was the beautiful instrument Watson first saw, a square-section mahogany tube on a wooden stand. The 7-foot telescope was Herschel's favorite instrument for carrying out his reviews or "sweeps" of the sky, as he called them.

The search for parallax and the effort to build larger aperture telescopes kept Herschel running from one task to another, and even sometimes giving music students the slip. Yet in 1781, he was spurred to new heights in the third of his main goals, the discovery of new nebulae and other unusual objects. In December of that year Watson gave him a catalog of nebulae prepared by the French astronomer Charles Messier. Messier, a comet-hunter, intended his catalog primarily as a list of comet look-alike objects that comet-hunters could safely ignore. The first item on his list, now known as M1 for "Messier No. 1" is the so-called Crab Nebula in Taurus, the aging remnant of a supernova explosion in the year 1054. Many of his items, such as M13, are actually clusters of stars that looked round and "fuzzy" to Messier.

Herschel's pride in his telescope's light-collecting aperture and high-power eyepiece magnification is evident in his account of his re-examination of Messier's objects. "As soon as the first of these volumes [of Messier's] came into my hands," he wrote in 1784, "I applied my former 20-feet reflector of 12 inches aperture to them; and saw, with the greatest pleasure, that most of the nebulae, which I had an opportunity of examining in proper situations, yielded to the force of my light and power, and were resolved into stars." He added that in many cases Messier had seen "only the more luminous part" of his nebulae.[26]

As a specific example of the difference between his view and Messier's, he compared their descriptions of the 53rd listed object, the globular cluster M53 (see chapter 2, figure 2.6 for an illustration of a globular cluster). Herschel quoted Messier in French: "Nébuleuse sans etoiles...ronde et apparente." ("Nebula without stars, round and prominent.") Herschel's own astronomical journal entry for this object ran: "A cluster of very close stars; one of the most beautiful objects I remember to have seen in the heavens. The cluster appears under the form of a solid ball, consisting of small stars, quite compressed into one blaze of light, with a great number of loose ones surrounding it, and distinctly visible in the general mass." For good measure, he appended a hand-drawn illustration of the globular cluster.[27]

These early observations of Messier's objects, in which the apparent nebulosity "yielded" to the force of Herschel's higher resolution and magnification, led him to believe for a long time that virtually all nebulae, no matter how cloudy in appearance, consisted of stars. The most cloudy or faint were simply the most distant. For example, in his enthusiasm for brushing aside earlier accounts of nebulae as consisting of some sort of luminous fluid, he described both M1, the gaseous Crab Nebula, and M3, an indistinctly seen globular cluster of stars, as showing "a mottled kind of nebulosity, which I shall call resolvable; so that I expect my present telescope [i.e., the second, larger aperture 20-foot, not yet used in the examination of all Messier objects] will, perhaps, render the stars visible of which I suppose them to be composed."[28]

Herschel evidently thought of these nebulae as island universes, comparable in nature to our own sidereal system, although clearly of a diversity of shapes and sizes. The contemporary novelist Fanny Burney, daughter of Herschel's friend, the physician Dr. Charles Burney, reported in the 1780s that Herschel told her he had discovered 1500 new universes.

With characteristic enthusiasm and energy, Herschel undertook the search for double stars and nebulae, the casting and endless polishing of mirrors, the construction of wooden stands to support his telescopes, and the reporting of his results in papers for the philosophical societies. The ceaseless round of activity taxed his resources to the utmost, however. His brother and helper Alexander had interests of his own and a musical career that took him to Bristol during the summers when Bath was quiet. Caroline noted in her memoir that these circumstances involved her more closely in her brother's work, though mostly, at this time, as his amanuensis:

"Alex was always very alert, assisting when anything new was going forward, but he wanted perseverance, and never liked to confine himself at home for many hours together. And so it happened that my brother William was obliged to make trial of my abilities in copying for him catalogues, tables, &c., and sometimes whole papers which were lent to him for his perusal. [...] When I found that a hand was sometimes wanted when any particular measures were to be made with the lamp micrometer, &c., or a fire to be kept up, or a dish of coffee necessary during a long night's watching, I undertook with pleasure what others might have thought a hardship," she wrote.[29]

In the spring of 1781, the extraordinary quality of Herschel's telescopes and his unusual assiduity in searching the heavens for new objects converged in allowing him to make a spectacular discovery. On the 13th of March, during the course of his systematic sweeping of the sky, Herschel noted the appearance of an unusual "nebulous star." The star seemed uncommonly large, and he suspected it of being a comet. On the 17th, he noted it had changed place with respect to the background stars, as a comet should. By the 19th he had further confirmed his impression, determining that the object moved in the ecliptic. That is, it traveled through the constellations of the zodiac like most solar system objects.

Other astronomers, including the Astronomer Royal at the Greenwich Observatory, Nevil Maskelyne, confirmed the object's motion. As news spread and astronomers checked their records, it became apparent that the object had been seen before, but had not attracted special attention. Herschel, with his superior optics and well-trained eye, had been the first to notice that this star-like object looked different, and he was therefore the first to track its position over the course of several nights.

Astronomers and mathematicians across Europe set to work calculating the orbit of the comet and comparing its predicted motion to its evolving place among the stars. By May the startling truth was beginning to sink in: the data only made sense if the object orbited the sun at about twice the distance of Saturn, in a nearly circular path more similar to that of a planet than that of a typical comet.

Herschel's "curious" object was in fact the first planet to be discovered since the dawn of recorded history. A new wandering star joined the ranks of those five known to the ancient Babylonians and Greeks: Mercury, Venus, Mars, Jupiter, and Saturn.

Strong opinions emerged on all sides about what to call the new planet. Herschel advocated that it be named in honor of

King George III, a fellow Hanoverian. This did not go over well on the continent. As Ireland's nineteenth-century Astronomer Royal, Sir Robert Ball, put it, European astronomers "considered that the British dominions, on which the sun never sets, were already quite large enough without further extensions to the celestial regions."[30] They in fact proposed the name Herschel, as well as Uranus, and until 1847, the planet went by three different names. The name Uranus at least was in keeping with tradition, because Uranus, in Greek myth, is the father of the character associated with the Roman god Saturn.

The Royal Society promptly elected Herschel a fellow and awarded him its Copely medal for his discovery of the new planet. Soon thereafter George III appointed Herschel as his personal astronomer: not the Astronomer Royal, but the King's Astronomer. A modest life pension freed Herschel from his now onerous music teaching duties. He and Caroline moved closer to Windsor Castle, to be available to entertain members of the court with views of the heavens.

Herschel's skill in the manual art of making telescopes, and his inclination to investigate and debate the nature of things he saw through them, had propelled him from musician-astronomer to full-time astronomer. With the pension from the crown he began filling his time as many modern astronomers do: conducting research, raising money to support his investigations (in Herschel's case, through the sale of telescopes and by applying for grants from the king), and writing up his work as he went along. He trained Caroline as his assistant, and supervised workmen who were constantly refurbishing or maintaining his instruments and the wooden structures for mounting the telescopes.

The construction of the heavens

With the move to the Windsor area—first to Datchet, then to Old Windsor, and finally to Slough, Herschel's permanent residence—Caroline found that once again, as was the case when she first moved to England, she must learn new skills to be useful to her beloved brother. William gave her a small telescope, with which he instructed her to "sweep for comets" and to note any other "remarkable appearances" she found along the way.[31] She started her own astronomical journal on 22 August 1782.

Gradually Caroline learned to use the bigger telescopes and to make some of the calculations required on the data collected. William taught her to re-measure with the micrometer the separations of the double stars he had observed, either as a check on his own observations or, perhaps, to monitor the separations for any change that might indicate a parallactic shift. She also jotted on pieces of paper notes on finding logarithms required in calculations, determining an observed object's coordinates and locating planets from their predicted positions in an almanac. Some of William's tutoring was evidently given at mealtimes, for she later recalled good-humoredly that he asked her to guess the angle of her slice of pudding at dinner, and made her "fall short" of her share if she gave the answer incorrectly.[32]

In the first year of her systematic scanning of the sky with the small telescope no comets materialized, but Caroline discovered about a dozen previously uncataloged nebulae and star clusters. However, her own observing, fruitful as it was, suffered continual interruptions as her brother requested help with his. When he trained her to look for new comets, Herschel evidently underestimated the amount of help he would need at his newest telescope, inaugurated shortly after he became King's Astronomer.

William and Alexander completed most of the optical work on William's "large 20 foot" telescope of 19 inches aperture during the summer of 1783. By late October, the telescope stood ready for use in the garden. Its mirror end rested near ground level; its top end, where the eyepiece was located, pointed up into the sky, suspended by a rope and pulley to a wooden frame shaped like a giant letter A in cross-section. The telescope was too unwieldy to be re-pointed frequently, so Herschel allowed the slow nightly revolution of the dome of the sky to bring objects to him. Perched on a wooden platform in the support structure, he gazed through the eyepiece at the stars and occasional nebulae as they drifted through his field of view. When an interesting object came along, he called to Caroline to note the time (for determining the object's coordinates) and to take a "memorandum" on its appearance.

The nightly routine was not without its dangers, as Herschel was often suspended 15 feet or more above the ground on a structure that once collapsed in a high wind. Caroline, hastening to answer her brother's call, once wounded her leg on an iron hook that projected from the machinery used to turn the telescope on its stand. The visiting astronomer Giuseppi Piazzi similarly tripped over a projecting beam and went home with "broken shins," according to the ever-worried Caroline.[33]

Several important studies with the new telescope occupied Herschel during the mid-1780s, including observations of Mars and the seasonal variation of its polar icecaps. His ground-breaking work on the shape and size of our stellar system (or galaxy) also began in this period. He called the topic of this investigation "the construction of the heavens."[34] This work tied together his observations of nebulae, which he believed at this time were clusters of stars, more or less distant, and his interest in the most profound philosophical questions pertaining to man's place in the universe.

Herschel no doubt evolved an interest in the three-dimensional structure of the starry system independently, but it is interesting to note that he owned a copy of Wright's *Original Theory*. From marginal notes in Herschel's hand, it appears that he read his copy cursorily sometime after 1781—and possibly before he began his star-gauging. However, even if his interest in the shape of the universe of stars was piqued by Wright's proposals and illustrations, it is safe to say that Herschel would have had no patience for Wright's approach. Though he was sometimes given to audacious speculation himself, Herschel took very seriously his duty to test his ideas with observation, and he disdained philosophical or metaphysical works on astronomy whose authors tried to "build a system upon hypotheses."[35]

Herschel presented his thoughts and observations on the construction of the heavens to the Royal Society in 1784 and 1785. The first of these papers is memorable for its depiction of the starry realm as crossed by extended layers or "strata" of stars and nebulae, analogous to the structures perceived by the geologist in the Earth's crust. He must have been inspired by advances in the study of the history of the Earth in this period, for in his introduction he predicted that the astronomer would soon have a three-dimensional view of the heavens, just as the geologist comprehended his terrain both in depth and in horizontal extent. He wrote in his introduction, "In future, therefore, we

100

shall look upon those regions into which we may now penetrate by means of such large telescopes, as a naturalist regards a rich extent of ground or chain of mountains, containing strata variously inclined and directed, as well as consisting of very different materials. A surface of a globe or map, therefore, will but ill delineate the interior parts of the heavens."[36]

Thanks to his deep familiarity with the night skies, Herschel had in fact uncovered patterns in the distribution of nebulae whose significance would not be understood until the middle of the twentieth century. He noticed that some nebulae were congregated in irregularly shaped and sometimes filamentary groups. In the course of "sweeping" the sky, when one or two nebulae appeared in his telescope, he had the impression of being "on nebulous ground," and often saw his suspicions confirmed when the next vertical swath of sky showed more nebulae in connected regions. One such "stratum" of nebulae he found we know as the Virgo Cluster or Virgo Supercluster of galaxies. Herschel correctly described the stratum as running more or less at right angles to the zone of the Milky Way, passing through the constellation of Ursa Major in the north, snaking through Coma Berenices and ending in the northern parts of Virgo. Indeed, as viewed through large telescopes, thousands of spiral and elliptical galaxies appear in this part of the sky, representing a real clustering of galaxies in space and the dominant structure in our neighborhood of the universe. The Virgo Supercluster lies some tens of thousands of light-years away. Even binoculars or small telescopes will reveal some of the fuzzy patches of light scattered along this "nebulous ground."[37]

Herschel was not always correct in the interpretation of his observations, and a serious problem with his 1784 and 1785 papers is his assumption that all, or virtually all, nebulae consist of clusters of stars. The galaxies in the Virgo cluster are, of course, themselves very distant clusters of stars, and the globular clusters consist of stars, but some of the nebulae he described are cloudy, truly gaseous objects in our own galaxy. But his description of nebular "strata" touched on a nugget of truth, even if he sometimes applied the idea too widely or included both types of nebula in one stratum. Galaxy-nebulae tend to be seen far from the plane of the Milky Way, where there is less obscuring dust, and are often found in clusters; globular clusters can be found in all directions but are most abundant in the direction of the center of our galaxy, near the constellation Scorpius; and even the gaseous nebulae that misled Herschel could be said to lie in strata, for they tend to occur in the spiral arms of our galaxy.

The idea of the Milky Way system itself as a stratum or layer of stars, also enunciated in the 1784 paper, was not new with Herschel, but he considered the implications of this interpretation more carefully than did Immanuel Kant and other predecessors. Herschel never glossed over the apparent dark rift in the Milky Way, visible particularly from Cygnus, through Sagittarius, and on to Centaurus (see chapter 2), which we now understand to be due to obscuring dust and gas. The "immense starry bed" of the Milky Way "is not of equal breadth or lustre in every part, nor runs on in one straight direction, but is curved and even divided into two streams along a very considerable portion of it," he wrote.[38] He conceived of the Milky Way as a stratum of stars, but in a rather un-geological way, as a vertically "branched" stratum. Figure 4.7 shows his view of the Milky

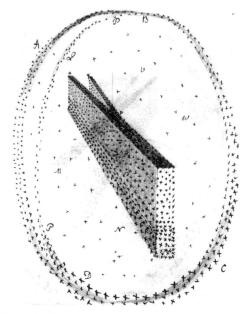

Figure 4.7 Herschel's branched-stratum Milky Way. (Credit: Royal Astronomical Society.)

Way as a layer, divided on one side like a partially split English muffin. The Sun, he reasoned, lay near the middle of the stratum. The branching of the stratum on one side gave rise to the divided-stream appearance of the Milky Way in the sky, as he showed in his figure with the double arc of stars.

The second paper, presented in 1785, contains a famous illustration of our galaxy as Herschel delineated it with his 20-foot telescope. Already in the first paper he had described a method of determining the position of our Sun within the Milky Way stratum, although he had yet to finish carrying out his investigation. This was the method he called the "Star-Gage." In the second paper, he described the fruits of that investigation, "an actual survey of the boundaries of our sidereal system."[39]

Herschel did not mean a trigonometric survey. Rather, he had used the space-penetrating power of his telescope as a kind of sounding line: he translated the average number of stars in a given direction to an estimate of the relative "depth" of the stellar system in that direction. He assumed that, generally speaking, the stars were uniformly distributed. In that case, the number of stars in a given direction must correlate with the depth of the system in that direction. He also had to assume that his telescopic view probed to the very edge of the system, and furthermore that no stars, either near or far, were hidden from view by some obscuring matter.

On a practical level, his technique began with choosing a given direction in the sky. He pointed the telescope in this direction and counted the stars in his field of view, some 15 minutes of arc in diameter. Sometimes he counted the stars in ten fields of view close together, and calculated an average number of stars for this gauge. For lines of sight into the Milky Way, a single field of view included hundreds of stars; his highest single gauge revealed 588. Some areas yielded no stars, or very few. He took 675 gauges, many of them averages of ten separate fields.

Next, Herschel calculated the depth of the heavens in the direction of his gauge—or more correctly, the relative depth, for the absolute distances to the stars were not known. The field with 588 stars must, according to his assumptions, represent the deepest extent of the sidereal system, and he estimated the

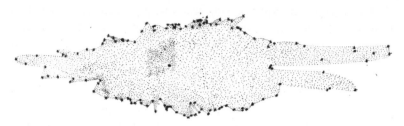

Figure 4.8   Milky Way from "star-gages." (Credit: Royal Astronomical Society.)

distance probed as "not less than 497 times the distance of Sirius from the sun."[40] A field with only 5 stars indicated that the Milky Way nebula in that direction could not extend to more than 100 times the distance of Sirius.

Herschel had dreamed of gauging the entire heavens this way and plotting the resulting surface, but he settled on a cross-section created from selected gauges. His cross-section formed a great circle passing through the poles of the Milky Way (i.e., at 90° from the plane of the bright zone) and intersecting the Milky Way at the point where it appears to branch or divide in two. The resulting plan of the starry system (see figure 4.8) is roughly oval, with very irregular edges and two thin "branches" extending from one side. He filled in the interior with stars, and showed the Sun as a dark dot near the middle.

The picture is startlingly like a rough cross-section of our galaxy, at first glance: the slab of stars is thin but wide. The correspondence is largely fortuitous, because, contrary to his belief, Herschel only probed a fraction of the distance to the edge of the system. Interstellar dust severely limited his view, particularly in the direction of the center of the Galaxy where he saw the dark rift in the Milky Way. And because he could not see the edge of the system, he naturally found that our Sun lies near the center. Nonetheless, this "map" of the heavens represented a radical advance in thinking about the form of our galaxy. His cross-section was the first of its kind and nothing was available to replace it for about 100 years. All future work on the shape of the Galaxy would pick up where Herschel left off.

Although he could not fix the absolute size of our Milky Way nebula, nor gauge the sizes or distances of the hundreds of other known nebulae, Herschel could not resist speculating about the nebulae as island universes. The Milky Way—"the stupendous sidereal system we inhabit, this extensive stratum and its secondary branch, consisting of many millions of stars"[41]—he believed must be a "detached nebula," or, to use a more modern term, an island universe, separated from other islands by a considerable distance. He estimated qualitatively the true size and shape of other island universes, based on their resolvability—whether he could see stars, or just an indistinct haze. He concluded that the Milky Way was "not indeed one of the least, but perhaps very far from the most considerable of those numberless clusters that enter into the construction of the heavens."[42] He correctly identified the Andromeda nebula as one of the nearest of the island universes.

The 40-foot telescope

After the publication of his two great papers on the construction of the heavens, Herschel busied himself on a more mundane level with the construction of his next great telescope, the so-called 40-foot telescope of 49 inches diameter. Already in the early 1780s he had been making smaller telescopes for sale to raise money for the large telescope he dreamed of. But he chafed at the loss of time to work on his own projects. As Caroline noted, "[H]e was then on the wrong side of forty-five, and felt how great an injustice he would be doing to himself and to the cause of Astronomy by giving up his time to making telescopes for other observers."[43] Sir Joseph Banks, President of the Royal Society, came to his rescue with the promise of a £2,000 grant, which was later followed by a second.

The second half of the decade brought moments of great satisfaction for both Caroline and William, in the midst of all their work on the great telescope. In 1786, while William and Alexander were in Germany on business, Caroline enjoyed some uninterrupted time at her own small telescope and discovered her first comet. This success was repeated in 1788 with a second, and in 1790 with two more. Altogether she discovered eight comets in her spare time, a number not far out of line with that of professional comet-hunters such as Messier, who in the course of a lifetime discovered (or independently co-discovered) 20. By the time she discovered her eighth comet in 1797, Caroline was so familiar with the night sky that she began her nightly searches with the naked eye, relying on her memory of the constellation patterns to bring anything unusual to her notice.[44]

In January 1787, Herschel tried a simple experiment with his 20-foot, 19-inch aperture telescope that was spectacularly successful. He removed the small secondary mirror in this Newtonian design instrument and arranged his eyepiece so he could view the light directly off the primary mirror (see figure 4.4, the "Herschelian"). By eliminating one of the reflections in the optical path, he decreased the loss of light that inevitably occurs, also. To make the new arrangement work, he tipped one side of the primary mirror, so that the light could be brought to a focus off to the side of the primary axis of the telescope. With this slightly improved light-gathering ability, the 20-foot telescope showed two previously unknown moons of "his" planet Uranus, moons now called Titania and Oberon.[45] The experiment encouraged Herschel to design his 40-foot telescope, still in the making, as a "front view." Coincidentally, one of the first observations he made with the 40-foot—in August 1789, before the telescope was completely finished—revealed two new moons of Saturn, Mimas and Enceladus.

Although the mirrors for the 40-foot telescope tarnished quickly, and the instrument as a whole required such time-consuming maintenance that Herschel often preferred to use the 20-foot, the completion of this behemoth must also count as one of Herschel's great rewards during the late 1780s (figure 4.9). Although by this time he had a small army of smiths, carpenters and polishers working for him, he never lost interest in the finest details of its construction. "There is not one screw-bolt about the whole apparatus but what was fixed under the immediate eye of my brother," Caroline noted.[46]

For William, the second half of the 1780s also brought a happy change in his personal circumstances. In 1788, he married a wealthy widow, Mary Pitt. Financial worries eased, and

Figure 4.9 Herschel's 40-foot focal length reflecting telescope, of aperture 49 inches. (Credit: Royal Astronomical Society.)

William began to allow himself more time for pleasurable activities such as hosting concerts at his house and taking his wife, and, later, his son John on vacations.

The change occasioned by William's marriage was not such a happy one for Caroline, now aged 38, for after 16 years, she was being displaced as a caregiver and companion to William. She mentions her brother's marriage in her memoir only in connection with how busy she was in the months preceding it: "[I]t may easily be supposed that I must have been fully employed (besides minding the heavens) to prepare everything as well as I could against the time I was to give up the place of a house-keeper, which was the 8th of May, 1788."[47]

The pages of her memoir and journal pertaining to the period 1788 to 1789 do not survive, a fact that her family members attribute to her initial bitterness toward her sister-in-law over her new situation, and later "calmer judgement which counselled the destruction of all record of what was likely to be painful to survivors."[48] But Caroline had adapted to her evolving opportunities as housekeeper, singer, and assistant astronomer, and she adapted once again to her new situation. She later explained to the wife of her nephew John how she came to be granted a salary from the crown to continue her astronomical work: "I refused my dear brother's proposal (at the time he resolved to enter the married state) of making me independent, and desired him to ask the king for a small salary to enable me to continue his assistant."[49] This salary was granted, making Caroline the first woman to be paid to do astronomical work, although payments proved to be irregular. Caroline moved to her own apartments in town, and commuted on foot to William's house to continue her astronomical duties.

The evolution of nebulae

In his two papers on the construction of the heavens in 1784 and 1785, Herschel echoed the language, if not the scientific approach, of physical geographers and geologists. He articulated a view of the Milky Way system as a branched "stratum" of stars, and the nebulae as variously shaped beds of stars, more or less distant in space. In two papers that appeared in 1789 and 1791, he borrowed his metaphors more from zoologists and botanists interested in the evolution of living things. Although he again over-simplified the picture presented by an array of observations, he deserves credit for emphasizing, as no one had before him, the dynamic nature of nebulae and star clusters, evolving with time under the influence of gravity.

In his 1789 paper, Herschel proposed that specimens of the many varieties he had noted among the nebulae—globular, double or triple, "narrow but much extended," comet-like, fan-shaped, etc.—would, if properly arranged, form a temporal sequence that would illuminate a basic organizing principle. Diffuse, irregularly shaped nebulae, he thought, were just beginning to draw themselves together, while globular clusters seemed to have settled into their symmetrical, centrally condensed shape through the action of a central power (which he tentatively equated with Newton's gravitational force) over a long period of time.

Although many of the nebulae he fit into this scheme did not belong in the same temporal sequence — he lumped galaxies and relatively small star-forming regions in our own galaxy together — he showed a keen intuition for the formation of globular clusters. More generally, he demonstrated an apt approach to a central problem in astronomy: that of inferring the course of the physical evolution of stars and stellar systems, which occurs on extremely long timescales, from observations recorded over the span of a human lifetime. His concluding paragraph in the 1789 paper is widely quoted for beautifully expressing this approach. "This method of viewing the heavens seems to throw them into a new kind of light," he wrote. "They are now seen to resemble a luxuriant garden, which contains the greatest variety of productions, in different flourishing beds; and one advantage we may reap from it is, that we can, as it were, extend the range of our experience to an immense duration. For, to continue the simile I have borrowed from the vegetable kingdom, is it not almost the same thing, whether we live successively to witness the germination, blooming, foliage, fecundity, fading, withering, and corruption of a plant, or whether a vast number of specimens, selected from every stage through which the plant passes in the course of its existence, be brought at once to our view?"[50]

About 18 months after he presented this paper to the Royal Society, on a cold, clear night in November 1790, Herschel came across an object in the constellation Taurus (now known by its catalog number NGC 1514) that completely changed his view of nebulae and prompted him to add a new twist to his ideas on the evolution of stars and stellar systems. He saw a star apparently embedded in a spherical cloud, like an extended atmosphere. The nebulous cloud seemed to be associated with the star, and therefore at the same distance. But the cloud itself would not resolve into a cluster of stars, as Herschel would have predicted.

"A most singular phaenomenon!" he commented in his observing notebook. If the round nebulosity was actually a collection of stars, but at a great distance, the central "star" of this object must be superlatively bright, and unlike any other known star. It seemed much more reasonable to admit, instead, the existence of a "shining fluid," here and probably elsewhere. Herschel could only guess at the nature of this self-luminous matter; he wondered if it might resemble "the electrical fluid in the aurora borealis [the northern lights]."[51]

No matter what the detailed nature of the shining fluid, Herschel understood at once that its existence might serve to "unravel" other mysterious phenomena. In particular, he had never been sure how to classify what he called "planetary nebulae." These objects looked a little like planets in his telescopes: large and round, but of uniform milky brightness, not brighter in the middle like globular clusters. He recognized that an orb of shining fluid surrounding a star, like the "singular phenomenon" he noticed in the constellation Taurus, accounted very well for the appearance of the planetary nebulae. Never one to curb his speculative impulses, he further surmised that the shining fluid might constitute a first stage in the formation of a star: "If [...] this matter is self-luminous, it seems more fit to produce a star by its condensation than to depend on the star for its existence," he wrote.[52]

As in so many instances before, Herschel had hit on kernels of truth but applied his theories too widely. A planetary nebula consists of a cloud of heated material ejected from a central star. But the gaseous material is not self-luminous as Herschel understood the term; it glows as a result of the intense ultraviolet heating it receives from the star. And planetary nebulae such as the one he saw in Taurus consist of stars in their later stages of evolution, not stars caught in the act of forming. The gaseous envelopes have come from the star and are not, as Herschel thought, condensing to create the central body.

These misunderstandings notwithstanding, in broad outline Herschel's theory anticipated our current view of star formation. A proto-stellar cloud of hot gas and dust, as we find in the Orion nebula and other emission or star-forming types of nebula, contracts and heats up further, until the proto-star reaches a high enough density to begin the thermonuclear processes that characterize a full-fledged star. And in Herschel's own time, his account of the evolution of nebulae was favorably viewed by the greatest exponent of the theory of nebular evolution, the French astronomer Pierre-Simon Laplace.

Laplace's nebular hypothesis

Laplace independently developed a theory about the origin of the solar system that resembled Herschel's sketchy theory of nebular evolution. He first brought out his thesis in 1796, in his *Exposition du Système du Monde* (contemporaneous English translation: *The System of the World*), a few years after Herschel's planetary nebula paper appeared. He enlarged on his theory in subsequent editions of his book, and it became widely known as "Laplace's nebular hypothesis" or "nebular theory."

Laplace's nebular hypothesis played a prominent role in the development of ideas about our solar and Milky Way systems. He envisioned the origin of the solar system as a giant cloud of nebular material, rotating due to some primordial impetus (which he did not explain). As the cloud contracted due to its own gravity, the rotation must, Laplace knew, speed up, just as a twirling ice-skater spins faster as she draws her arms in toward her body. The rotation would gradually flatten the nebular cloud so that it took the shape of a rotating disk.

Laplace claimed that as the disk continued to contract and spin up, rings of gas or nebular material would be left behind, separated from the main disk. Several rings might be spun off successively, in this theory. He further claimed that the material in each ring would gradually draw together to form the planets. The Sun would be left as a remnant of the original spinning cloud, just as, in Herschel's theory, the Sun would form from the inward condensation of an extended nebula.

Laplace's theory to some extent eclipsed Herschel's, although the two men, who met during a visit by Herschel to Paris in 1802, did not consider themselves rivals. Laplace acknowledged their agreement on the basic tenets of the nebular hypothesis in a letter to Herschel in 1814, and incorporated Herschel's views in the 1835 edition of the *Exposition du Système du Monde*. There he noted that Herschel had "descended" from a consideration of nebulae in various stages of condensation to the idea that one or more stars might form from an initially amorphous, diffuse nebula; while he, Laplace, had "ascended" from a consideration of the probable history of our own solar system to a similar conception of a star-forming nebula.

111

Variants of Laplace's theory were widely accepted until modern times to account for the structure of the solar system and its planets. As we shall see in subsequent chapters, his theory also shaped astronomers' interpretations of the appearance of some nebulae as flattened systems in apparent rotation.

Infra-red radiation and binary stars

At the turn of the nineteenth century, Herschel, by then a naturalized English citizen, celebrated his 62nd birthday. He had come to astronomy only in his late thirties, following a respectable career in music, but he had devoted so much time and energy to his new calling that in the next dozen years he had made several historic discoveries. He had revealed previously unknown planetary bodies in the solar system, catalogued thousands of double stars in a difficult search for stellar parallax, brought to light some 2500 nebulae, proposed a first-of-its-kind map of the Milky Way system, and advanced the idea that different kinds of nebulae might be related in a temporal sequence. His creative period was far from over, however.

Two of his discoveries from the early years of the new century relate to subsequent developments by other astronomers profiled in this book. The first is his discovery of the infra-red part of the electromagnetic spectrum. He was observing the Sun and its sunspots with darkened glass filters of different colors when he noticed that through some of the filters he felt a sensation of heat, but saw little light, while through other filters he saw light but felt no heat. Intrigued, he temporarily abandoned his study of sunspots and set up an experiment to see if he could find some explanation. He dispersed the light from the Sun with a prism and set out an array of thermometers to measure the temperature rise produced by the different colors. In this way he found that the rays of light beyond the red portion of the visible spectrum — we would now think of these as longer-wavelength emissions — did not produce any illumination, but did produce heat. Radiant heat, he speculated, must consist of "invisible light."[53] In other words, he found that the rainbow of visible colors from the sun continued on the red side, or what we would now call the infra-red portion of the spectrum, with invisible radiation that was efficient at heating. Telescopes sensitive to infra-red and other portions of the spectrum are now an important part of the astronomer's tools to explore the universe, as we shall see in chapter 10.

The second of his noted achievements in the early part of the nineteenth century is his discovery, in 1802, that some of the double stars he had cataloged are not chance alignments of nearby and more distant stars, but binary systems — pairs of stars in orbit around each other. While searching for parallactic shifts, Herschel found that the separations of some stars changed, but not in the way he would expect from a seasonal variation in Earth's vantage point. The orbital periods of some of these binary systems were too long for him to observe a complete orbit — one has a period of about 200 years — but he saw enough to deduce that the shift in separation was due to the pair's binary nature. On the face of it, this finding represented a setback, because it precluded the use of those stars for Galileo's method of searching for parallax. However, astronomers welcomed this news as possible evidence that Newton's laws of gravity applied to bodies outside the solar system.

As indicated earlier, Herschel never did find parallax, at least in part because his equipment was optimized for the discovery of faint sources and not for the precise determination of stellar positions or the measurement of very small shifts. The parallax of the nearest star is less than 1 arcsecond, and detecting a shift of 1 arcsecond is equivalent to measuring the diameter of a penny at a distance of more than 1 mile. Most of the stars are much farther away and have a correspondingly smaller parallax. As we shall see in the next chapter, the measurement of parallactic shifts did not occur until almost the middle of the 1800s.

The unfathomed Milky Way

Herschel continued his observations with his 20- and 40-foot telescopes through the first decade of the 1800s, and continued to try to fit his observations of nebulae into a comprehensive theory of nebular evolution. Alexander helped with the mirror polishing and other duties. Caroline continued to assist William and to care for the family—when the irresponsible younger brother Dietrich showed up again, penniless, in 1809, she complained she had "not a day's respite from accumulated trouble and anxiety" until he left for Hanover four years later.[54] Meanwhile William's son John (figure 4.10) grew up and entered St. John's college at Cambridge University, where, as an undergraduate, he distinguished himself in mathematics.

In 1816, Alexander, then 71 years old, retired to Hanover, supported by an annuity from William. Perhaps his departure, and the loss of his assistance to William, precipitated some crisis at Slough, for in this year John decided to give up his

Figure 4.10 John Frederick William Herschel (1792–1871). (Credit: National Portrait Gallery, UK.)

114

graduate studies in law and go home to take up his father's "star-gazing."[55]

Herschel read one of his last major papers to the Royal Society in 1817, when he was 78 years old. He had reluctantly given up on finding the absolute distances to the stars: he knew that the parallax of even the nearest stars must be less than one arcsecond, and he had no hope of obtaining any satisfactory result from this method of triangulation. But he still sought to map the positions of the stars in three dimensions, or what he called "longitude, latitude, and Profundity."[56] Accordingly, he devoted much time in his later years to developing a quantitative method of comparing the relative distances to the stars. The 1817 paper describes his method, and his application of the method to the problem of "the construction and extent of the milky way."[57] His final conclusions on the subject are embodied in a sketch of the Milky Way, replacing the diagram he had earlier drawn from his star gauges.

To begin with, Herschel cast aside the notion of "big" and "small" stars, even though he commonly used such terms in his own observing notebooks. His method required him to admit "a certain physical generic size and brightness" of the stars, and besides, he felt, "we cannot really mean to affirm that the stars of the 5th, 6th, and 7th magnitudes are really smaller than those of the 1st, 2d, or 3d."[58] In that case, Herschel noted, the difference in the apparent magnitudes of the stars must be due to their different distances, rather than their different sizes. This overly simplistic but very common assumption has been called by a modern astronomer the "faintness means farness" axiom; as we shall see in chapters 8 and 9, astronomers continued to rely on it until the early years of the twentieth century, and it still may be used in some cases as an approximation to the truth if stars of one particular type are considered.[59] In fact, the intrinsic luminosity of a star may range from more than 10 000 times greater than that of the Sun, down to 1/100 times or less, so the assumption that faint stars are distant ones and bright stars are near, is a very poor one.

Next, Herschel pointed out, correctly, that the amount of light from a star is inversely proportional to the square of its distance. That is, a star of a given brightness at a distance of one unit (say, one light-year) will appear to be only a quarter as bright if removed to a distance of two units (or two light-years). The fact that this property was known meant that Herschel could put his relative distances on a quantitative basis.

To actually find the relative brightnesses and hence the relative distances, Herschel used two similar telescopes side-by-side to compare target stars with standard stars in what he called the method of "the equalisation of starlight." He equipped the telescope to be used on standard stars with an optical diaphragm (like the kind used in a modern camera) so that its aperture could be stopped down or fixed to smaller sizes. The telescope to be used on target stars was left "unconfined."[60]

Herschel pointed the first telescope to a chosen standard such as the bright star Arcturus in the constellation Bootes. He stopped down the aperture to a quarter, and viewed Arcturus this way. Then he searched through the second telescope for stars that seemed equally bright as the dimmed-down Arcturus. This search required him to move many times between the two telescopes, comparing the two views. He rejected many target stars that seemed slightly dimmer or brighter than the dimmed-down Arcturus, but finally found that the star Alpha Andromeda seemed to be a perfect match. Thus he established that Alpha Andromeda is twice as far as Arcturus, and by repeating this laborious equalization method with other targets and standards, he established the relative distances to other, more distant stars. A star of the "second order of distances" such as Alpha Andromeda could itself be used as a standard star. So, for example, Alpha Andromeda stopped down to a quarter of its brightness was found to be equal to the star Mu Pegasi.[61]

The "equalization" technique allowed Herschel to assign a "space penetrating power" to each of his larger telescopes. This in turn allowed him to measure, in a crude way, the "profundity" of the Milky Way system in various directions. For example, he turned one of his smaller telescopes to a spot of the Milky Way in Perseus, and saw stars which he knew must be of the 12th to the 24th order of distances (i.e. 12 to 24 times as far as his standard star). With a somewhat larger telescope, more stars became visible, and a whitish background appeared. These stars he estimated were at the 48th to 96th order of distance. With the 7-foot telescope he saw yet more stars, up to what he estimated was the 204th order of distances, and so on.

In this way he mapped out the relative depth of the stars on the celestial sphere in different directions. In some directions of the sky, perpendicular to the plane of the Milky Way, he thought he reached the limit of the stars, but within the Milky Way itself some parts seemed out of reach. "[T]he utmost stretch of the space-penetrating power of the 20 feet telescope could not fathom the Profundity of the milky way, and the stars which were beyond its reach must have been farther from us than the 900^{dth} order of distances," he wrote.[62] He had not yet used the temperamental 40-foot telescope in such a study, he wrote, but he predicted that even this great aperture instrument would not reach the end of the stars. "From the great diameter of the mirror of the 40 feet telescope we have reason to believe, that a review of the milky way with this instrument would carry the extent of this brilliant arrangement of stars as far into space as its penetrating power can reach, which would be up to the 2300^{dth} order of distances, and that it would then probably leave us again in the same uncertainty as the 20 feet telescope," he wrote.[63]

Herschel illustrated his new conception of the Milky Way with a simple diagram (see figure 4.11). The circle in the middle represents a sphere containing stars visible to the naked eye, that is, in his notation, to the 12th order of distances. The parallel lines show the stratum of the Milky Way. From the center of the circle where we sit to the top or bottom edge of the stratum, Herschel believed, is the 39th order of distance. For the purposes of this sketch he called the depth of the Milky Way along its greatest extent the 900th order of distances — but his illustration of the stratum is open, to show that the greatest depths have yet to be fathomed.

His illustration appears so tentative and imprecise, it is easy to overlook as the sum of his life's work on the construction of the heavens. He must have been acutely disappointed that he could not provide a better map of our stellar system after years of building and designing ever larger telescopes, counting the stars in laborious "star gages," and, late in life, starting afresh with the "equalization" technique of determining relative distances. Yet he retained to the very end a youthful and infectious enthusiasm for his subject.

Herschel's friend, the poet Thomas Campbell, said after a visit in 1813, when Herschel was 76, "[H]is simplicity, his

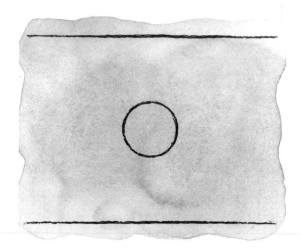

Figure 4.11 Herschel's "Unfathomable" Milky Way. The circle represents the limit to which the naked eye can see; the parallel lines delimit the Milky Way system as Herschel conceived it late in life, when he believed the system extended to an unknown distance in breadth.

kindness, his anecdotes, his readiness to explain, and make perfectly perspicuous too, his own sublime conceptions of the Universe, are indescribably charming.... [A]nything you ask, he labours with a sort of boyish earnestness to explain."[64]

Herschel died in August 1822, at the age of 83. Caroline, who was then 72, was despondent. She made a hasty and much-regretted decision to return to Hanover and live with her brother Dietrich. She always expected to live only a short time more herself, but endured another 26 years as a local celebrity, complaining all the time about incompetent servants and the "useless" life she led. In 1828 the Royal Society awarded her its Gold Medal for her arrangement of a catalog of nebulae and for her lifelong assistance to her brother.

John Herschel, who was destined to become one of the most prominent English astronomers of the nineteenth century, continued his father's work on double stars—now interesting in their own right as probes of the law of gravitation—and verified his father's observations of nebulae. He helped establish the Royal Astronomical Society, which met first in 1821, shortly before William Herschel's death. After his mother Mary's death, John and his wife and children sailed for the Cape of Good Hope, where John swept the skies for southern hemisphere nebulae and double stars.

5

WILHELM STRUVE:
SEEKER OF
PARALLAX

"The barrier has begun to yield."

John Herschel, 1841,
on parallax measurements by Bessel, Struve, and Henderson,
which breached the "hitherto impassable barrier" to a
knowledge of the stellar distances.[1]

Wilhelm Struve, a tall and athletic youth who looked older than his 13 or 14 years, was striding purposefully down a street in the outskirts of Hamburg one day in 1806 or 1807. Suddenly, recruiting officers for Napoleon's army seized him and bundled him off to a nearby house, where they locked him in a room on the upper floor of a two-story house.

Struve did not lack daring or self-reliance, two qualities that would characterize his personal and professional style. He escaped by taking a risky drop out the window, and ran back home to neighboring Altona.

Struve found safety at home, and later at a university campus far from his province, although political upheavals would punctuate his life and career. His escape was surely a turning point in his life and a providential accident for the sciences of astronomy and geodesy. Instead of fighting under the Emperor Bonaparte, he measured the exact size and shape of the Earth and found the distances to the nearer stars—a feat that had eluded astronomers for centuries. Twice he commissioned the largest and finest telescopes of his day. And, late in his career, he made a serious attempt to extend William Herschel's study of the construction of the heavens.

120

A Classical Education

Wilhelm Struve—or, to use his full name, Friedrich Georg Wilhelm Struve (see figure 5.1)—was born on 15 April 1793 in Altona, near Hamburg. An independent port city until Hamburg annexed it in the twentieth century, Altona was then part of the German, though Danish-ruled, province of Schleswig-Holstein, which lies between the North and Baltic Seas. Baltic Sea countries claimed Struve as one of their own all his life; although he spent his youth under German and Danish influences, he is known as a forefather of both Estonian and Russian astronomy. (See the map on the inside front cover of the book to locate many of the northern European cities mentioned in this chapter.)

Struve's family included four brothers and two sisters. Wilhelm became particularly close to his eldest brother Karl, a future teacher, and a younger brother, Ludwig, who was to study medicine. Not much is known about the Struve children's early life. We learn from a family history written later by

Figure 5.1 Friedrich Georg Wilhelm Struve (1793–1864). (Credit: Institute of Astronomy, Cambridge, UK.)

Wilhelm's son Otto that the household Wilhelm grew up in was well-regulated, to say the least; every hour had its appointed task. The children went on long walks "with gymnastic exercises" in summer, and skated in winter. Wilhelm's father once said, "We Struves cannot live happily without unceasing work, since from earliest youth we have been convinced that it is the finest and most valuable spice of human life."[2]

Struve's education and family background granted him all the advantages William Herschel's father had struggled to provide for the Herschel children about a half-century earlier in Hanover. Struve received a comprehensive classical education, the best his province had to offer. His father Jacob taught classical languages, dogmatics, and exegesis in Altona's *Gymnasium*, a university preparatory school. Indeed, Jacob Struve must have been a brilliant as well as educated man, and could have fostered his children's talents in any academic field. Though an expert in philology, the study of the historical development of languages, he filled his private notebooks with problems in statistics, probability, and number theory. Jacob even wrote a mathematics textbook that students praised for its clarity, and received an honorary doctorate for his mathematical work from the province's main university in Kiel.

Struve enrolled in the most junior class of the *Gymnasium* at the age of six, and began the final two-year program of university preparation at 14. He never completed the program, however, as a result of his run-in with Napoleon's press gang.

The incident, which Struve later recounted to his son Otto without recalling the precise date, probably occurred shortly after the famous battle of Jena in October 1806. Napoleon was then near the height of his powers, and had completely routed the Prussian forces. To rejuvenate his own forces, Napoleon had ordered the confederation of subordinate German states, such as Hamburg, to draw up fresh armies. His conscription rate in France at this time was on the order of 100 000 men a year.

French troops would not have pursued Struve to Altona because it belonged to the Danish crown, which was then playing a neutral role in Europe's complex political affairs. Struve's father, however, took no chances. He wrote to his elder son Karl, who taught at a *Gymnasium* in the city of Dorpat, and arranged for Wilhelm to apply to the new university there as soon as it was safe to travel through war-torn Prussia.

Dorpat is the old German name for Tartu in modern Estonia. In the nineteenth century, Czar Alexander I counted it as part of the Russian empire. Although Struve's family could guess that Napoleon planned to take on the Russian army, Dorpat seemed far enough off Napoleon's probable invasion route to be out of harm's way. Struve made his way there in late summer 1808 by a difficult journey lasting over a month. He carried a Danish passport, which his parents hoped would provide a cloak of political neutrality, and a letter of recommendation from his *Gymnasium* professors.

The day after his arrival, Struve climbed up the hill over-looking the Emajogi river. All around him workmen hammered away at the university's new buildings—so many of them in the neo-classical style that later campus denizens jokingly referred to it as "Athens on the Emajogi."[3] Struve undoubtedly saw them at work on the university's observatory; they had laid the cornerstone only a few weeks before.

The university admitted him at once. Luckily for the 15-year-old Struve, he would not have to learn a new language to enroll. Dorpat was the only university in the Russian empire to conduct classes in German.

Struve lived at first with his brother's family, but soon realized that Karl could not afford to support him as well as two small children. He began working as live-in tutor to the children of well-to-do townfolk almost immediately. At the beginning of his second semester he accepted what would become a long-term position in the household of a nobleman, Count von Berg, devoting 34 hours a week to the education of the count's four sons. This position might easily have derailed Struve's own studies. The von Bergs spent all but the coldest winter months on a family estate far from town, preventing Struve from attending many lectures, and they liked to show off their private tutor at dances and hunts.

As if the life of a teacher–student were not difficult enough, Struve also had to contend with the fact that he was stuck in the wrong academic field. At the end of his first semester, he wrote home to complain that his philosophy and philology courses were too abstruse, and to plead for permission to devote himself to mathematics. Unfortunately his father, despite his own passion for mathematics, thought the humanities would serve Struve better, both from the moral perspective and in his later attempts to find work. Jacob refused to be swayed and insisted that Struve stay with his classical studies.

Struve made the best of his lot. An astronomy course from J. W. A. Pfaff, the professor who had urged the construction of the university's observatory, helped sustain his interest in the natural sciences. At length his final degree examination in philology, which he passed with honors in 1810, established his independence. He declined a job offer from the Dorpat *Gymnasium* to teach history and quit teaching the von Berg boys, although he continued to live with them as a member of the household. With support from the university's rector, the physicist Georg Parrot, he embarked on a graduate program of his own design in physics, mathematics and astronomy.

Struve's first astronomy professor, Pfaff, left the university in 1809 and his replacement suffered from poor health, leaving the observatory understaffed. As a graduate student, Struve turned this situation to his advantage. He found a transit telescope, a popular nineteenth-century instrument that could be used in the determination of longitude, still packed in the crate in which it had been delivered to the observatory. Struve would have to teach himself the art of astronomical observation, but the use of this small telescope would furnish an excellent doctoral dissertation topic. Precise longitude determinations still held great interest for mapmakers and astronomers.

The special feature of a transit telescope lies with its mounting—how it is attached to its support structure—rather than its aperture or light-gathering power. A typical transit telescope (figure 5.2) rotates around a horizontal axis and can point only along a north–south line. The observer can view stars or planets only when they move east to west across his local meridian in the course of the night. In other words, it is possible to observe any object in the sky, but each object has its appointed viewing time, when it transits or crosses the meridian. The trade-off for limited range of movement is the superior accuracy and precision of pointing, which is vital for the kind of observations Struve wished to make. In the nineteenth century, the user of a transit telescope relied on a micrometer, a set of threads or wire reticles visible through the eyepiece, to establish precise star positions (see chapter 4, figure 4.6).

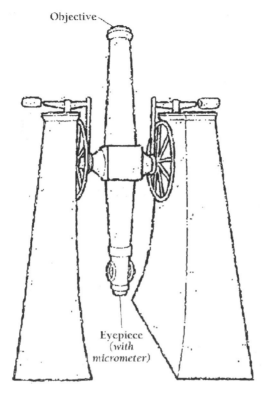

Objective

Eyepiece
(with
micrometer)

Figure 5.2 The transit telescope. A typical transit telescope is supported between two heavy piers, and can rotate only on one axis to show objects on the observer's meridian. The transit telescope was designed for the measurement of accurate stellar or planetary positions. The angle of elevation of the telescope, and the time at which the object transited the observer's meridian (as seen by the passage of the object across the wire in the micrometer) gave its declination and right ascension. (Credit: Layne Lundström.)

A typical observation with such an instrument proceeded by the "eye and ear" method. First, the observer consulted a star catalog or astronomical almanac to point the telescope to the correct position along the north–south direction, ready to catch sight of the desired star as the Earth's rotation brought it into view. As the star appeared in the eyepiece and approached the

125

wire or set of wires in the eyepiece micrometer, the observer would glance at a clock, or call to an assistant to do so, and note the time. Then the observer would pay close attention to the sound of the clock, counting the ticks while watching the star edge closer and closer to the wire. The micrometer often included several wires parallel to that marking the position of the meridian, so that an average of a set of measurements could be made.

A star might require two seconds to cross a given wire. As we now know, the moment of crossing of bright stars was often recorded a little too early and the crossing of faint stars a little too late, because of characteristics of human visual perception. The observer's estimate of the exact time the star crossed the meridian therefore incorporated his subjective perception of the passage of time and of the moment of the star's crossing the wire, in addition to instrumental biases, distortions inherent in observing through the Earth's atmosphere, and other effects. After the observer had accounted for all known sources of distortion, he would compare the time at which the object crossed his local meridian to the time it crossed the meridian at an observatory in Paris, London or some other reference point. The time difference gave the observer's longitude east or west of that reference.

Before he could begin his observations, Struve had to arrange for the transit telescope's final mounting. Workers had erected granite pillars in the new observatory to support the transit telescope and hold it steady, but Struve had to chisel holes in the pillars himself, to hold the posts attaching the telescope to the pillars. Later, his son recalled that Struve liked to tell the story of the "great time and toil" he had expended to get the telescope mounted.[4]

When he was not teaching himself astronomy or digging holes in stone pillars, the indefatigable Struve often picked up surveying instruments. Almost a hundred years after Thomas Wright found it a useful source of income, surveying continued to form, during Struve's lifetime, part and parcel of an astronomer's work. An important difference was that, during the Napoleonic wars, walking around with telescopes or surveying sextants tended to arouse suspicion. On a summer day, Struve was out surveying the von Berg's country estate when members of a Russian patrol picked him up, certain they had found a

French spy preparing the way for Napoleon's invasion. Struve was able to clear up the misunderstanding before a military judge, but the episode interrupted his work for a week.

Between 1810 and 1813, Struve completed his timing observations with the transit telescope at the Dorpat observatory and combined them with timing data from other observatories in Europe. This allowed him to determine the longitude of Dorpat with unprecedented accuracy — surpassing the achievements of two former astronomy professors at Dorpat.

He also determined the observatory's latitude; this is a simpler question of accurately measuring the height or altitude of a star above the horizon, as it crosses the meridian. The pole star provides a convenient indicator for observers in the northern hemisphere, because its altitude is roughly equal to an observer's latitude. An observer at 40° latitude, for example, will see Polaris reach a maximum height of 40° above the northern horizon, while an observer at the north pole, latitude +90°, will find it overhead, or 90° from the horizon. Struve used the same principle to determine the latitude of the Dorpat observatory.

The faculty committee supervising Struve's combined master's and doctoral work scheduled his thesis defense for a day in late October 1813. At the last minute the exam had to be postponed, for a reason that probably brought Struve particular relief, considering his brushes with the French and Russian armies. News arrived, by mail, that Napoleon's army had suffered its first major defeat in Central Europe. Spontaneous city-wide celebrations took precedence, for a day, over university business.

The university's natural science faculty members could not bring themselves to say goodbye to their industrious prodigy. After the 20-year-old Struve passed the exam they offered him a faculty position in mathematics and astronomy, with the understanding that he would direct the observatory as well. Struve's transition from student of philology to professional astronomer was finally complete.

Bessel's influence

Struve accepted the faculty position at Dorpat, but arranged for a leave of absence in the summer of 1814 to return to Altona and see

Figure 5.3 Friedrich Wilhelm Bessel (1784–1846). (Credit: Institute of Astronomy, Cambridge, UK.)

his family. On this long-awaited vacation, he met two people who would become very important to him: Emilie Wall, the teenage daughter of family friends and his future wife, and Friedrich Wilhelm Bessel, director of the Königsberg observatory (figure 5.3). Bessel's work paralleled Struve's in many ways, and he would become Struve's friendly rival in the race to nail one of astronomy's toughest problems, the establishment of stellar distances by parallax.

Bessel, older than Struve by about 10 years, had left school at 14, uninspired by academic subjects. He found work as a commercial accountant at a shipping company in Bremen, and there developed an interest in many practical aspects of the import/export business. He taught himself languages, geography and navigation; eventually the problem of determining longitude at sea led him to study astronomy and mathematics.

In 1804 Bessel wrote a paper showing how he had calculated and refined the orbit of Halley's comet, based on data from the comet's 1607 apparition. He then contrived to meet a leading German astronomer, Heinrich Olbers, on the street, and to raise the subject of his orbit calculation. The kindly Olbers recognized Bessel's exceptional talent in this work, and acted as Bessel's mentor until he made the transition to professional astronomer. At the age of 26, Bessel accepted the responsibility of guiding the construction of the observatory in the Baltic port city of Königsberg (now Kaliningrad, in Russia). Bessel directed that institution for the rest of his life, while a professor of astronomy at the University of Königsberg.

According to Struve's son's biography of his father, Bessel and Struve met for the first time on Struve's trip back from Altona to Dorpat via Königsberg. The recently completed observatory rose on one of the city's highest hilltops. Struve learned that during its construction, Bessel had embarked on some data analysis that would allow him to quantify the amount of "bending" or refraction that light rays exhibit as they pass through the Earth's atmosphere, a phenomenon of interest to all astronomers, and particularly those engaged in determining precise stellar positions.

A concern for precision was, indeed, the hallmark of Bessel's perspective on astronomy. Earnest and rather uncompromising by nature, Bessel considered obtaining precise positions and orbits of celestial bodies to be the only true aim of the astronomer. In contrast to William Herschel, who had investigated the surface of Mars and the orbits of double stars with equal enthusiasm, Bessel believed that besides positions and orbits, "Everything else that one may learn about the objects, for example their appearance and the constitution of their surfaces, is not unworthy of attention, but is not the proper concern of astronomy."[5] Struve was not such a purist, but in making precision observations a large part of his life's work, he may have been inspired by Bessel's zeal in that direction.

Refraction is by far the largest factor an astronomer must correct for in establishing a celestial object's true coordinates (see figure 5.4). Refraction occurs as light travels from a region of low density — such as interplanetary space — to one of higher density, such as Earth's atmosphere. The amount of refraction

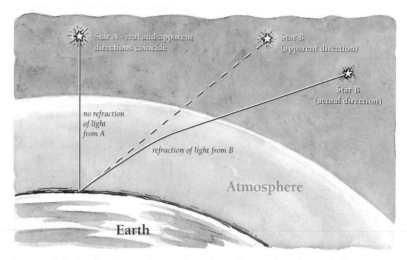

Figure 5.4 Refraction or "bending" of light passing through the atmosphere. As light passes through the Earth's atmosphere, its path deviates from a straight line according to the angle at which it enters the atmosphere and the amount of dense material it encounters. The light from star B ("actual direction" on the diagram) follows a curved path upon traversing the atmosphere. The observer's eye cannot make any accommodation for curved rays of light, however, so the apparent direction of the star is given by a backward extrapolation of the ray of light as it enters the eye. Thus the star appears to be higher in the sky (closer to the zenith, and farther from the horizon) than it actually is. Light from a star actually at the zenith, however, suffers no refraction. The real and apparent positions of the star coincide. (Credit: Layne Lundström.)

an observer has to contend with depends on the atmospheric conditions of temperature and pressure. It also depends on the height of the star or other object above the horizon, because the object's height above the horizon relates to the thickness of the atmospheric layer that the light must pass through. At the horizon, where refraction is strongest, an object often appears displaced higher in the sky than it actually is by as much as 34 arcminutes, or 2040 arcseconds. This is a large displacement, equivalent to the diameter of the full moon. The same object might not show any effect of refraction when it is at the observer's zenith.

William Herschel had not worried very much about refraction in his efforts to find parallactic shifts, because he sought this shift among pairs of stars that appeared close together in the sky, and so probably experienced the same amount of distortion in their apparent positions. If, say, he had seen two stars separated by 5 arcseconds in January and 3 arcseconds in June, he might have assumed that the shift was due to the changing line of sight to the nearer star, while the more remote star appeared at rest. Since he was observing both stars through the same patch of atmosphere, refraction could not act differently on each one to cause an apparent change in their separation. Bessel, however, with his characteristic thoroughness, sought a complete understanding of factors affecting stellar positions and the possible evolution of those positions in time.

Bessel took the lead among nineteenth-century astronomers in probing not only refraction, but also precession, nutation, and aberration. Struve's visit to Königsberg may have encouraged him to follow in his friend's footsteps, for he began conducting his own research on these phenomena.

Precession is the slow gyration of the Earth's spin axis, caused by the gravitational pull of the Sun, Moon, and planets on the unevenly distributed mass of the Earth (see figure 5.5). If the Earth's spin axis were a visible line extending into space, it would currently point close to Polaris, our "pole star" in the north. Over a period of 26 000 years, precession causes the axis to delineate a circle in the sky, such that different stars near that circle assume the role of Earth's pole star. The star Al Deramin will do so around AD 7500, then Deneb, Vega, and Alpha Draconis at successive points around the circle.

As the Earth's spin axis gyrates in space, the celestial equator necessarily rotates around too. That has implications for stellar coordinates; the zero-point or reference mark of right ascension is at the intersection of the celestial equator and the plane of the solar system, so as the celestial equator rotates, the intersection point also moves and keeps stars' coordinates on the move. Bessel worked on determining the precise rate of this precession because he wanted to compare the current positions of stars with those measured by astronomers in centuries past. Even though precession manifests itself in a cycle of about 26 000 years, a few hundred years is enough time for precession to cause a noticeable change in a star's coordinates. In fact, the coordinates of some stars are affected by as much as 50 arcseconds a year.

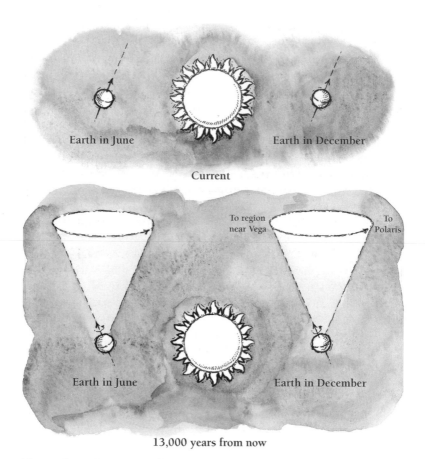

Figure 5.5 Precession of the Earth's spin axis. The Earth's spin axis currently points in the direction of the star Polaris, but the axis slowly "wobbles" as the Earth spins. Top panel: Over the short time span required for the Earth to orbit the Sun, the spin axis maintains its orientation. The axis points the same direction in June and in December. Bottom panel: Some 13 000 years from now, the axis will have rotated and will point to the bright star Vega. (Credit: Layne Lundström.)

Nutation is a "nodding" or wobbling of the Earth's spin axis and is also a result of forces on the Earth's uneven distribution of matter. Its effect on star positions is smaller than precession's — at most, a change of 10 arcseconds a year. Nutation is a more complex phenomenon than precession, and took astronomers much longer to appreciate. While precession was known to the Greeks in the second century BCE and explained by Isaac Newton in the eighteenth century, nutation wasn't even suspected until the eighteenth century.

Aberration is another apparent displacement of the objects observed. It arises from the combined motion of the light-emitting object, the observer on the surface of the Earth, and the finite speed of light. An analogy for aberration on a more familiar scale can be observed from inside a car on a snowy or rainy day. When the car is in motion, falling snowflakes or rain-drops appear to emanate from a point in front of the windshield, rather than up in the sky. The apparent deviation of the path of the falling snowflake or raindrop is like the apparent bending of starlight due to aberration.

In 1728, the English astronomer and parallax-seeker James Bradley first demonstrated the existence of aberration in repeated observations of the star Gamma Draconis. Over the course of the year, as the Earth revolved around the Sun, the position of this "fixed" star appeared to trace out an ellipse in the sky, simply due to the effect of aberration. The effect is quite large; the aberra-tion displacement amounted to about 20 arcseconds, while that star's displacement due to parallax was completely undetectable to Bradley, at 0.03 arcseconds.

Struve and Bessel were among the first to understand that stellar parallaxes were likely to be very small, and that the effects of refraction, precession, nutation, and aberration would mask the stars' small parallactic shifts, unless properly accounted for. In retrospect, their patient, methodical pursuit of factors affecting stellar positions, by up to tens of arcseconds, contrasts with the hasty and overly optimistic claims of scores of early nineteenth-century astronomers who claimed to have measured stellar parallaxes. The Italian astronomer Giuseppi Piazzi, for example, the unfortunate visitor to the Herschel house who "broke his shins" on the telescope's framework, believed at this time that the star Vega exhibited a parallax of about 2 arcseconds, and Sirius the enormous parallax of 4 arcseconds, which would put it at a distance of only 0.8 light-years. Bessel, Struve, and other astronomers greeted these results with skepticism, but, in the early decades of the nineteenth century, could not yet offer anything better.

Struve's career at Dorpat: geodesy, double stars, and the search for parallax

In 1815, shortly after Napoleon's final defeat at Waterloo, Struve married Emilie Wall in Altona and brought her to Dorpat. Little is known about Emilie's personality or her perspective on her new life with Struve, but her experiences were at least physically taxing. In 19 years, she carried 12 babies to term, beginning with Gustav, born in 1816, Alfred, born in 1817, and Otto, born in 1819.

Struve tried to reserve Sundays for family gatherings, but nights at the telescope and work-related travel often kept him away from home. He embarked on two main research projects that would dominate his 25-year tenure as a professor at Dorpat. The first of these, which he worked on mainly during the summers, was a series of land surveys on a scale large enough to take into account the Earth's curvature. The second — well suited to take advantage of long winter nights — was his study of double stars and stellar parallaxes, leading to his *magnum opus* on stellar astronomy and the distribution of stars in the galaxy, the *Etudes d'Astronomie Stellaire* [*Studies in Stellar Astronomy*].

In 1815, an organization called the Livland Public Utility and Economic Society asked Struve to survey the Russian province of Livland, territory that comprises the modern nations of Estonia and Latvia. It was an important project, as the survey formed the basis of new maps showing the precise location of hundreds of landmarks, and included information on elevation above the Baltic sea level, essential for planning public works in this area of extensive wetlands and swamp forests.

Upon completing the Livland survey in 1819, Struve proposed to the Russian Czar Alexander I, through university intermediaries, an even more substantial geodetic survey. He would measure the length of an arc of longitude stretching from an island in the Gulf of Finland to a town in present-day Latvia, considerably longer than the span available from the Livland survey. The project was approved, and this work kept Struve busy until about 1851, as he linked the initial survey to others in progress and, after 1825, petitioned the new Czar Nicholas I for funds to extend that work also. He finally came to supervise a multinational effort to measure an arc from the shores of the Barents Sea in extreme northern Norway to the city of Izmail (Ukraine) near the Black Sea. This arc, covering a distance of some 1800 miles (3000 km), is sometimes known today as the Struve Arc.

Struve's involvement in geodetic surveys followed naturally, if not inevitably, from his technical expertise, his passion for pursuing fundamental questions in science, and his position in Russia's most westernized university. All western powers carried out large-scale surveys in the nineteenth century, the better to assess the extent of their territories or overseas empires and to plan for their commercial exploitation. The British embarked on a trigonometric survey of India in 1802; the Americans established the precursor to their Coast and Geodetic Survey agency in 1807, beginning work on their transcontinental arc in 1871; Bessel himself carried out a triangulation of East Prussia in the 1830s; the Dutch mapped Indonesia in the 1860s; the French produced topographic maps of Southeast Asia in the 1880s. Struve's surveys helped link the westernmost provinces of the Russian empire to their European neighbors, physically and politically.

The geodesist's instruments for measuring positions and angles are essentially the same as the astronomer's, so governments usually tapped observatory directors for national surveys. For Bessel and Struve, the surveys also carried the allure of big questions in science. In the late 1600s and early 1700s, a set of French surveys showed a curious thing: the length of a degree of longitude measured north of Paris appeared to be shorter than a degree south of Paris, suggesting that the Earth is slightly egg-shaped. If the Earth were perfectly round, a degree of longitude would be the same length everywhere; if, as Isaac Newton had convincingly argued, the Earth were rather flattened at the poles, a degree should be slightly longer north of Paris, not shorter. Indeed, Newton's explanation of precession depended on the Earth's having an equatorial bulge and polar flattening.

British scientists vehemently dismissed the French data suggesting an egg-shaped Earth. To settle the question, the French Academy of Sciences dispatched geodesists to the equatorial regions of South America and to Lapland, to measure the lengths of arcs in these two extreme positions. The results of these eighteenth-century expeditions clearly vindicated Newton's theory, and stimulated more interest in the exact shape or "figure" of the Earth.

Selection of target stars and instruments

When he wasn't out measuring the Earth or teaching his many classes, Struve liked to sweep the sky with the transit instrument at Dorpat or with a later acquisition, a $3\frac{1}{2}$-inch aperture refracting telescope equipped with a micrometer. As William Herschel had done before him, Struve spent a great part of his early astronomical career looking for double stars. Indeed, Struve consciously took many cues from Herschel, not only cataloging double stars—something few other astronomers devoted so much time to—but also following up on specific observing projects Herschel had begun. For example, Struve confirmed Herschel's contention that the bright star Castor in the constellation Gemini actually consists of a physical binary system, with one star in orbit around the other.

Struve knew that many of the double stars he observed would eventually reveal themselves to be physical binaries, or even triple stars in mutual orbit, and so not suitable for the double-star approach to parallax determinations. Nevertheless, Struve approved of Herschel's plan (actually, Galileo's method) to seek distance measurements among those doubles that were not members of physical binary systems—and especially among those doubles in which one star was much brighter than the other, suggesting that one star was near and the other, far. Struve tried more than one scheme for measuring parallax, but his efforts to compile double star catalogs testify to his belief that double stars would ultimately yield a rich harvest of parallax measurements, and that he might succeed where Herschel had failed.

In 1818, Struve began a multi-year attempt to determine parallax angles for a number of stars near the north celestial pole, including some in Ursa Minor, the "Little Bear." In the same year Bessel published a monumental work, *Fundamenta Astronomiae*, that would set the stage for his own search for parallax. *Fundamenta Astronomiae* lists the positions of stars that Bradley had observed in the mid-1700s, corrected for instrumental effects and for distorting effects such as refraction and precession. This work by Bessel allowed later generations of astronomers to compare their observations with Bradley's, giving them a long time span over which to note changes and proper motions of stars. Bessel himself concluded from his review of Bradley's star catalog that parallactic shifts must be less than one arcsecond — a stiff challenge for seekers of parallax.

Bessel had also by then chosen his favorite target for parallax measurements, a star in the constellation Cygnus, known in catalogs as 61 Cygni. He had even published preliminary results in 1815 and 1816. The race was on: for the first time, two astronomers who understood the difficulty of parallax measurements and who perceived the futility of earlier efforts set out to do it right. In these early years of the race, however, both Struve and Bessel concentrated their efforts on laying the groundwork — Struve cataloging double stars and both Struve and Bessel gauging effects that might distort their measurements — knowing that their data on specific stars might have to be corrected later.

The star Bessel chose, 61 Cygni, is actually a binary system about 30 arcseconds wide. Its components, two orange stars of similar brightness that are not hard to see and separate with today's binoculars, orbit each other with a period of several hundred years. Bessel did not rule out 61 Cygni for parallax measurements on account of its binary nature, but treated the binary star system as a whole, and measured its position with respect to a number of widely-spaced background stars.

Bessel cast his lot with 61 Cygni because he had some clues, independent of parallax measurements, that it might be one of the nearer stars. In 1806, Piazzi had drawn the astronomical community's attention to the fact that 61 Cygni, also known as the "flying star," moves very rapidly across the sky, exhibiting a proper motion like the moving stars Edmond Halley had noticed in the early 1700s (see chapter 3). This proper motion is easy to distinguish from the shift due to parallax or aberration: a star with proper motion keeps moving in one direction, instead of wobbling back and forth against a field of background stars as the observer's line of sight changes with the Earth's orbit around the Sun. In the case of 61 Cygni, the proper motion is at the unusually high rate of more than 5 arcseconds a year. Piazzi suggested this might be an indication of the star system's proximity. All stars, he reasoned, might possess proper motion to some degree, and simple geometric arguments indicate that, on average, the nearest stars or star systems would exhibit the highest proper motion. In the same way, cars in an adjacent lane on the highway appear to move quickly, while cars on a distant highway appear to cross one's field of view more slowly.

A second clue that 61 Cygni might prove to be a good target in the quest for parallax lay buried in Bradley's mid-1700s observations of the system. Bradley had determined the approximate period of rotation of the two component stars about their mutual center of gravity. By applying basic principles of physics in the form of the laws of planetary motion (Kepler's third law, in particular), Bessel could deduce the approximate distance between the two stars. A comparison of this linear distance with the apparent separation of the two stars in the sky suggested the pair must be somewhat more than 7 light-years distant, with a parallax somewhat less than half an arcsecond. A small angle, to be sure, but perhaps measurable.

In devoting most of his effort to 61 Cygni, which he thought for independent reasons was relatively close, Bessel cleverly maximized his chances in what might turn out to be a cosmic wild goose chase. Careful astronomers all knew that parallax angles might prove to be too small for even the best observers to measure. Indeed, Bessel's first effort with 61 Cygni gave a negative number for the parallax angle, clearly indicating that his measurements and corrections were not sufficiently refined. Struve, initially, didn't even try to do more than set upper limits on the parallax for his set of polar stars.

In 1820, Struve took a step that would improve his own odds in the search for parallax. On a trip to Munich to order geodetic survey equipment, he looked in on Joseph von Fraunhofer, Europe's pre-eminent glass manufacturer and maker of optical instruments. Fraunhofer's help would be critical to both Struve and Bessel in their efforts to make astronomical observations of very high precision.

Fraunhofer, orphaned at a young age, had worked under harsh conditions as an apprentice mirror-maker and lens-grinder. In 1801, his master's house suddenly collapsed, killing his master's wife, but the young Fraunhofer was pulled alive from the rubble four hours later. The accident brought him to the attention of a wealthy civil servant and entrepreneur, Joseph von Utzschneider, who encouraged Fraunhofer's subsequent technical education and eventually—recognizing Fraunhofer's great talent—formed a partnership with him, the optical firm of Utzschneider and Fraunhofer.

Around 1814, while studying the refraction of light through glass in an effort to improve telescope optics, Fraunhofer made the discovery that he is best known for. He dispersed the Sun's light through prisms and found that the resulting spectrum did not form a continuous rainbow of color, but was crossed by some 500 dark lines. The lines, which are now known as the Fraunhofer lines, proved to be indicators of chemical elements in the solar atmosphere. The discovery of these lines stimulated the development of the science of spectroscopy, which in turn revolutionized astronomy, as we shall see in the next chapter. Fraunhofer, however, was singlemindedly dedicated to refining telescope lenses, and did not inquire in depth into the nature of his solar lines.

At the time of Struve's visit in 1820, Fraunhofer was working on a lens 9.6 inches in diameter (9 "Paris inches"), engineered to minimize the color distortions or "chromatic aberration" that glass lenses tended to produce in the telescopic images of stars. Chromatic aberration is more than a cosmetic problem: images blur when the different colors composing white light focus differently. To diminish the appearance of colored rings around the images of stars, Fraunhofer combined two types of glass, known as flint glass and crown glass, side by side. The chromatic aberrations produced by each lens individually could be made to cancel out in a so-called achromatic lens, producing a visually sharp image.

The challenge for the glass industry was to produce large samples of crown and flint glass without imperfections such as streaks and bubbles. Utzschneider, dissatisfied with the glass available on the market, had set up his own workshop in Benediktbeuern, some 30 miles [50 kilometers] from Munich, to supply his scientific instruments company. There he employed a Swiss artisan, Pierre Louis Guinand, who had developed a technique to make superior glass. Guinand eventually returned to Switzerland, but not before confiding—or being forced to divulge—his trade secrets to Fraunhofer.

139

A telescope made with Fraunhofer's 9.6-inch achromatic lens would be the largest refracting or lens-based telescope in the world, allowing for the clearest distinction between close double stars and producing crisp, uncolored images. Properly mounted, it would also allow for the most precise position determinations. Of course, in light-gathering power, Herschel's large mirror-based telescopes had the advantage: his 20-foot telescope had an aperture of 12 inches, and the flawed 40-foot telescope had an aperture of 49 inches. However, reflecting telescopes of the eighteenth and early nineteenth centuries produced poor quality images. As Herschel knew from experience, their metal surfaces were more difficult to figure accurately than comparable glass surfaces and they tarnished easily, requiring constant maintenance. Instrument makers such as Utzschneider and Fraunhofer were eager to solve problems of glass production because they knew that refractors promised to form the basis of more maintenance-free, reliable products.

Struve, who was more interested in precision observations than in discovering exceedingly dim sources such as nebulae, took a keen interest in Fraunhofer's prototype large refractor telescope. He resolved to secure it for the observatory at Dorpat if he possibly could. On his way home from Altona, he stopped in Königsberg and discussed Fraunhofer's innovation with Bessel. As soon as he reached Dorpat he submitted a proposal to the university's rector, offering to sacrifice some of his other research expenditures to make room for the telescope in the university's budget. The proposal wended its way up to the chancellor of the university, and Struve, to his elation, received a positive response.

Dorpat's great refractor

Fraunhofer's workshop took several years to complete what everyone referred to as the "great refractor." Struve, while waiting, pressed ahead with his program of observations with existing instruments. In 1822, he published a result for a single star, Delta Ursae Minoris, the second star away from Polaris in the handle of the Little Dipper asterism. He claimed to have measured its parallax angle (defined as half the star's annual shift in angular position) as 0.163 arcseconds, with a "probable error," a measure of his uncertainty, of 0.026 arcseconds. This parallax angle corresponds to a distance of 20 light-years. Measuring the parallactic shift of 0.163 arcseconds, as Struve claimed to have done, is equivalent to measuring the diameter of a penny seen 8 miles away.

Textbook authors tend not to mention this result because it is incorrect (the modern value is 0.001 arcseconds); the result didn't arouse any strong feelings among Struve's peers that he had convincingly solved the problem, and was later repudiated by Struve himself. Yet at the time of publication, Struve must have believed that he had found the first reliable distance to a star. He must have been impatient to receive the new, larger telescope from Fraunhofer to extend these important parallax measurements to other stars. The search for parallax had become something of a cottage industry among amateur astronomers, and Struve was eager to lay inferior and misguided claims to rest.

In the fall of 1824, Fraunhofer's firm sent the long-awaited "great refractor" to Dorpat. Struve's own description of the event—including his assembly of the telescope parts without the benefit of instructions—tells the story best. The *Astronomische Nachrichten*, a scientific journal for reporting astronomical research, had been circulating since 1823; Struve wrote in a letter to the editor as follows:

> On the 10th November last [1824] this immense telescope arrived here, packed up in 22 boxes, weighing altogether 5,000 pounds, Russian weight. On opening the boxes, it was found that the land carriage of more than 300 German miles [about 940 km] had not produced the smallest injury to the instrument, the parts of which were most excellently secured. All the bolts and stops, for instance, which served to secure the different parts, were lined or covered with velvet; and the most expensive part (the object-glass) occupied a large box by itself, in the center of which it was so sustained by springs, that even a fall of the box from a considerable height could not have injured it.
>
> Considering the great number of small pieces, the putting together again of the instrument seemed to be no easy task, and the difficulty was increased by the great weight of some of them; and unfortunately the maker had forgotten to send the direction for doing it. However, after some consideration of the parts, and guided

by a drawing in my possession, I set to work on the 11th, and was so fortunate as to accomplish the putting up of the instrument by the 15th; and on the 16th (it being a clear morning) I had the satisfaction of having the first look through it at the Moon and some double stars.

I stood astonished before this beautiful instrument, undetermined which to admire most, the beauty and elegance of the workmanship in its most minute parts, the propriety of its construction, the ingenious mechanism for moving it, or the incomparable optical power of the telescope and the precision with which objects are defined.[6]

The telescope represented a point of national pride, and Struve must have felt some pressure to produce great results from it. Czar Alexander I gave both Fraunhofer and Struve diamond rings for their roles in creating this "great refractor" and bringing it to Dorpat. Utzschneider, Fraunhofer's partner, informed the university that the firm had absorbed most of the cost over-runs, anticipating that the telescope's reputation would bring new business.

Struve appears to have taken with ease to his high profile position. Already at this time his colleagues thought of him as a "powerful, almost Goethean character" on campus.[7] His subordinates knew him as a tough man, genial but always dignified and a bit formal, sometimes demanding and autocratic. Some years later, when Struve directed the prime observatory of the Russian empire, Pulkovo, it became obvious that his style was rather patriarchal. His friend, the explorer and scientist Alexander von Humboldt, speaking in confidence and perhaps somewhat jokingly to Bessel, once referred to Struve as "the tyrant of Pulkovo."[8]

Working alone, for the most part, Struve set to work on a catalog of all double stars north of declination −15°. On clear nights he systematically swept the telescope along the north–south direction, pausing at every double star to estimate the magnitude of each component, the separation between the components, and the position of the double system. One out of every 35 or so stars turned out to be double. Even allowing for time to record his observations, he found he could examine more than 400 stars an hour this way. He ultimately listed about 3000 items, mostly doubles with some triples included.

With more time, Struve could have recorded precise positions and separations. His aim, for the moment, was simply to compile a list of all double stars brighter than ninth magnitude, adding his to those that the Herschels and other observers had found. The astronomical community evidently valued this contribution, for when the resulting work, *Catalogus Novus*, came out in the summer of 1827, the Royal Astronomical Society rewarded him with its prestigious Gold Medal prize. Fraunhofer, unfortunately, did not live to see the use Struve made of his great instrument; he died of tuberculosis in 1826, at the age of 39.

Struve intended the catalog to support a search for measurable stellar parallax, and in a lengthy introduction, he took the trouble to help other astronomers make the best use of it. The main problem for seekers of parallax was to select optical doubles, as opposed to physical binaries. Struve explained, using a statistical analysis of his data, that closely-spaced doubles were more numerous than anticipated and were probably stars in orbit around each other, while the more widely-spaced double stars were more likely to be chance alignments of near and distant stars, suitable for parallax tests on the brighter of the two in the pair. He generously pointed out to his readers two specific examples of widely-spaced double stars that he thought held the most promise for parallax measurements. One was Alpha Andromedae, the brightest star in the constellation Andromeda. The second, which he was to work on extensively himself, was the bright star Vega, otherwise known as Alpha Lyrae, with its companion 42 arcseconds away.

Professional contacts

The next logical step in Struve's astronomical career — and the next lap in the race to establish stellar distances in a convincing way — was for Struve to determine precise positions for the double stars listed in his *Catalogus Novus* and to target the most likely stars for ongoing parallax measurements. His projects came to fruition only after many years' delay, however. Between 1828 and 1834, his geodetic field work, travel, new administrative responsibilities, and family life all contributed to postpone progress on the parallax effort.

In 1828, Struve lost three family members. His younger brother Ludwig, who had lived in Dorpat for the past four years, died rather suddenly. Gustav, the Struves' oldest child, and Alexandra, aged three, died in an epidemic of typhoid fever. Other children, including Otto, were ill. The same year, Struve broke his leg in an accident with the very heavy "great refractor" telescope.

More difficult times yet lay in store, but Struve's professional contacts during this period buoyed him. Among the most important of these was his friendship with Alexander von Humboldt, the explorer and scientist who later coined the term "island universe" for what we now call galaxies.

Humboldt, a generation older than Struve, made a famous research voyage to Central and South America between 1799 and 1804. He surveyed the Spanish territory there and collected a wealth of geographical, meteorological, and botanical data, which he described in a lively account, *Personal Narrative of Travels to the Equinoctal Regions of the New Continent during the years 1799 to 1801*. Charles Darwin said of his influence, "I shall never forget that my whole course of life is due to having read and re-read as a youth his *Personal Narrative*."[9]

Humboldt, Struve and Bessel shared a written correspondence and exchanged visits. The three men made an unlikely trio, yet they complemented each other in character and inspired one another intellectually. Humboldt, cosmopolitan and witty, came from an aristocratic family and returned from his South American voyages to serve as personal advisor to the Prussian king. His scientific interests were broad; only he could have attempted to write everything known about the Earth, as he did in his monumental and encyclopedic work, *Cosmos*, the first volume of which appeared in 1845. Of the three men, Humboldt had perhaps the greatest originality and broadest scope, Bessel the greatest genius in mathematics, and Struve the purest focus on uncovering the construction of the heavens.

Humboldt visited Dorpat in 1829, at the outset of an expedition to the Urals and central Asia, and consulted with Struve on a challenging project to measure the difference in elevation between the Caspian and Black seas. He was one of several distinguished visitors to Dorpat during Struve's tenure as director. Struve also cemented his relationship with his professional

144

contacts in Europe by traveling west for astronomical conferences or meetings with instrument makers.

In 1830, Struve made one such trip, to Germany, France, and England. Struve stopped in Königsberg, as he usually did on his voyages, to see Bessel. That these visits stimulated both parties can be seen from a letter Bessel sent afterwards to Humboldt. Bessel and Struve were to participate together in a surveying project: in 1831 and 1832, Bessel directed geodetical measurements of meridian arcs in East Prussia, to link up with Struve's surveys in the Baltic states. Bessel wrote, "[T]he warmth which he shows for this project seized me too, so that the sacrifice which I must thus make of other tasks [in order to participate in the project] really seems smaller than it did before." He added that he counted himself lucky to have had first the astronomer Johann Encke, then Struve stay with him, for each in his own way exemplified how to live the "astronomical way of life" (*das astronomische Leben*).[10]

Struve's stop in Paris involved another of his oddly frequent brushes with Napoleonic wars and civil disturbances. Struve happened to pay a call to his colleagues at the Paris observatory on the morning of 27 July 1830. By the time he was ready to return to his hotel, a riotous crowd of students and workers had barricaded the streets, protesting repressive government decrees. France had returned to monarchy after Napoleon's defeat at Waterloo, and King Charles X and his cabinet had alienated the country with limits on suffrage and freedom of the press. For three days, during what came to be known as the 1830 revolution, Struve camped out at the observatory.

During a less agitated sojourn in London, Struve visited the Herschel house in Slough. The 40-foot telescope still rested in its wooden frame in the garden, a monument to the ambitions of its maker. William Herschel had died eight years earlier, in 1822; his widow Mary still lived. As we saw in chapter 4, William's sister and assistant Caroline had gone to live with a surviving brother in Hanover.

At the time of Struve's visit to London, William's son John Herschel, recently married, was reviewing the nebulae and clusters that his father had observed. He had already received his own Gold Medal for his work on double stars; along with the English astronomer James South, he was one of Struve's

145

few competitors in this field. He had published papers on methods of determining parallax, had recently finished one of his most famous works, *Discourse on Natural Philosophy*, and was tackling the mathematical problem of determining the motions of double stars in orbit around a mutual center of gravity.

The highlight of Struve's trip occurred when John Herschel presented him with a complete set of William Herschel's papers, annotated by the author. Struve treasured these papers all his life. He believed they gave him essential insight in his own studies. Later they formed the nucleus of a world-class library of historical astronomical texts that Struve amassed at the Pulkovo Observatory.

Bessel's heliometer

While Struve traveled and recovered from the loss of his brother and two children, his sympathetic friend Bessel enjoyed some productive years. Beginning in 1821, he had begun to compile a catalog of the tens of thousands of stars brighter than ninth magnitude, lying between declination $-15°$ and $+45°$. He had continued to pursue sources of error in observing star positions, and had refined his studies of the rate of Earth's precession. As a foundation for his own and others' future work on parallax and proper motion, he had established a reference system for the positions of stars. All this work culminated in 1830 with the appearance of *Tabulae Regiomontanae* (the Latin name translates as King's mountain, or Königsberg in German). The *Tabulae* includes a listing of the positions of 38 exceptionally well-studied stars, which serve as fixed points in the stellar reference system, and rules for systematically treating data on stellar positions. Generations of astronomers would come to rely on this work of Bessel in making comparisons for parallax or proper motion determinations, or in accurately measuring the positions of the Sun, Moon, and planets.

In 1829, Bessel finally laid his hands on a new instrument that he had ordered from Fraunhofer almost a decade earlier. Bessel had helped Fraunhofer design the 6-inch telescope, of a type known as a heliometer. Utzschneider's workshop completed the heliometer after Fraunhofer's untimely death from tuberculosis.

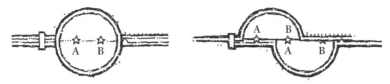

Figure 5.6 The heliometer type telescope. Originally conceived as an instrument to measure the width of the disk of the Sun (*helios* in Greek), the heliometer was used by Bessel, in place of a wire micrometer, to measure the separation of stars. The two halves of the objective lens are slid apart until the image of star A lies directly under the image of star B. The amount of displacement of the two halves of the lens gives the angular separation of the stars. (Credit: Layne Lundström.)

This instrument would allow Bessel to implement his own plan to measure parallax, which, unlike Struve's plan, did not call for using double stars.

The heliometer first emerged in the mid-1700s as a device to measure the width of the disk of the Sun. (Hence its name—*helios* is Greek for the Sun.) It consists of a round lens split along a diameter. The two semicircles can slide past each other along the cut (figure 5.6). Each half of the lens produces its own image, so the heliometer is like a double-duty telescope. The astronomer who wanted to measure the width of the solar diameter slid the two halves of the lens apart until the top and bottom images of the Sun moved apart. When the left edge of the solar image on the bottom lined up with the right edge of the solar image on top, he read off a scale the linear displacement of the two halves of the lens. This reflected the width of the solar disk.

Bessel eventually thought of another use for the heliometer. Given stars A and B in the field of view, he could turn a screw to slide the two halves of the lens apart until the lower image of A and B moved by an amount equal to the stars' separation—that is, until A in the lower image lay under B in the upper image, or vice versa. This would allow him to measure, with unprecedented accuracy, the distance between two stars separated by up to 2°, or 7200 arcseconds. This range is far greater than that available with wire micrometers. He would not be limited to measuring the parallax of one star with respect

147

to its companion in an optical pair; he could measure the apparent displacement of a star, or star system, with respect to scattered "field stars" in the same part of the sky. Astronomers still use his principle today. Making measurements with respect to a number of field stars limits the effects of any small parallactic shifts in the field stars themselves.

Bessel did not immediately apply the heliometer to the study of 61 Cygni, but renewed his attack on the parallax of that system in 1834. Even then, Struve was still occupied with new duties and problems. The race for parallax took a long time to heat up.

Crises and new directions for Struve

Shortly after Struve's return to Dorpat from Paris and London in 1830, Czar Nicholas I called on him to evaluate the state of the observatory at St. Petersburg. That institution owned, among other instruments, a 20-foot telescope that William Herschel had made for sale, but the building housing the telescope stood too close to the smoky city to permit good quality observations. Struve gave his frank opinion that the observatory served little purpose. From that moment, Struve played a leading role in selecting the site and planning for the new observatory the Czar wished to build.

In 1834, just as the commission that Struve sat on finalized the plans for the new observatory and received funds for the building, Struve's family life reached a crisis. Emilie was pregnant with the couple's 12th child; the 11th child died that year. Alfred, the oldest surviving son, lay bedridden, stricken by a diseased hip, with Emilie acting as his nurse.

On 1 January 1835, Emilie gave birth to a daughter, also named Emilie. The next day, Alfred died. Emilie recovered from the delivery of her daughter with difficulty, and three weeks later, as her son Otto described it, "a kind spasms with fierce pains in the back and sides suddenly appeared."[11]

Sensing that her end was near, 37-year-old Emilie called Struve and urged him to marry again after her passing — specifically, according to Otto, to marry Johanna Bartels, the daughter of one of Struve's colleagues. A few days later, on 1 February, Emilie died.

Otto writes in his biography of his father that he bore his losses "manfully," and that caring for his flock of children helped numb the pain.[12] Some of his zest for life returned with new astronomical charges. About 10 weeks after Emilie's death, Struve and Sergei Uvarov, Minister of Public Instruction and President of the Imperial Academy of Sciences, appeared before the Czar again. Nicholas designated Struve as the director of the new observatory, which was to be built at Pulkovo, some 20 kilometers from St. Petersburg on one of the few hills rising above the marshy plains.

Nicholas asked Struve to commission the best possible instruments to equip the observatory as soon as possible, so in June 1835, Struve departed for Germany and Austria, where the most renowned instrument-makers had their workshops. He took Otto, his oldest surviving son, with him, leaving the rest of his household somewhat in disarray. Otto, 15, had just finished his studies at the Dorpat *Gymnasium*.

Struve's most important appointment awaited him in Munich with Utzschneider's firm, which since Fraunhofer's death had been directed by Georg Merz. The artisans there had just finished a refractor of 11.2 inches aperture for the Munich Observatory, which they proudly offered for Struve's inspection. Struve pressed them to try to make an even bigger objective, despite the fact that the Czar's commission's original plans only called for this size.

Father and son visited Bessel on both outbound and inbound legs of their journey, and showed him the detailed plans for the new observatory at Pulkovo. The Struves also stopped to see Alexander von Humboldt in Berlin. They doubtless visited Berlin's new observatory, which Johann Encke directed but which owed its existence to Humboldt's strong support.

Parallax

By the spring of 1835, Struve's affairs were settling and he could look forward to a resumption of his studies of double stars and the search for parallax. He had married Johanna Bartels, and she expected the birth of their first child in the fall. Johanna, about 28 years old, restored order to the large household. Astronomers who knew her later described her as quiet and pious, but well-loved and respected. Otto described her as "the refuge of all who needed help and council."[13]

Beginning in 1835 and continuing through 1836, Struve resumed his work on double stars, following up on the *Catalogus Novus* work with measurements of the separations of a promising set of 17 double stars, including the bright star Vega and a nearby companion star. Civil construction had begun on the new observatory at Pulkovo, and Struve was trying to accomplish as much as possible at Dorpat before the inevitable disruption of the move. Bessel, meanwhile, had begun monitoring the position of 61 Cygni in September 1834, using a standard micrometer; he had not yet started using the heliometer for this purpose. He was trying to use two 11th-magnitude stars as references, but these stars were so faint in the heliometer, which had an aperture of only 6 inches, that he found their positions difficult to measure unless atmospheric conditions were outstandingly good. Bad weather made for slow progress on the parallax effort, and in 1835, geodetic field work claimed his attention, too. Thus in 1836, Struve had the opportunity to concentrate on his effort to measure parallax, while Bessel was somewhat diverted both by unrelated work and by difficulties with his approach. Struve's correspondence indicates that he had not yet mentioned to Bessel his intensive effort to find the parallax of Vega.

In January 1837, Struve had some news. He gave a preview of his forthcoming results in a letter to the Permanent Secretary of the Russian Imperial Academy of Sciences, Paul Fuss.

"I trust I have been able to show that observations are now incomparably more exact, and parallax cannot escape us even if it consists of as little as 1/10th of a second of arc," he wrote. He added that a special series of observations of the star Vega (otherwise known as Alpha Lyrae) were not yet absolutely decisive, but that calculations gave a parallax of between 0.10 and 0.18 arcseconds. He concluded by explaining his result in units of "solar distance," the Earth–Sun distance: "If it turns out, as I hope, that further calculations confirm this result, this would constitute the important discovery that Alpha Lyrae is at a distance from the solar system of 1 million solar distances."[14] Such vast distances to the stars had been suspected by some, but Struve's parallax measurements provided welcome hard evidence.

Struve then completed his review of the separations of double stars he had earlier cataloged. In 1837, he published the results, to wide acclaim, as *Mensurae Micrometriae*. In the introduction to this work Struve laid out his updated thoughts on opportunities for parallax measurements. He considered — incorrectly, as we now know — that the brightest stars are likely to be the nearest, although he admitted that this might not be true in all cases. He noted — perhaps as a result of discussions with Bessel, who had articulated the idea as early as 1812 — that a star's high proper motion probably indicated nearness. This, of course, was Bessel's criterion in selecting 61 Cygni.

The main thrust of *Mensurae Micrometriae* is Struve's effort to measure stellar distances for stars that he considered the most amenable to the parallax method. He quoted the parallax for Vega as 1/8 of an arcsecond (0.125 arcseconds), with a somewhat large "probable error" or range beyond this value of about 1/20 of an arcsecond (0.055 arcseconds). This parallax for Vega compares well with the modern value of 0.123 arcseconds.

Astronomers all over Europe took note of Struve's progress in parallax determinations, the first successful application of Galileo's double-star method. They admired Struve's results despite the large uncertainty associated with the quoted value. James South, the double-star observer in England, remarked to a visitor, "Struve has reaped the golden harvest among the double stars, and there is little now for me to hope or expect."[15] Indeed, he became so bitter over his own failure in this regard that, in a fit of pique, he destroyed an expensive lens he had ordered for his observations.

Bessel, who had a greater personal fondness for Struve and didn't incline toward fits of pique anyway, praised the book enthusiastically. Struve had advised him of its imminent publication, and of his still inconclusive work on Vega, in a letter dated 25 July 1837.[16] A month later, on 26 August, Bessel wrote back to Struve to praise the *Mensurae Micrometriae*, which he had by then received, and to say that Struve's attempt to measure the parallax of Vega stimulated him to do the same with 61 Cygni and another parallax candidate, Alpha Bootis. He noted, however, that Struve's final parallax results would likely emerge first, since (as Bessel had learned from Struve's letter) Struve had already been collecting data for a year. "Forever the honor of having tried this method first belongs to you," Bessel wrote.[17]

To Humboldt, Bessel wrote in September that Struve's "great work" had given him much pleasure, and that some of Struve's comments had provoked him to scrutinize his own methods again. He did not consider differences between Struve's measurements and his to be pertinent for public comment, but, Bessel wrote, he had satisfied himself that specific disagreements arose either from technical difficulties in the use of the micrometer or from Struve's inability to compensate precisely for the unwanted movement of his telescope due to the clock drive mechanism, whose function was to keep the telescope pointed at the stars as they rose and set.[18]

In October, Bessel wrote to his mentor Olbers, acknowledging that he saw himself in competition with Struve. "I think Struve has taken the lead, for he has made an attempt which, though not yet a complete success, nevertheless seems to offer good prospects," he wrote.[19]

Spurred and perhaps worried by Struve's progress, Bessel resumed his effort on 61 Cygni with renewed vigor. As he later wrote to John Herschel, various things, including work on the Earth's gravitational effect on timekeeping devices, had distracted him earlier:

"In the year 1835, researches on the length of the pendulum at Berlin took me away for three months from the Observatory; and when I returned, Halley's comet had made its appearance and claimed all clear nights. In 1836, I was too much occupied with the calculations of the measurement of a degree in this country, and with editing my work on the subject, to be able to prosecute the observations of α Cygni [he meant, 61 Cygni] so uninterruptedly as was necessary, in my opinion, in order that they might afford an unequivocal result. But in 1837 these obstacles were removed, and I, thereupon, resumed the hope that I should be led to the same result which Struve grounded on his observations of Alpha Lyrae, by similar observations of 61 Cygni."[20]

During the course of his new observations in the winter of 1837–38, Bessel abandoned work on Alpha Bootis and focused on securing the parallax of 61 Cygni. He stopped using the two 11th-magnitude reference stars he had had difficulty with and found some brighter ones, magnitudes 9 and 10, farther away from 61 Cygni. Now he applied the heliometer to measuring these distances. The comparison stars were about 8 and 12 arcminutes away, far outside the range of a conventional micrometer but well within the range of the heliometer. In March 1838, Bessel described this progress to Struve, adding that Struve would be the first to learn of the results, since they were encouraged by Struve's observations of Vega. Struve responded that measurements of Vega continued to accumulate, but that he would wait to process all his data until November of that year.[21]

In October 1838, just before Struve's anticipated new results on Vega, Bessel announced his own results to the editor of the *Astronomische Nachrichten* and to John Herschel at the Royal Astronomical Society. He found 61 Cygni's annual shift to be about 1/3 of an arcsecond (0.314 arcseconds); the results might be in error by 1/50 of an arcsecond (0.020 arcseconds). The parallax angle for 61 Cygni is larger than that Struve found for Vega, and indicates that Bessel's target was closer, at a distance of about 10 light-years. Bessel's result is not far from the modern value for the parallax of 61 Cygni, 0.287 arcseconds, determined with the use of instruments on board the *Hipparcos* satellite.

Neither Bessel nor Struve considered the race to measure stellar distances to be over. Both continued to refine their results on the target stars they had chosen. However, textbooks usually cite this parallax determination of Bessel's as the first measurement of the distance to a star. His measurement was more precise: he found the distance to 61 Cygni to be 10.4 light-years, plus or minus 0.6 light-years. Struve, on the other hand, found that the distance to Vega lay in the range 18 to 47 light-years.

Bessel's result appeared in the *Astronomische Nachrichten* of December 1838. Struve's reaction is not recorded in his correspondence, although he must have noted Bessel's superior result. This year too was a difficult year for Struve, with or without the competition with Bessel. His son Friedrich, the second child with Johanna Bartels, had recently died in infancy. Johanna's father, Struve's close friend and colleague in the mathematics faculty, died. Paul, the couple's third child, was born in 1838, but he did not survive infancy either, leaving only the first child by this marriage, Karl.

Yet again, Struve could not allow himself to be distracted by family misfortunes, on account of the impending move to Pulkovo. In the late summer of 1838, he took Otto back to Munich to purchase books for the new library he intended to create, and to inspect the instruments that were nearing completion. The firm of Merz and Mahler, the successor to that of Utzschneider and Fraunhofer, had succeeded in making a lens of 15 inches, significantly bigger than the 10.5 inches that the Czar's commission had originally proposed. During this time, Struve also advised assistants on surveys to determine the levels of the Black and Caspian seas.

Unexpectedly, in January 1839 a newcomer joined the parallax contest, if rather belatedly and timidly. Thomas Henderson had been stationed at the British observatory at the Cape of Good Hope in what is now South Africa. Upon returning to his home in Scotland with his data, he discovered that his observations of the star Alpha Centauri would lend themselves to parallax determinations. Henderson had stumbled on what is in fact the closest star, with a parallax of three-quarters of an arcsecond. He published his result of 1 arcsecond, with an uncertainly of 0.75 arcseconds. The result struck the astronomical community as rather tentative, given that he quoted such a large estimated uncertainty. Henderson's paper must have sounded a kind of warning bell for Struve, however, signaling that other observers deemed parallax measurements within range of their instruments.

Struve gathered as much additional data on Vega as he could at Dorpat, because the precision required for parallax measurements would not allow him to combine data from two different instrumental set-ups at Dorpat and Pulkovo. But he did not have time to analyze his data before the move, which came in May 1839.

Johanna presumably took care of moving the extended — and still growing — family into their new quarters, a two-story apartment connected to the main building. (See figure 5.7 for a nineteenth-century view of Pulkovo.) Between 1839 and 1842, Wilhelm and Johanna had three more children, Anna, Ernst, and Nikolai. Johanna took responsibility for educating the Struve children, too, as there was no school at Pulkovo. The observatory lay too far for the family and visiting astronomers

Figure 5.7   The Pulkovo Observatory at the end of the nineteenth century. Fraunhofer's 15-inch Great Refractor occupied the space under the central dome. (Credit: Institute of Astronomy, Cambridge, UK.)

to have regular contact with St. Petersburg society. They even grew some of their food on the observatory grounds.

The various telescopes, clocks, and accessories that Struve had ordered from Germany and England, including the "15-inch great refractor" telescope, arrived in parts in 102 crates a few months after the family moved. Once again, as in Dorpat, Struve could not wait for company representatives to take charge of the assembly. He had taken a keen interest in the details of the instruments' construction — Merz and Mahler's assistants might have called it an overbearing interest, as legend has it that Struve vetted every screw. Armed with this knowledge, he and the new Pulkovo employees erected the new telescopes by themselves in six weeks.

Having assembled Pulkovo's instruments with all possible speed, Struve returned to analyzing his Dorpat data on the parallax of Vega as soon as he could. He put Otto in charge of the "15-inch great refractor." Struve himself did not observe as much at Pulkovo as he had at Dorpat. However, in conjunction with his attempt to refine the parallax measurement on Vega, he made more observations pertaining to the strength of aberration and nutation constants.

His new value for the parallax shift of Vega came out in 1840: an angle of 0.2619 arcseconds, with an uncertainty of 0.0254 arcseconds. With hindsight, it appears that Struve's attempt at refining the value of parallax failed, for his first value was more nearly correct. It is Bessel's 1838 result for 61 Cygni, not Struve's work on Vega, that history has viewed as the crossing of the finish line in the race to measure the distances to the stars by parallax. Even in his day, Struve's colleagues must have felt their confidence shaken by the fact that the new value was almost twice the old. Nevertheless his peers recognized it as a step forward. A small change Struve made in his method of collecting data even prompted Bessel to revise his estimate of the parallax of 61 Cygni.

Bessel's health began to decline about this time, due to cancer. From a professional standpoint, however, he continued to reap the rewards for a life of exceptional industry and accomplishment. In 1841, the Royal Astronomical Society presented him with a Gold Medal for his determination of the parallax of 61 Cygni.

John Herschel, as President of the Royal Astronomical Society, gave the address on the occasion of Bessel's Gold Medal. He alluded to Struve's and Henderson's work also, praising all three results as "among the fairest flowers of civilization." The measurement of stellar distances had seemed a "great and hitherto impassable barrier to our excursions into the sidereal universe," he said, adding that the barrier's being "almost simultaneously overleaped at three different points" was "the greatest and most glorious triumph which practical astronomy has ever witnessed."[22]

Certainly the search for parallax had proved to be frustratingly difficult. In the late 1600s, even the parallax of Mars, our neighbor in the solar system, challenged the most sophisticated observers. In the mid-1700s, Bradley, the English astronomer who discovered aberration while searching for parallax, could only conclude that parallax was a smaller effect than many supposed, and that the stars must lie at about 400 000 times the Sun's distance from Earth. About a century of technological progress elapsed after Bradley's laborious attempt to find parallax before Bessel, Struve, and Henderson showed that the nearest stars are truly as remote as Bradley had suspected.

Perhaps because John Herschel took so much care to refer to Struve's and Henderson's accomplishments in his address for Bessel's Gold Medal, Struve made no public complaint about the relatively small fuss made over him. Perhaps it was his affection for Bessel that mollified any bitterness. Bessel died in 1846; two years later, in a pamphlet describing the Pulkovo observatory, Struve remarked without acrimony that Bessel's measurement of the annual parallax of 61 Cygni was "one of the greatest discoveries of our century."[23]

The *Etudes*

Considering the great effort Struve and Bessel expended to measure the distances to only a small number of stars through parallax, they and other astronomers of the mid-nineteenth century could well have given up on charting the relative positions of stars in all directions and uncovering the structure of the sidereal system. Struve did not forget this ultimate objective set by his champion William Herschel, however. In a report Struve wrote in French for the French-speaking Russian Imperial Academy of Sciences in 1847, *Etudes d'Astronomie Stellaire*, he returned to this important and still elusive goal.

"The preparation of this report has given me an opportunity to return to the study of the Milky Way," Struve wrote. "This entity is so puzzling, at first glance, that one is almost tempted to give up on a satisfactory explanation. However, the man of science must never retreat, neither when faced with the cryptic nature of a phenomenon, nor with the difficulty of an inquiry. Let him procure earlier studies, let him set out to increase knowledge of the phenomenon through new, precise observations; and he can be sure of a measure of success in his studies, if he employs a calm speculation, without giving himself over to the influences of an excited and predisposed imagination."[24]

Struve began his *Etudes* with an historical review of philosophical and astronomical views on the nature of the Milky Way and the distribution of stars in the stellar system. Drawing on the manuscripts he had received from John Herschel on his trip to London, he then examined the evolution of the elder Herschel's thoughts on the construction of the heavens. Struve described in detail Herschel's "star-gages," which explored a section of the stellar system and which suggested that the Milky Way consists of a thin but broad layer of stars.

To bring Herschel's 1785 drawing of the stellar system into a modern framework, Struve derived an approximate scale in light of recent parallax determinations. At the time Struve wrote the *Etudes*, 35 estimates of stellar distances were available. Assuming that stars all have about the same intrinsic brightness, Struve computed that the average distance to the first magnitude stars in the sample amounted to about 1 million Earth–Sun distances. He noted that Herschel's 1785 system, which Herschel had described in units of the distance to the bright star Sirius, stretched some 817 million Earth–Sun distances at its widest point, or almost 13 000 light-years, according to the approximate scale.

Struve reminded his readers that Herschel himself had reconsidered the assumptions on which the 1785 system was based. Herschel had come to recognize that the stars are not distributed uniformly and, more importantly, that his telescopes did not permit him to see to the edge of the stellar system. Still, Struve noted, the outdated drawing retained its status as a standard representation of the Milky Way system. "As to the explanation of the Milky Way, science has remained approximately at a standstill since the passing of Sir W. Herschel," Struve wrote. "We may ask, why have astronomers generally championed the old doctrine on the Milky Way, articulated in 1785, despite the fact that it was entirely abandoned by the author himself?"[25] Struve would probably be surprised to know that many modern textbooks still show Herschel's early drawing as representative of his life's work without mentioning that Herschel became aware of its limitations.

The next sections of Struve's report offered some new thoughts on the stellar system, picking up where Herschel left off. First, Struve insisted that we must consider the Milky Way to be "fathomless," as Herschel had suspected. We cannot see the edge of our system in any direction, Struve asserted. He continued, "It follows that if we consider all the fixed stars that surround the Sun as forming a large system, that of the Milky Way, we are in perfect ignorance of its extent, and we have not the least idea about the external shape of this immense system."[26] Struve was quite right in that the best telescopes of his day could not penetrate to the edges of our galaxy.

The fact that he thought the stars extended far beyond the limits of available telescopes did not dissuade Struve from trying to glean what information he could from a study of the distribution of stars in different parts of the sky. His approach was statistical—that is, he used samples of data gleaned from different directions in the sky to infer properties of the entire stellar system. He combined Herschel's star-gauges and more recent data to model the density of stars in different directions. The data suggested to him that the Sun is located near the center of the disk of the Milky Way. He envisioned the Milky Way system as a set of thin layers of stars stacked vertically. Within each layer, the stars were distributed uniformly with a particular density. With increasing distance from the central plane—the disk of the Milky Way galaxy—the density of stars decreased in a way that Struve could describe mathematically (see figure 5.8). In other words, Struve could not define the size of the disk in the horizontal direction or vertically, perpendicular to the plane, because he considered it fathomless, but he tried to express quantitatively how the density of stars decreased with distance in the vertical direction.

This model of Struve's stirred up controversy, because not all astronomers agreed that the distribution of stars could be fit to a mathematical function. But the most difficult section of the *Etudes*, for Struve's peers, was yet to come. In one of the final sections of his report Struve claimed, again based on star counts by Herschel and later researchers, that space is permeated with some absorbing material that diminishes our view of the universe.

Struve used statistical arguments to show that Herschel's telescope did not "penetrate" the layers of stars as deeply as one might expect, given its light-collecting power, and argued that the shortfall was due to some absorbing material. Struve made the common and useful—but inaccurate—assumption that all stars are of intrinsically the same brightness. If that were the case, the brightest stars would be the nearest ones, and the apparent brightness of a star would be directly related to the star's distance. Then the number of stars in each magnitude category would increase in a calculable way, e.g., there would be about four times as many second-magnitude stars as first-magnitude stars, and so on. The greater the distance probed, the greater the space corresponding to that distance, and the greater the number of faint stars of the corresponding magnitude limit.

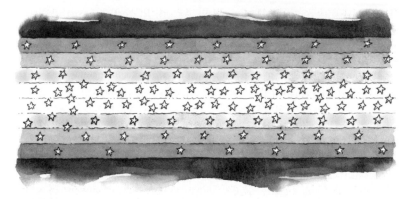

Figure 5.8 Struve's model of the Milky Way, as described verbally in his *Etudes d'Astronomie Stellaire* (1847). The Milky Way system of stars, in his conception, is thin in one direction but extends to unknown reaches in the other direction. Struve described the distribution of stars mathematically, envisioning them as very densely packed in a thin central layer, surrounded by layers of decreasing density. (Credit: Layne Lundström.)

Struve noted that Herschel's star gauges did not reveal as many faint stars as he should have found, if the assumptions about uniform distribution of stars and uniform brightness applied. For example, he calculated that Herschel should have counted about 3000 stars when probing the deepest field of the Milky Way, yet the largest number of stars Herschel found in one of his gauges was 588. The light from the missing faint stars, Struve believed, was too much absorbed en route, and the stars remained essentially invisible.

Struve normally shied away from speculation or bold hypothesis in astronomy, and his suggestion that space is not perfectly transparent marks a departure from his usual style. He was not the first to discuss the possible presence of absorbing material; indeed, he revived arguments for such a phenomenon made by Bessel's mentor, Olbers, and by other astronomers. But he was both correct and ahead of his time in proposing that visible light is partially absorbed in its passage through space. His reasoning, though not strictly correct, even gave him a fair idea of the amount of absorption. He calculated that the light from a first-magnitude star loses about 1/100 of its intensity over the distance it travels. Knowing that Struve believed first-magnitude stars were about 16 light-years away, we can convert his formulation of the amount of absorption into modern terms and compare with our modern value. He got a decrease of about 2 magnitudes per kiloparsec, or 2 magnitudes decrease over a distance of about 3300 light-years; that is a factor of 2 larger than the canonical value quoted today.

Struve sent a copy of *Etudes* to the main astronomical journal of his day, the *Astronomische Nachrichten*. He knew that many of his readers would take a stern view of his attempt to gain a broad understanding of the stellar universe and to make simplifying assumptions and statistical arguments for the sake of illuminating the "big picture." He even admitted, at the beginning of the section in which he discussed interstellar absorption, that even the soberest of analysis and speculation sometimes led to "unexpected" conclusions.[27] As the reviews of his work appeared in subsequent issues of the journal and elsewhere, Struve learned that his apprehensions were well founded. His colleague Johann Encke, director of the Berlin Observatory, wrote in outrage that Struve's authority would "secure immediate entry" for his results in all writing on the subject. Encke stressed, "It appears to me of importance for astronomy that the assumptions and parallaxes of the *Etudes* do not get into our astronomical and popular writings."[28]

The *Etudes* also received some praise. John Herschel took Struve's conclusions seriously, holding them up for examination in his widely-read textbook, *Outlines of Astronomy*. George Airy, Astronomer Royal at the Greenwich observatory and a friend of the Struve family, praised the work's importance and the "ingenuity" of the mathematical arguments in a carefully-worded review. Privately he told Struve he thought Encke had missed Struve's main points. Still, Encke's criticisms became well known. In condemning Struve's work, he also, perhaps not advisedly, condemned the whole statistical approach.

Struve's methods and conclusions were, in fact, ahead of their time. His statistical approach was particularly important because only a few hundred stellar distances could be found from parallax shifts at the end of the nineteenth century, not enough for astronomers to reconstruct the shape and scale of the Galaxy from direct distance probes. Struve's basic approach, adapted and refined by others, proved to be necessary far into the future. But this final great work of his appeared when he was no longer at the peak of his career, and it did not receive the attention that it probably deserved.

Final years

After 1841, Struve saw his position at the forefront of astronomy increasingly eroded by progress around the world. In Ireland, William Parsons, third Earl of Rosse, constructed a reflecting or mirror-based telescope of 6 feet (72 inches) aperture. From 1845 until 1917, this remained the world's largest aperture telescope. Perhaps most galling for Struve, Harvard University in the United States ordered from Merz and Mahler a replica of Pulkovo's 15-inch great refractor. Pulkovo no longer boasted the world's best instruments.

Beginning in the mid-1800s, photography grew in importance as a way of gathering data in astronomy. In 1845, the French physicists Leon Foucault and Armand Fizeau showed off the first daguerrotype of the Sun, showing sunspots and other features of interest to astronomers and physicists. From then on, photography conferred important advantages to those astronomers who learned the new techniques.

Not only were techniques and instruments evolving; the important questions in astronomy were changing, too. Precise observations of stellar positions and planetary orbits no longer dominated the field by the latter part of Wilhelm's career, replaced to some extent by questions concerning the physical nature of the stars and planets, which could only be answered using the new science of analyzing light, spectroscopy. No doubt, Struve felt the sands shifting under him.

After he published *Etudes d'Astronomie Stellaire*, Struve's professional life became less active and he spent more time with his family. He prepared previous work for publication; *Positiones Mediae*, the positions of the double stars he had listed in *Catalogus Novus*, appeared in 1852, and thereafter he applied himself to the first volume of *Arc du Meridien*, his major work on the geodetic surveys he had carried out or supervised. His letters to friends and colleagues related the family chronicles — his older children's marriages, and the efforts he and the other adults in the family made, in the absence of a school at Pulkovo, to educate the younger children by his marriage to Johanna, along with Otto's children.

In 1858, Struve, who had already been ill with infected sores behind his ears, developed a serious infection from a swelling on his neck. As the illness progressed, he became feeble, and could not recall anything from the past 20 years. Otto wrote to the Astronomer Royal George Airy of the sad state his father's health:

"When I came to his bed one morning, he spoke with me as if I were a complete stranger. On my remarking, 'Father, don't you know me? I am Otto' he first looked at me fixedly, then drawing me to himself with the words, 'Otto, my old comrade, I did not know you, that is terrible!' he broke out into a stream of tears — an appearance that we had never before known in him. From this moment the memory for recent times began to come back, but remained very weak."[29]

Over the next few years, Struve did recover enough physically to take a restful family holiday in warmer climes, but he never returned to work. Otto and a former colleague had to complete the *Arc du Meridien* for him. In 1864, at the age of 71, he died of pneumonia. He lies buried on the grounds of Pulkovo, amid a grove of birch trees that he planted himself.

Struve's son Otto Wilhelm Struve remained at the helm of Pulkovo another 27 years after Wilhelm's death. Thereafter the directorship of Pulkovo passed out of the Struve family, but the Struve astronomical dynasty and tradition of leading observatories continued. Otto's son Hermann eventually took Bessel's place at the Königsberg Observatory, then was tapped to direct the observatory that Alexander von Humboldt had helped establish in Berlin. Hermann's son Georg became an astronomer, and, like his great-grandfather Wilhelm and grandfather Otto, enjoyed a reputation as an excellent observer. Georg had two sons, one of whom studied astronomy but did not make a career of it.

Another of Otto's sons, Ludwig, followed even more closely than his older brother Hermann in the Struve family footsteps. Ludwig studied at his grandfather's alma mater, the University of Dorpat, and spent some time there as a professional astronomer, before being appointed director of the observatory at the University of Kharkov.

Ludwig's son, named Otto, was destined to carry on the Struve tradition in the United States. In the 1920s, the second Otto Struve earned a PhD in astrophysics at the University of Chicago and assumed the directorship of Chicago's Yerkes Observatory. During his tenure at Chicago he negotiated an agreement that led to the founding of the McDonald Observatory in Texas. The 82-inch aperture telescope there, the Otto Struve telescope, is still used most clear nights.

Otto would certainly have made his great-grandfather Wilhelm proud. Like each of his astronomer forebears, he received a Gold Medal from the Royal Astronomical Society in London. He edited the *Astrophysical Journal*; presided over the American Astronomical Society; chaired the department of astronomy at the University of California, Berkeley, and directed its Leuschner Observatory; and directed the National Science Foundation's first national observatory, the National Radio Astronomy Observatory in Greenbank, West Virginia. The Struve astronomical dynasty came to an end when he died, married but childless, in 1963.

6

WILLIAM HUGGINS: PIONEER OF THE NEW ASTRONOMY

"If we were to go to the sun, and to bring away some portions of it and analyze them in our laboratories, we could not examine them more accurately than we can by this new mode of spectrum analysis."

Warren De La Rue, 1861[1]

On a fine morning in May 1851, London's first international exhibition of works of art and industry threw open its doors to the public. The Crystal Palace, a glass-walled building resembling a giant greenhouse, sprawled over 18 acres in Hyde Park. Inside, a dizzying array of manufactured goods, machines, and raw materials invited visitors to take stock of participating countries' mineral resources and innovations in design and technology. The famous Koh-i-Noor diamond, which the British East India Company had recently presented to Queen Victoria, occupied a central showcase.

Every day except Sunday for five and a half months, tens of thousands of visitors jostled each other to admire silks and velvets, musical instruments, carriages, false teeth, electroplated coffee pots and creamers, ornate wooden cabinets, and other housewares crowding the display cabinets and galleries. Working models of industrial and agricultural machines, powered by a steam engine located outside the main building, provided an overview of the industrial revolution, which by then had reached its peak. Electric telegraph instruments, various types of batteries, and exhibits on iron and steel manufacturing processes hinted at further transformations to come.

The "Great Exhibition," as it was called, expressed Victorian society's intense drive to achieve scientific and industrial progress. The world it reflected—the world in which William Huggins, then in his late twenties, began his scientific career—was full of new or newly improved tools and instruments. Microscopes first drew Huggins to independent research and the philosophical societies, but he made his mark with telescopes and spectroscopes, instruments to analyze light that came into prominence in the 1860s. Beginning in the 1870s, he and his wife Margaret were among the first to apply the art of photography—publicly presented for the first time at the "Great Exhibition"—to astronomy and spectroscopy.

One of Huggins's most important accomplishments was to apply spectroscopy to the nebulae, extending one of William Herschel's lines of inquiry. Herschel had explored two main questions: first, what is the shape and size of our stellar system, and second, what is the nature of the nebulae, and what role does nebulous matter play in the evolution of starry systems. Recall that Herschel at first thought that all nebulae were distant aggregates of stars, some of them unresolved even with the best available telescopes. But later, after examining the "singular phenomenon" of the planetary nebula NGC 1514 in Taurus, Herschel changed his mind. He began speaking of "nebulous stars, properly so-called." He even speculated that stars might evolve from nebular condensations.

Herschel's follower Wilhelm Struve had mainly taken him up on the first question, charting the shape and scale of our stellar system. Huggins, although he worked on a variety of astronomical topics, drew attention back to the question of the nebulae. The shift in focus during the latter half of the nineteenth century stemmed in large part from the advent of spectroscopy, which promised unprecedented insight into the nature of nebulae. Indeed, by allowing astronomers to gauge the chemical and physical constitution of stars and nebulae, spectroscopy transformed astronomy into "astrophysics." Huggins's career coincided with an era during which, for the first time, astronomers could fully appreciate the great variety of stars and nebulae, and categorize them according to their chemical and physical characteristics. In the excitement of such "spectrum analysis," the measurement of parallax and other work on the scale of the galaxy temporarily lost some of its urgency.

An informal education

William Huggins (figure 6.1) was born on 7 February 1824 in London. He grew up as an only child; his parents, who ran a silk mercery and draper's shop on the busy commercial street of Gracechurch, had lost an earlier infant.

A biography of Huggins that his wife sketched out after his death and shortly before hers includes a few anecdotes about his childhood.[2] It relates that Huggins showed a talent for building instruments at a young age: when he was not much older than six, he constructed one of the static electricity generators that

Figure 6.1 William Huggins (1824–1910). (Credit: National Portrait Gallery, UK.)

167

were popular amusements at the time. Such a machine would commonly have involved brushing a piece of fur against a turning wheel embedded with amber; after an electric charge had built up on the wheel, the experimenter drew sparks from it using metal wires. Young William announced his success with his machine by running through the house shouting, "I've had a shock, I've had a shock!"[3]

Neither Huggins nor his contemporary biographers precisely documented the course of his education. The biographical sketch, which friends of the family edited and published after Margaret Huggins's death, tends to dramatize the hardships Huggins faced and his dedication to self-improvement. It is clear, in any event, that he received much of his education informally, from tutors and from attendance at public lectures in chemistry and physics.[4]

The family's circumstances precluded Huggins's attendance at any of England's prestigious public schools. In 1837, however, the City of London school opened to serve the needs of the upper middle class, and Huggins, 13, was among the first to enroll. The school offered lessons not only in English grammar, history, mathematics, and the classical languages, but also in less traditional subjects related to the trades, such as French, bookkeeping, and experimental philosophy, which we would now call applied science. Huggins also studied violin. This formal schooling, for some unknown reason, turned out to be but a relatively brief interlude. The school records and history show that he left two years later, in the first term of 1839, although he remained close to his mathematics teacher there. For a few years after he left the school, he apparently studied with tutors again.

According to the biographical sketch, Huggins adopted photography as a hobby almost as soon as the technology became widely available. In 1838, when he was 14, his mother took him on holiday to Paris. He came back with apparatus for taking daguerrotypes, a form of photograph.

Taking daguerrotypes meant lugging around 110 pounds [50 kg] of equipment and waiting patiently through long exposures. Louis Daguerre's official kits for amateurs — which could not have been available before Daguerre perfected his process in 1837, the year prior to Huggins's trip to Paris — included a wooden box camera with a 3-inch [8-cm] lens in front and, at the rear of the box, a ground-glass screen on which to focus the image.[5] The photographer sensitized a silvered copper plate by exposing it to the vapor of iodine in an iodizing box, before placing it in the focal plane of the camera. Even the brightest outdoor subjects required a plate exposure of about 15 minutes. Then the photographer would develop the image by exposing the plate to mercury vapor. The mercury attached itself to the iodide in the brightest, most sunlit areas of the image. The photographer fixed the image by washing the plate in "hypo," or sodium thiosulphate, which left the mercury deposits intact in the brightly lit areas but washed away the silver iodide from the shadowed areas. If Huggins mastered the use of all the chemicals, heating lamps, and plate polishing materials in his kit, his daguerrotypes must have been among the first created in England, as claimed in the biographical sketch.[6]

The few examples of Huggins' non-scientific writing that survive provide slim clues to his personality. A tone of ironic detachment, which may have been an early form of the pompousness he displayed later in life, characterizes his style. In a depiction of a sea-voyage with friends in 1824 he wrote, humorously: "As the vessel leaves the harbour, what a change steals over the smiling face of many of our companions. The relaxed features contract. The blush of health departs like the glow of evening, leaving the paleness of night upon the cheek. At the same time an intense interest is experienced to look over the vessel down into the sea."

In the same journal, the short-statured Huggins exhibited a certain aloofness on the subject of female companionship. "I cannot describe the ladies, as I never notice them," he wrote. After visiting a spa, he wrote, "It is a very fashionable bathing place and much renowned for diseases of various kinds, especially those of the skin. Even celibacy, says our witty guide book, is often cured here. *I* much doubt whether the remedy is not found much worse to bear than the disease itself."[7]

Attraction to astronomy

Huggins bought his first telescope around 1842, when he was about 15 years old. He probably had made his first acquaintance with optical instruments in his parents' shop, where his father or shop assistants would have used a magnifying glass to count the number of threads per inch in cloth samples. His parents encouraged his scientific interests by giving him a microscope sometime in his youth. He used the microscope to study plant and animal specimens, apparently with some squeamishness over the animal dissections. The biographical sketch goes so far as to suggest he eventually chose astronomy over biology because of his distaste for dissection: "[A]lthough he recognised the lawfulness of such experiments as were necessary for the benefit of mankind, he realized that he was too sensitive to do some of the work inevitably demanded by biological research."[8]

Astronomy-related news in the next decade would certainly have fueled Huggins' interest in astronomy, and would have encouraged him to pursue it as an amateur. In March and April 1843, the London *Times* published almost daily accounts of the changing appearance of the "Great Comet of 1843," a comet so bright, it could be seen in daylight. The comet, whose tail extended over an arc of 40° in the sky, attracted more than usual interest, both in Europe and the United States. In fact, it was by capitalizing on public interest in this comet that Harvard College raised money for its best telescope, the replica of Struve's 15-inch Great Refractor in Pulkovo.

In 1845—by which time Huggins would have had enough experience with his own small telescope to start dreaming of larger ones—William Parsons, the third Earl of Rosse, made the astronomical discovery for which he is still remembered today, using a telescope that was for a long time the largest reflecting telescope in the world. In the late 1830s, Parsons set out on a quest to see if the nebulae that had eluded earlier attempts to resolve them into stars would be resolved by larger instruments than Herschel's. He built first a 3-foot aperture telescope, and then, in 1845, unveiled a truly gigantic telescope of 6 feet diameter. Shortly after he began using this monster or "Leviathan" telescope, as it came to be known, he announced the important discovery of spiral structure in some nebulae, illustrating his finding with a drawing (figure 6.2) of what we would now call the spiral galaxy M51. Parsons also became well known for his assertion, widely accepted at the time but now known to be

Figure 6.2 Drawing of the spiral galaxy M51 by William Parsons, third Earl of Rosse. Parsons constructed the largest reflecting telescope of his day (6 feet in diameter), and was the first to discern spiral structure in some of the so-called nebulae. His drawing of M51, the "whirlpool galaxy," made in 1845, accurately shows that galaxy's spiral arms. (Compare with the Hubble Space Telescope photograph of the same object, figure 10.4a.) The blob at the end of one of the arms is, in fact, a smaller companion galaxy, catalog number NGC 5195, interacting gravitationally with M51. Three years later, Parsons noted spiral structure in the galaxy known as M99. (Credit: Birr Scientific and Heritage Foundation, courtesy of The Earl of Rosse.)

spurious, that he had succeeded in resolving the Orion nebula into stars.

In 1846, an amateur astronomer shared the spotlight with a professional in an outstanding piece of news, the discovery of Neptune, the eighth planet of our solar system, and its system of satellites. Neptune is not all that difficult to see; even Galileo may have seen it through his primitive telescope, according to historical research into his observing notebooks. The difficulty was to recognize it as a planet, slowly moving against the

171

background stars in its stately 165-year orbit around the Sun, and not to mistake it for an eighth-magnitude star.

In the nineteenth century, Bessel and a number of other astronomers had predicted the existence of a massive planet beyond the orbit of Uranus, based on observed irregularities in Uranus's orbital motion. However, pinpointing the hypothetical planet's location from these perturbations had proved to be an extremely difficult problem in celestial mechanics. The French astronomer Urbain Le Verrier, at the Paris Observatory, solved the problem and was the first to have his predicted position for the planet checked. On 23 September 1846, astronomers at the Berlin observatory found Le Verrier's planet, the newest member of the solar system, approximately where Le Verrier said it would be. Just a few days after the *Times* of London reported this historic discovery, William Lassell, an English brewer and well-known amateur astronomer, found Neptune's largest satellite, Triton.

Even when professional astronomers led the way, amateur astronomers were often quick to follow their innovations. William Cranch Bond, the first director of the Harvard College Observatory, contributed stunning daguerrotypes of the Moon to the photography display at the Crystal Palace Great Exhibition. These photographic records of a celestial object elicited great enthusiasm from the public and prompted several amateur astronomers in London, including Huggins, to experiment with astro-photography.[9]

Whatever finally inspired Huggins — no clues emerge from the biographical sketch or his own writings — he committed himself to independent scientific research within a few years of the Great Exhibition. He became a fellow of the Royal Microscopical Society in 1852. In 1853, he bought a telescope of 5 inches aperture from the well-known maker, John Dollond, and the next year was elected to the Royal Astronomical Society, an organization that John Herschel had helped establish in 1820. The Royal Society itself, the mother institution to the more specialized societies, remained out of reach for Huggins; in 1846, its members had begun to limit membership to scientists who had already demonstrated significant accomplishment.

In the same post-Exhibition period when Huggins began his association with the Microscopical and Astronomical societies, his father fell seriously ill. Rather than take over the family business, Huggins sold it and moved himself and his parents to a house in a new development in Lambeth, south of London. Thereafter, Huggins appears to have drawn a modest income from rents. His father died shortly after the move, and Huggins occupied himself with taking care of his mother and pursuing his passion for astronomy.

Huggins' new house on Upper Tulse Hill included a large garden in the back where he could set up his telescope. But Huggins did more than trundle the instrument outside in fine weather. He hired a local carpenter to build an observatory with a 12-foot diameter dome. The observatory floor sat on columns stretching 16 feet above the ground, so that he could see above the trees and have access to the room from the second story of his house. Beginning in 1856 with a description of this observatory and its instruments, Huggins regularly offered news from "Mr. Huggins' Observatory" to readers of the Royal Astronomical Society's *Monthly Notices*.[10]

During his early years as an independent researcher, Huggins found a mentor in William Dawes, a well-respected amateur astronomer and friend of Lassell. In 1850, just a few years after Lassell's discovery of Neptune's satellite Triton, Dawes had discovered a previously overlooked ring in Saturn's system, a thin, translucent ring interior to the two brighter rings, known as the "crepe" ring for its textured appearance. Around 1858, Dawes sold Huggins a telescope with one of the best 8-inch lenses available, which Dawes himself had purchased from the American lens maker Alvan Clark. More importantly, Huggins appears to have learned from Dawes how to carry out a research program. He began documenting his observations with drawings and carrying out regular observations of objects he thought were of scientific interest.

In the early 1860s, less than 10 years after moving to Upper Tulse Hill and devoting himself to astronomy, Huggins embarked on an ambitious project that turned out successfully and catapulted him to prominence in the astronomical community. The turning point came in 1862 when he initiated a discussion on spectroscopy with his neighbor Dr. Miller. Credit for setting Huggins and other astronomers down a very productive path, however, belongs to the founders of spectroscopy, Gustav Kirchhoff and Robert Bunsen.

The beginnings of spectroscopy

Kirchhoff and Bunsen began publishing accounts of their epoch-making experiments in 1859. The experiments consisted of vaporizing various chemical compounds using a gas burner (known to high school chemistry students everywhere as the Bunsen burner), and analyzing the light the compounds gave off. In the case of sodium chloride, for example, they would deposit a little pellet of salt on a platinum wire loop, and then hold the loop in the gas flame. As the salt compound glowed in the flame, the experimenters pointed a spectroscope at the flame to capture the substance's spectrum.

At the heart of Kirchhoff and Bunsen's spectroscope lay a prism. A prism will spread or disperse a ray of sunlight into a rainbow. The prism bends the path of incoming light of different wavelengths by different amounts, so that each color (or wavelength) composing the apparently white beam of sunlight emerges from the prism at its own angle, forming a spectrum.

When Kirchhoff and Bunsen analyzed the light that the heated compound or metal gave off by dispersing it with prisms, they noted that instead of a continuous spectrum or rainbow of colors, a few isolated bright lines appeared, of various colors. What excited them was to find that every element seemed to produce its own characteristic lines: sodium, for example, produces a pair of bright, closely-spaced lines in the yellow part of the spectrum, as well as fainter lines elsewhere. Magnesium produces a distinctive bright triplet of lines in the blue/green part of the spectrum, as well as some strong lines in the green and yellow.

Kirchhoff and Bunsen could not yet explain why or how these lines occurred, but the fact that each element corresponded to a unique set of lines suggested an entirely new way to perform chemical analysis. And because the method was so sensitive—they noted that very small amounts of sodium contamination of a sample would produce the characteristic yellow doublet, for example, in addition to the lines of the intended sample—they predicted that spectrum analysis would enable chemists to find rare elements that traditional chemical analysis had missed.

A second facet of Kirchhoff and Bunsen's experiment, and one that their contemporaries found particularly hard to understand, was to show that the bright colored lines they saw when they heated their samples could be, in a sense, reversed, in another kind of spectrum. When a continuous spectrum of light passed through a sodium flame, for example, the resulting "absorption spectrum" showed the rainbow of colors, with a pair of *dark* lines in the yellow, exactly where the sodium doublet would be. Thus there are three types of spectrum: the continuous spectrum, the emission or bright-line spectrum, and the absorption spectrum. The continuous spectrum arises from any incandescent solid or liquid, and reveals no clues about chemical composition. The emission spectrum of colored lines against a dark background arises from hot vaporized substances, and reveals the unique signature of any chemical element in the pattern of lines. The absorption spectrum of dark lines in a "rainbow" of color arises when a viewer sees a continuous spectrum of light illuminating a vapor from behind, and (as we now understand) the vapor selectively absorbs the light at certain wavelengths instead of emitting at those wavelengths, as in the emission spectrum.

Kirchhoff and Bunsen concluded one of their seminal papers on spectrum analysis, in 1860, with some comments on Fraunhofer's lines and on the potential applications of spectrum analysis in astronomy. Fraunhofer, the lens maker whose firm had built the "Great Refractors" for Wilhelm Struve and the heliometer for Bessel, had noted in his experiments with glass prisms that unexplained dark lines appeared in the solar spectrum (figure 6.3). The effect is not readily apparent when sunlight passes through only one prism, because the dark lines are very fine, but when sunlight passes through a series of prisms and the light is dispersed as widely as possible, hundreds of unequally spaced dark lines appear. Fraunhofer labeled the more prominent of the dark lines A through H, running from the red end of the spectrum toward the blue. Kirchhoff and Bunsen suggested a connection between Fraunhofer's lines and their observations on absorption spectra.

Spectrum analysis, Kirchhoff and Bunsen said, offers a simple means for detecting even small traces of elements in laboratory samples, and more; it "opens to chemical research a hitherto completely closed region extending far beyond the limits of the Earth and even of the solar system. Since in this

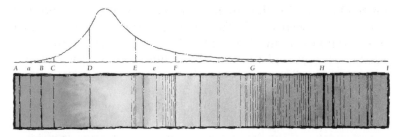

Figure 6.3   Fraunhofer's solar spectrum. The optical instrument maker, Joseph Fraunhofer, dispersed sunlight through prisms and detected the presence of hundreds of fine dark lines amid the colors of the rainbow. Top panel: The intensity of light from the Sun is strongest in the yellow part of the spectrum, as shown by the peak of the curve. Bottom panel: Fraunhofer labeled the more prominent lines with letters of the alphabet. The lines were later shown to relate to the presence of various elements in the Sun. For example, the C and F lines are due to hydrogen, A and B are due to oxygen, and D is due to sodium. (Credit: Adapted from the original by Layne Lundström.)

analytical method it is sufficient to see the glowing gas to be analyzed, it can easily be applied to the atmosphere of the sun and the bright stars."[11]

The only difference between the familiar laboratory spectra and the spectra of the Sun, they noted, was that the solar spectral lines appeared "reversed" as dark lines. Today we know that the Fraunhofer solar spectrum is an absorption spectrum because the temperature of the Sun decreases with height. (The temperature increases in the extremely tenuous outer layer called the corona, but that rise does not affect the Fraunhofer spectrum.) The deeper layers act as a heater, producing a continuum of emission, and the cooler, more elevated layers act as a "vapor," selectively blocking some of the light from below. Kirchhoff and Bunsen thought of the Sun as having a surface and an atmosphere, so they wrote that "the spectrum of the sun with its dark lines is just a reversal of the spectrum which the atmosphere of the sun would show by itself."[12]

Kirchhoff and Bunsen concluded with a kind of call to arms to laboratory chemists: "The chemical analysis of the sun's atmosphere requires only the search for those substances that produce the bright lines [in the laboratory] that coincide with the dark lines of the spectrum."[13] The problem turned out to be much more complex than they made it sound, due to the large number of lines and the special conditions of temperature and pressure prevailing in the Sun's atmosphere; by 1895, some 14 000 lines were known in the Fraunhofer spectrum, but in 1924, more than 6000 of these lines had yet to be identified with their corresponding elements.[14] However, as scientists learned over the course of the next century, the investigations Kirchhoff and Bunsen urged held just as much promise as they thought. In effect, they paved the way to an understanding that some philosophers of science thought was impossible: a knowledge of the chemical composition of the stars.[15]

The scientific societies in England, which were already in a state of excitement over Charles Darwin's 1859 publication of the *Origin of Species*, eagerly took up discussions of Kirchhoff and Bunsen's papers. Warren De La Rue, a prominent astrophotographer who was, like Huggins, a member of both the Microscopical and Royal Astronomical Societies, wrote in 1861, "The physicist and the chemist have brought before us a means of analysis so wonderfully exact that, as Dr. Faraday recently said, if we were to go to the sun, and to bring away some portions of it and analyze them in our laboratories, we could not examine them more accurately than we can by this new mode of spectrum analysis."[16]

Collaboration with Miller

More than one scientific society called on Huggins' neighbor, William Allen Miller, to explain the latest developments in spectrum analysis. A founding member of the Chemical Society; Treasurer and Vice President of the Royal Society; and an eminent professor of chemistry at King's College, London; Miller had been working on spectrum analysis before it was fashionable to do so, although without achieving a breakthrough as Kirchhoff and Bunsen did. In 1861, he was photographing the spectra of pure metal vapors, building up a database of the spectral line patterns. Sometime early in 1862, Huggins approached him and proposed a collaboration to investigate the spectra of celestial objects. Miller apparently hesitated, daunted by the faintness of stars. As Huggins himself liked to point out, to emphasize the difficulty of stellar spectroscopy, "the light from Vega is one forty thousand millionth part of light from the sun."[17] Nevertheless, after visiting Huggins' observatory, Miller agreed to share his expertise, and the two embarked on a pioneering effort in astronomical spectroscopy.

The first challenge Huggins and Miller faced—and there were many—was to devise an appropriate spectroscope and attach it to the observing end of Huggins' 8-inch aperture, 10-foot telescope. They needed to disperse the light coming down the telescope with one or more prisms. The more dispersed the spectrum, the easier it would be to see any lines. On the other hand, the faint intensity of the light meant that it would not remain visible if it were spread *too* much. For examining the spectrum of bright stars, they eventually settled on a combination of two glass prisms.

The stellar spectroscope as a whole consisted of a fine slit at the eyepiece end of the telescope, to admit a narrow beam of light from the star or planet under investigation; a tube to funnel the light from the slit toward the prisms; the series of prisms; and a small telescope—part of the spectroscope—about 7 inches long and with an aperture less than 1 inch, mounted on a rotating base, for observing the light that emerged at different angles from the second prism (see figure 6.4 for a simplified but comparable set-up). The person using the spectroscope would record the small telescope's position against a finely divided scale, and note the aspect of the spectrum at this position; that is, he would note whether or not lines appeared. Then with the turn of a screw, he would advance the small telescope by a slight increment, bringing light of a different wavelength into view, and repeat the observation of any lines. The scale recording the position of the small telescope indicated some 1800 divisions between the Fraunhofer A and H lines of the spectrum.[18]

The second challenge was to record the spectral features in a reliable way. Each element has its own characteristic spectrum, but the position of one or more lines in one element's spectrum might appear in nearly the same positions as the lines in another element's spectrum. For example, both calcium and lithium have strong lines that appear to be the same shade of green. Because

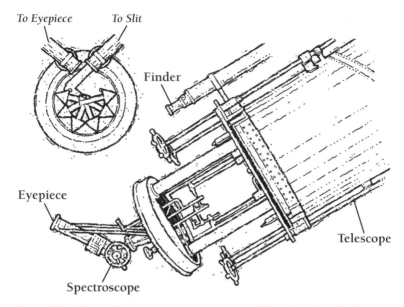

Figure 6.4 The astronomical spectroscope of Huggins' day. Light from the main telescope is directed to a spectroscope attached at the eyepiece end. Inset: The spectroscope admits light from the main telescope through a slit and disperses it with prisms. The observer puts his eye to the eyepiece of a small telescope (part of the spectroscope) to observe spectral lines. A micrometer screw allows the position of the small telescope to be read with precision, giving the location of spectral lines within the spectrum. (Credit: Layne Lundström.)

each color in the spectrum corresponds to a particular wavelength of light, the modern way to phrase this is to say that both elements have a line at about 610 nanometers [1 nanometer is 10^{-9} meters]. Huggins and Miller came up with a solution to this problem that was simple in concept, but tricky to execute: they found a way to observe simultaneously the celestial spectrum and a comparison spectrum, created in the laboratory. The apparatus for vaporizing metal samples sat on a rolling cart. A mirror reflected the laboratory or terrestrial spectrum generated on this rolling cart into the spectroscope. The observer saw the bright laboratory spectrum directly below the fainter stellar or planetary spectrum, like two lines of music in a score.

This allowed him to note any coincidence of lines — or a lack of exact coincidence — or to measure the position of lines in the celestial spectrum with reference to the laboratory standard. Huggins and Miller undoubtedly went to greater lengths than their competitors to calibrate their line measurements, and Huggins would later reap the benefits of their care and persistence.

The third challenge, central to their objective of identifying the chemical constituents of other stars, was to catalog the lines corresponding to the known elements. Early on in their experiments, Huggins and Miller accordingly allowed themselves to digress from their work on stellar spectra to improve existing maps or charts of the lines produced by metals in the laboratory. Miller had already photographed the lines of more than a dozen metals in the ultraviolet part of the electromagnetic spectrum. Huggins extended this work, documenting the lines of the metals in the visible part of the spectrum — that part of the spectrum where light from stars similar to the Sun is most intense. Among those metals he studied were common ones such as sodium and potassium, and newer elements, such as thallium, which chemists had only recently discovered through spectrum analysis.

In 1863, during the course of their earliest work on stellar spectra, Huggins and Miller attempted to add a fourth layer of complexity: they tried to photograph the spectra emerging from the prisms instead of manually recording the positions of the lines. By this time Daguerre's photographic process had been superseded by others with improved sensitivity, although no method gave as good results in the visible part of the spectrum as in the ultraviolet. Huggins and Miller experimented with the so-called wet collodion process. This was a messy enterprise involving coating glass plates with a sticky, light-sensitive substance that became progressively less sensitive as it dried. They tried it on the spectra of two of the brightest stars in the sky, Sirius near the foot of Orion and Capella, the sixth-brightest of all stars. They succeeded in obtaining photographs of the spectra, but noted that the lines were not distinct enough to measure. They suggested in their papers that they intended to resume experiments with photography later.

In fact, Huggins and Miller could not afford to hold up their investigation of stellar spectra while they worked out the difficulties of photographing the spectra. Competitors, even other amateurs, nipped at their heels. In 1863, both Italian and American scientists actually bested them and published data on the spectra of the brightest stars. This spurred Huggins and Miller to present a preliminary sample of their results for three stars.

Their schematic drawings of the spectra were, as they had promised their readers, more detailed than those of any other observers. They included more lines and had taken care to indicate not only the positions, but also the relative width or "strength" of the lines. Some lines appeared as very fine dark traces; others blotted out a more sizable band in the spectrum. The reasons for the diverse widths of the lines had yet to be uncovered, but Huggins and Miller rightly assumed that any such spectral features might prove significant, and should be recorded.

In May 1864, Huggins and Miller published their most complete round-up of results to date in the journal *Philosophical Transactions of the Royal Society*. Their introductory paragraphs remind us, as modern readers, how little was known about the stars before the advent of spectroscopy less than 150 years ago. They noted that since the discovery that some stars orbit each other in binary systems—William Herschel's discovery—demonstrating the applicability of Newton's laws beyond the solar system, astronomers had made virtually no progress in elucidating the nature of stars. Most stars were too remote to allow even for an estimate of their distances from parallax, and hence their true brightnesses were unknown. Spectroscopy promised to change all that; the most detailed analysis possible on Earth could be extended to the stars, no matter how distant. A knowledge of the construction of the universe, Huggins and Miller suggested, had finally become a reasonable goal: the success of spectrum analysis as applied to determining the nature of some of the Sun's constituent elements "rendered it obvious that it would be an investigation of the highest interest, in its relations to our knowledge of the general plan and structure of the visible universe, to endeavor to apply this new method of analysis to the light which reaches the earth from the fixed stars."[19]

For this article, Huggins and Miller examined the spectra of about 50 stars, and fully measured the lines in a small subset of these. Aldebaran, the red star marking the eye of the bull in Taurus, served as a target for one of their more detailed investigations. Many hours of work had yielded the positions of about 70 lines in the visible range of its spectrum. Some lines appeared to be the same as those in the solar spectrum. Nine sets of lines coincided with lines in laboratory spectra of sodium, magnesium, hydrogen, calcium, iron, bismuth, tellurium, antimony, and mercury. The rest of the lines did not match any of the known laboratory spectra.

The authors had firmly established their authority in the matter of resolution, or the level of detail they could see. While other astro-spectroscopists had stated, for example, that the spectrum of Rigel, the bright bluish star in Orion's shoulder, was perfectly continuous, Huggins and Miller had found it to be full of innumerable very fine lines. Indeed, they found no spectra without lines, indicating the ubiquity of stellar atmospheres. "The stars admit of no such broad distinctions of classification" on the basis of the presence or absence of lines, they wrote. "Star differs from star alone in the grouping and arrangement of the numerous fine lines by which their spectra are crossed."[20] The sample of stars that they examined in full showed that no stellar spectrum was likely to be less complex than that of the Sun, although it might display a different pattern of lines.

In their concluding remarks, Huggins and Miller discussed the application of their results to a number of important hypotheses. They noted that their findings might require some modifications or refinements to Laplace's nebular hypothesis, according to which the stars and planets condensed out of a primordial cloud of nebular material (see chapter 4). Huggins and Miller believed that the diversity of spectra among the stars implied that the stars were made up of different mixtures of materials, and therefore that "the composition of the nebulous material must have differed at different points."[21] Their concern was slightly misplaced; we know today that all stars are made largely of the same material, hydrogen and helium, and that the variation in stellar spectra arises from variations in proportions of constituents that exist only in trace amounts.

In another speculative vein, Huggins and Miller were inspired by discussions of Darwin's work to consider the "great plan" of the universe and the "design" of the stars. Darwin's

Figure 1.1. The "Hubble Deep Field"—a view taken with the Wide Field and Planetary Camera 2 on board the Hubble Space Telescope. As described in the text, most of the objects seen here are distant galaxies. A foreground star, within our own galaxy, has "rays" extending from it—an artifact of the imaging system. The view is actually a synthesis of separate images in red, green, and blue light. (Credit: Jeff Hester and Paul Scowen (Arizona State University), and NASA.)

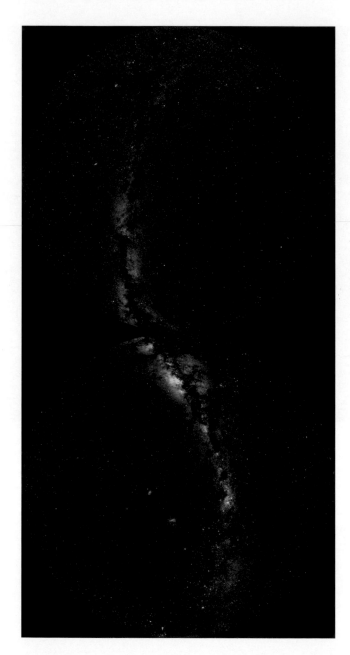

Figure 1.2. The Milky Way in the northern and southern hemispheres (left and right panels, respectively). Mosaic assembled by Axel Mellinger from 51 wide-angle photographs taken over the course of three years. (Credit: Axel Mellinger. Reprinted with permission.)

Figure 2.5. The Pleiades. Only about half a dozen stars in the Pleiades open cluster are visible to the naked eye, but many more appear through telescopes or in long-exposure photographs. In telescopic views, one can see a blue veil near some of the brighter stars—a reflection nebula caused by dust. (Copyright Anglo-Australian Observatory/Royal Observatory, Edinburgh. Photograph from UK Schmidt plates by David Malin. Reproduced with permission of David Malin Images.)

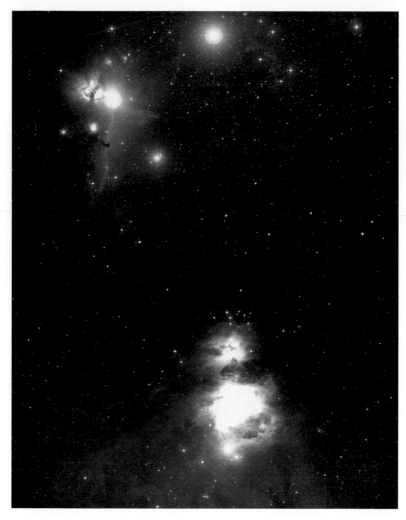

Figure 2.7. The Orion nebula, the most famous example of a diffuse emission nebula. The nebula shrouds from view a stellar nursery, where stars are condensing out of hydrogen and helium gas. (Copyright Anglo-Australian Observatory/Royal Observatory, Edinburgh. Photograph from UK Schmidt plates by David Malin. Reproduced with permission of David Malin Images.)

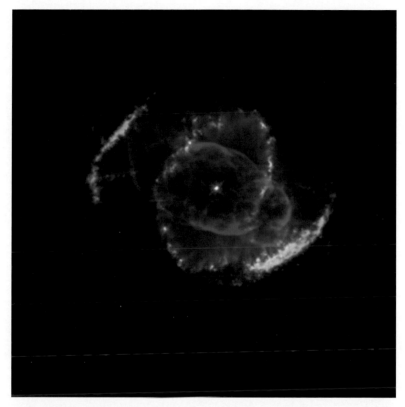

Figure 2.8. The Cat's Eye nebula, an example of a planetary nebula. The name comes from the fact that the gas and dust surrounding the star looks, through a small telescope at least, like a disk or planet. (Credit: J.P. Harrington and K.J. Borkowski (University of Maryland), and NASA.)

Figure 2.9. The Horsehead nebula in Orion. The Horsehead nebula is a dark nebula, silhouetted against the brighter light from the emission nebula in Orion. (Credit: NASA, NOAO, ESA and The Hubble Heritage Team. STScI/AURA.)

Figure 10.4. "Grand Design" galaxy. The spiral arms are prominent and extend in an unbroken line from the center to the extremities in this example of a so-called "grand design" galaxy, M51, the "Whirlpool" galaxy. (The bright round patch shown in Lord Rosse's drawing at the end of one of the spiral arms (figure 6.2) is just off the top edge of this image.)

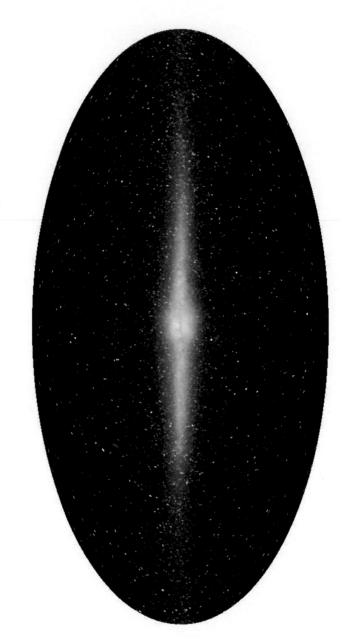

Figure 10.7. The Milky Way galaxy from inside. This view of our galaxy, taken in the infra-red portion of the spectrum using instruments on the Cosmic Background Explorer (COBE) satellite, shows our galaxy's thin disk of stars, and dust (which appears red or orange in this image) within the disk. Infra-red light penetrates dust and gas much better than visible light, so the image reveals much more of the central swath of the Galaxy than we could see with traditional telescopes. (Copyright Edward L. Wright. Used with permission.)

evidence for a common structure among species related by evolution and natural selection had led some philosophers to suggest that a similar relationship might exist in the inorganic world, as well. Huggins and Miller wrote that, just as all vertebrates showed "a unity of plan observable amongst the multiform varieties," their observations suggested that "a similar unity of operation extends through the universe as far as light enables us to have cognizance of material objects." They inferred that "the stars, while differing the one from the other in the kinds of matter of which they consist, are all constructed upon the same plan as our sun, and are composed of matter identical, at least in part, with the materials of our system."[22] In other words, the basic structure of the stars was similar to that of the Sun, and the elements in the stars, such as sodium, magnesium, and calcium, were the same as those found in the Sun and on the Earth, and only the relative proportions of the constituent elements varied.

Not only were the stars basically similar to the Sun, Huggins and Miller proposed; their spectral investigations lent credence to the idea of other planetary systems similar to ours. They wrote "There is . . . a probability that these stars, which are analogous to our sun in structure, fulfil an analogous purpose, and are, like our sun, surrounded by planets, which they by their attraction uphold, and by their radiation illuminate and energise. And if matter identical with that upon the earth exists in the stars, the same matter would also probably be present in the planets genetically connected with them, as is the case in our solar system."[23]

One can easily imagine Huggins' and Miller's enthusiasm over their illumination of some of the mysteries of the universe with the simple application of the spectroscope to the light from the stars. They had identified terrestrial elements in the stars, found some unknown lines that might represent new elements, discovered a challenge to Laplace's nebular theory, and given new life to speculation about other planetary systems. They had uncovered new information about the stars that seemed to make them, for the first time, amenable to classification and study just like the plants and creatures of Darwin's organic world. Traditional astronomy, with its concern for planetary orbits and accurate stellar positions, must have seemed quite staid in comparison.

Spectra of the nebulae

The 1864 paper on the spectra of the fixed stars established Huggins and Miller as authorities on the technique of spectroscopy and put them well ahead of their competition. But, having surveyed the spectra of the brightest stars and ascertained that a full record of the lines of even one star might take years to compile, Huggins was impatient to break new ground again. Working alone, he turned his telescope and spectroscope to the nebulae.

His first observation was almost bound to be an historic one, considering the controversy that still swirled around the nature of these objects. Would all nebulae turn out to be groups of stars, resolvable with bigger telescopes, and perhaps showing the spiral structure Lord Rosse had found? Or were some intrinsically fuzzy, composed of some "shining fluid" as William Herschel had supposed?

For his first observation, Huggins chose the so-called Cat's Eye nebula in the constellation Draco, visible year-round from the northern hemisphere (see chapter 2, figure 2.8). The nebula is not visible to the unaided eye, but through small telescopes it looks like a slightly out-of-focus blue disk. Its disk-like appearance puts it in the class Herschel named "planetary nebulae."

"On August 29, 1864, I directed the telescope armed with the spectrum apparatus to this nebula," Huggins reported. "At first I suspected some derangement of the instrument had taken place; for no spectrum was seen, but only a short line of light perpendicular to the direction of dispersion [i.e., an emission line]. I then found that the light of this nebula, unlike any other ex-terrestrial light which had yet been subjected by me to prismatic analysis, was not composed of light of different refrangibilities [i.e., wavelengths] and therefore could not form a spectrum."[24]

What Huggins had found, he realized, was the first example of an emission spectrum among the celestial objects, the same type of spectrum produced by vaporized chemical compounds in the laboratory. This nebula, at least, was no agglomeration of stars, but a hot gas.

Huggins was taken aback by the purity of this nebular light, all concentrated in one narrow green band of the spectrum. If he harbored any suspicions that nebulae represented clouds of vaporous material about to condense into stars, he might have expected to see emission spectra consisting of *hundreds* of bright lines, corresponding to the hundreds of absorption lines in the spectra of fully formed stars — not just one line.

On further examination he saw that there were, in fact, two other lines, fainter than the first one. The brighter line became known as the "chief nebular line."

Huggins turned his telescope and spectroscope to another nebula, then another. The planetary nebula all showed similar spectra: the chief nebular line in the green, which Huggins tentatively — and mistakenly — identified with one of the nitrogen lines, and two others. One planetary nebula showed a fourth faint line, so not all planetary nebulae were exactly the same.

The other nebulae, those that had definitely been resolved into stars by larger telescopes than Huggins', should have a different sort of spectrum. Indeed, when Huggins examined them he found that the spectra of two globular clusters in Hercules fit the puzzle perfectly. Though too faint to show absorption lines, the spectra were distinctly star-like — more like continuous spectra. Finally Huggins turned his telescope to the Great Nebula in Andromeda, M31 in Messier's catalog, and its nearby companion, known as M32. Although the light from each was so faint that Huggins could see only the middle part of the visible spectrum, he had no difficulty discerning that both spectra were star-like, with no sign of emission lines.

Huggins could see no way to relate the simple, three- or four-line emission spectra of the gaseous nebulae with the complex absorption spectra of the stars and star clusters. If the "nebulous fluid" he was seeing was the raw material for stars, it should give rise to as many bright lines as there are dark lines in a typical stellar spectrum. Furthermore, by surveying a number of nebulae he should have come across some in a more advanced state of development, and, as he noted in a follow-up paper in 1865, "an advance in the process of formation into stars would have been indicated by a more complex spectrum."[25] But only one nebula had more than three lines. Huggins gave the opinion that the nebulae did not fit the apparent "unity of plan" in the visible universe that he and Miller had previously noted: "[T]he nebulae which give a gaseous spectrum are systems possessing a structure, and a purpose in relation to the universe, altogether distinct and of another order from the great group of cosmical bodies to which our sun and the fixed stars belong," he wrote.[26]

Huggins was trying to make sense of his observations long before it was possible to do so. The dominant line in the spectrum of the gaseous nebulae is so bright that, in his experimental set-up, it overshadowed the presence of other lines.[27] These fainter lines, if he had been able to see them, would have indicated to Huggins a more complex chemical composition of planetary and other nebulae. He was right, however, to question Herschel's idea about the origin of the nebulae. Stars do form from nebular clouds similar to those Herschel and Laplace spoke of, but there are many kinds of nebulae besides star-forming regions. The Cat's Eye nebula and other planetary nebulae consist of material ejected from the atmosphere of a star near the end of its life-cycle.

In the year or two following his discovery of emission lines in some nebular spectra, Huggins might well have put forward some hypothesis of his own to account for the existence of both starry and nebulous bodies. But Huggins was by character more cautious than, say, Herschel had been about advancing new ideas. He would only go so far as to say that the emission-line nebulae belonged in a class by themselves as far as distances were concerned. Since they were composed of truly nebulous material, it was not necessary to assume that they appeared nebulous because of their great distance. "The opinions which have been entertained of the enormous distances of the nebulae, since these have been founded upon the supposed extent of remoteness at which stars of considerable brightness would cease to be separately visible in the telescope, must now be given up," at least for those he had observed to be gaseous, Huggins wrote in his rather stilted style.[28] In general, Huggins preferred to wait patiently for the truth to emerge from new data. "It would be easy to speculate," he said in a lecture to the British Association for the Advancement of Science in 1866, "but it appears to me that it would not be philosophical to dogmatise at present on a subject of which we know so very little. Our views of the universe are undergoing important changes; let us wait for more facts, with minds unfettered by any dogmatic theory, and therefore free to receive the obvious teaching, whatever it may be, of new observations."[29] Unfortunately, as we shall see, the teachings of new observations were rarely obvious.

Both the Royal and Royal Astronomical Societies applauded Huggins' and Miller's work as an example of British ingenuity in developing the new science of spectroscopy. Astronomers particularly appreciated Huggins' clarification of the status of nebulae, with some fitting into the category of starry agglomerations and some gaseous. The Royal Society elected Huggins as one of 15 new Fellows in June 1865, and the next year presented him with its Royal Medal. Miller, an officer in the society, was not eligible to be named co-recipient. The Royal Astronomical Society, however, presented Huggins and Miller with the first jointly-awarded Gold Medal.

Famous visitors came to see Huggins' observatory. Otto Struve, son of Wilhelm Struve and current director of the Pulkovo observatory, stopped by in the summer of 1865. So did the gentleman-astronomer William Parsons, Lord Rosse, whose claims of resolving nebulae into stars with his giant 6-foot telescope were partly vitiated by Huggins' observations of the emission spectra in some nebulae.

Huggins savored the accolades. At least one of his friends, in fact, remarked on the egotism he displayed at this stage in his career. In 1870, Huggins led an expedition to Oran, Algeria, to observe a solar eclipse. One of his fellow travelers, the chemist William Crookes, noted that Huggins' pompousness put him at odds with his companions. Crookes wrote in his diary, "Little Huggins's bumptiousness is most amusing. He appears to be so puffed up with his own importance as to be blind to the very offensive manner in which he dictates to the gentlemen who are co-operating with him, whilst the fulsome manner in which he toadies to Tyndall [a prominent physicist] must be as offensive to him (Tyndall) as it is disgusting to all who witness it. I half fancy there will be a mutiny against his officiousness." Later in the trip, in reporting on a series of violent storms at sea, Crookes gleefully singled out Huggins' misadventures: "Eating was almost impossible, for nearly all one's attention was required to keep the meal, &c., on the plate, and ourselves on the benches. Huggins being small and not very careful, disappeared once, plate and all, under the table."[30]

The motions of the stars

Huggins may have been overly anxious to garner the respect of prominent scientists, but having made his mark in his mid-forties, he had no intention of resting on his laurels. Rather, his success with the delicate operations of the spectroscope spurred him on to tackle other problems, particularly those requiring new or improved instrumentation. In the late 1860s, after some dabbling in research on the surface features of the Sun, he enlarged his research agenda, attempting to build an entirely different kind of instrument for his telescope, one that would record the heat emitted by celestial bodies. If this ambitious effort had been successful, it might have permitted him to quantitatively relate the temperature of stars to their spectra. This project has been eclipsed, however, by what some astronomers regard as Huggins' greatest legacy beyond his investigation of stellar and nebular spectra—his demonstration, in 1868, that spectroscopic data could be used to find the speed of stars in motion toward or away from the Earth.

When Edmond Halley discovered the proper motion of stars in 1718 (see chapter 3), he was observing the displacement of stars *across* the sky, or perpendicular to the lines of sight to the stars. If any had motion *along* the lines of sight, toward or away from him, this was undetectable. Yet such a "radial motion," as it came to be called, was almost inevitably present; the random movement of any given star would likely be composed of motion both radial and transverse to the observer. This radial motion, Huggins learned, must be reflected in the appearance of stellar spectra.

In 1867, Huggins began a correspondence with James Clerk Maxwell, a fellow of the Royal Society and a brilliant physicist, soon to become a faculty member in physics at Cambridge University. Huggins may have met Maxwell through his neighbor and collaborator Miller, for both Miller and Maxwell held faculty positions at King's College, London, until Maxwell left in 1865.

Huggins wrote to Maxwell to ask about the effect a star's motion might have on its spectrum. In response, Maxwell explained that a moving source of light, such as a star or nebula, will produce a spectrum in which the lines—whether they are absorption or emission lines—are shifted from the positions they would occupy if the source were at rest with respect to the viewer. The effect is analogous to the Doppler shift in sound: as a whistling train roars past a bystander on a platform, the person hears the whistle apparently rise in pitch as the train approaches, then flatten in pitch as the train recedes. In the case of a stellar spectrum, a given pattern of lines, such as the triplet produced by magnesium, will appear shifted toward the blue end of the spectrum if the Earth and star are in relative motion toward each other. If the Earth and star are moving apart, the star's lines, whose positions can be compared to those of a laboratory sample at rest, will be shifted toward the red end of the spectrum. If the star moves neither toward or away from the Earth, but moves perpendicularly to the line joining the Earth and star, no line shift should appear (see figure 6.5).

Maxwell showed Huggins how to calculate the amount of displacement he might expect to find in the spectrum of a star. The results were not encouraging. Even the Earth's rapid motion in orbit around the Sun, which causes some apparent

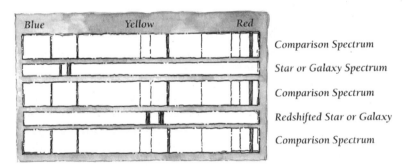

Figure 6.5 Shift of spectral lines due to motion of the source. The motion of a star or galaxy, either away from or toward the observer, may be inferred from the shift in position of characteristic spectral lines. The top, third, and fifth panels in the figure show the simplified spectrum of a laboratory sample, used as a reference. The second panel shows the spectrum of an astronomical source (in this simplified example, a single pair of lines) in the case where the source does not have motion in the line of sight. The fourth panel shows the same astronomical spectrum shifted to the red, as would be the case for an object moving away from the observer. The amount of shift gives the velocity of the source. (Credit: Layne Lundström.)

motion toward and away from any point in the sky, would produce an extremely small shift, very difficult for Huggins to measure. But Huggins reckoned that the stars with high proper motion, like the relatively nearby one that Bessel had used for his parallax measurement, might have unsuspected motion in the line of sight, too.

Once again Huggins experimented with his observing equipment and technique, pushing the limits of precision in his measurement of the line positions. He added prisms to his spectroscope to disperse the light even more and to accentuate the appearance of any shift in the stellar lines compared to those in the laboratory spectra. This meant that only the brightest stars provided enough light to work with, so he concentrated much of his effort on the spectrum of Sirius. Even with these precautions, he doubted his own initial results. He believed he saw a shift, but would his estimate convince readers of his scientific papers?

Huggins realized that the eye is very sensitive to discontinuities or jags in otherwise straight lines. His genius was to employ this sensitivity through his experimental set-up. He rearranged his apparatus so that the view through the spectroscope showed two duplicate laboratory spectra, one directly adjacent to and on top of the stellar spectrum and one below. Lines present in both the laboratory and stellar spectra would cut straight through all three spectra if no shift occurred. A slight shift of a stellar line to the left or right produced a crooked line, all the more jagged in appearance because of the double discontinuity.

A further complication to get around was that the line he chose for detailed comparison, a prominent line in the spectrum of hydrogen, was narrower in the stellar spectrum than in the laboratory spectrum. Huggins knew this discrepancy arose from differences in temperature and pressure in the vapors generating the hydrogen spectrum in the star and in the laboratory; the higher the pressure, the broader the line. Undaunted, he incorporated a vacuum tube into his laboratory set-up so he could produce a finer hydrogen line for comparison with the stellar spectra.

By the spring of 1868, Huggins felt he had convincing evidence of a shift in Sirius's spectrum—proof of its motion. Other stars he had examined were simply too faint to be studied with his current experimental set-up, and the natural unsteadiness of the atmosphere blurred the spectra, but he did not want to wait for more results and risk being "scooped." He submitted an article to the *Philosophical Transactions*, acknowledging the help from the well-known Maxwell, stressing the difficulty he had had with the experimental set-up, and announcing his results.

The shift he measured between the laboratory hydrogen line and the center of the stellar line amounted to an apparent change in the wavelength of the light emitted of 0.109 millionth of a millimeter, or about 1000 nanometers in modern units. Translating this shift according to the formula he had learned from Maxwell, he derived "a motion of recession existing between the Earth and the star of 41.4 miles per second." Of this motion, some part was due to the Earth's motion in space; Huggins calculated that at the time he made his observations, considering the position of the Earth in its orbit around the Sun, the Earth was moving away from the star with a velocity of about 12 miles per second. He noted, "There remains unaccounted for a motion of recession from the Earth amounting to 29.4 miles per second, which we appear to be entitled to attribute to Sirius."[31]

This paper — essentially correct in its approach, although not in the details of Sirius's motion — was to be Huggins' first and last on the subject for several years, but it secured him a reputation as a pioneer in measuring motion from spectra. His work on line shifts was so innovative that few of his peers even understood it. The principle is the same as that commonly used today to study the motion of galaxies, and the technique was used for this purpose by Hubble in the 1920s (see chapter 9). In the decades following Huggins' measurement of the radial velocity of Sirius, astronomers began compiling radial velocities in a routine way, along with other spectral data and proper motion transverse to the line of sight.

The 15-inch telescope

A few months after the publication of the paper on stellar motion in the line of sight, Huggins was deeply saddened by the death of his mother, whom he had been caring for. He felt himself quite unable to work for several months. He even gave up competing against astronomers in the solar physics community; he had hoped to be the first to find a method to observe the Sun's outer atmosphere without the benefit of an eclipse, but both English and French astronomers beat him to the punch.

For a time, nothing seemed to be going Huggins' way. He sold his 8-inch aperture telescope and was in negotiations to buy a larger one from Thomas Cooke, then the premier telescope maker in England, when Cooke died. At the same time, Huggins served as a consultant to the Australian government, which was planning the construction of a new telescope and spectroscope to be installed in Melbourne. Though he was happy to be sought out for his advice, Huggins knew that the Melbourne observatory would soon be competing with his own.

Huggins' friend Thomas Romney Robinson, director of the Armagh Observatory in Ireland and an influential member of the Royal Society, offered a solution: the Royal Society would purchase a larger telescope to be installed in Huggins' own observatory, and Huggins would have full use of it.

"Mr. Huggins has done wonders with the means at his disposal; but any one who is familiar with this kind of work must know that his 8″ object-glass cannot go much beyond what it has already revealed to him, and must regret that one so highly gifted for these investigations should not be enabled to pursue them to the greatest possible extent," Robinson told the Council of the Royal Society, meeting to consider his proposal.[32]

Huggins, still not feeling fully recovered from his loss, hesitated. "I am suffering from my nervous system having been shaken so that I am rather nervous about having a large instrument," he wrote to Robinson. "I fear I shall not be able to do all that the society might reasonably expect."[33] He was also worried about the changes that a larger telescope might entail. A larger telescope would require a larger dome over his observatory, and, as he confessed to Robinson, he feared that moving the dome by himself would tire him out before he even began his nightly observations. On the other hand, he was loathe to make himself dependent on an assistant, whom he imagined would not be sufficiently flexible to be available at moment's notice. "It would cripple me much to need an assistant. I should lose so many opportunities for observations," he confided.[34]

In the end, Huggins accepted the offer and the highly respected firm of Thomas Grubb built him a 15-inch aperture telescope. It arrived at Upper Tulse Hill in 1870. Huggins had the use of it until the end of his working life but, as he had foreseen, he was dogged by complaints from rival astronomers, who thought that he was not making good enough use of this valuable resource.

A curious aspect of Huggins' mode of operation during these years, considering his youthful hobby of taking daguerrotypes, is his apparent lack of interest in photography. Solar astronomers incorporated photography in their routine operations as a way of recording data, and stellar astronomers in the United States were beginning to do so. Henry Draper, a rival of Huggins in the United States, was the first to photograph the spectrum of a star in 1872. Another of Huggins' rivals, the solar astronomer J. Norman Lockyer, even asserted in a public lecture in 1873 that "we have in photography not only a tremendous ally of the spectroscope, but a part of the spectroscope itself."[35] Huggins and his friend and mentor Miller abandoned their early efforts to photograph stellar spectra in the early 1860s, relying instead on hand-drawn charts of spectra, and although they hinted that they would return to photography, they never did. Huggins went to great lengths to circumvent the need for photography during the eclipse expedition to Oran; he planned to use a pin to prick a paper card at the location of a spectral line. (In the end, he didn't use this system because clouds obscured the scientific team's view of the Sun at the last moment.) But just as his failure to keep abreast of developments in photography was about to become a serious impediment, a woman came into his life who brought the requisite skills into his observatory.

Margaret

Huggins' personal and professional life changed radically in the mid-1870s. Sometime in the early 1870s, probably in the home of mutual friends in London, he met Margaret Murray (figure 6.6). She had an interest in astronomy and an unusual—and no doubt latent—aptitude for laboratory work. He married her in

Figure 6.6   Margaret Lindsay Murray Huggins (1848–1916). (Courtesy of Wellesley College Archives.)

1875, when he was 51 and she was about 27. She became his devoted companion both inside and outside the observatory.

Although both she and Huggins referred to her as his assistant, Margaret was much more than a passive trainee. Recent research by historian of science Barbara Becker, for her doctoral dissertation, has shown that Margaret not only assisted Huggins, but played a key role in getting him to use photography. She had input on the research agenda at Upper Tulse Hill, optimized the instrumentation, and occasionally challenged her husband on the interpretation of data.

Margaret was born in 1828 near Dublin. Her father, a solicitor, and her mother were both of Scottish ancestry, but she grew up with her older and younger siblings in what is now the town of Dun Laoghaire, Ireland.[36]

194

Her mother died in 1857. According to some sources, she subsequently spent some time with her wealthy paternal grandfather, and was sent for a while to a private girl's school in the south of England. She liked to draw, and shared with her future husband an interest in music as well as astronomy.

After Margaret's death in 1915, her friends offered different accounts of how she came to be interested in astronomy and photography. Some thought her grandfather had acquainted her with the telescope, and others said that she had, in her teens, experimented with prisms after reading a popular article on spectroscopy, and had seen Fraunhofer's lines in the solar spectrum for herself. One friend added that Margaret had taught herself photography as a child or adolescent. The claim is not far-fetched; in the years following the Great Exhibition, photography grew rapidly in popularity, especially among the artistically inclined. No social conventions barred women from experimenting with photography as an amusement or even from working as professional photographers' assistants.

After her marriage to Huggins, Margaret picked up scientific research with alacrity. Within a few months of moving to Upper Tulse Hill, she began experimenting with photography. She took over the task of maintaining the laboratory notebooks in March 1876. It is not clear whether the remarks she made during the first months reflect her own thoughts or her husband's, but by December 1876, her use of "I" and "we" in her notes, and later her use of "W" to denote her husband William's pronouncements, leave no doubt that she conducted independent work, and that she made photographic techniques her special domain.

For example, Margaret wrote in June 1876: "I had a new and much smaller camera made to use in connection with the above described apparatus.... I was occupied upon all favourable days in testing and adjusting this photographic apparatus upon the solar spectrum: at the same time testing different photographic methods with a view to finding, relatively to different parts of the spectrum the most sensitive, and relatively to the whole spectrum the quickest method for star spectra."[37] Many years later, in 1893, she was confident enough to record her disagreement with Huggins on a matter of procedure: "I was... unable to be in the Observatory but W insisted on working alone. Again tried Messier 15, giving exposure from 6.10 to

195

9 p.m.... Developed next day and delighted to find a spectrum good enough to tell us something. It is not however as strong as I should have liked & I regret much that W would not take my counsel & have left the plate in so that it might have had continued exposure the next fine night."[38] These samples of her laboratory notes show that Margaret was much more than an assistant, even in the early days of her marriage.

Competition

Huggins served two years as president of the Royal Astronomical Society, 1876–1878. The work that he would be best known for was behind him: his collaboration with Miller—who had died suddenly in 1870—exploring stellar spectra, his discovery of the emission spectra from planetary nebulae, and his bold application of theory to develop a method of measuring stellar motion in the line of sight. He had reached the pinnacle of his scientific career.

Thanks in part to Margaret's assistance, Huggins was able to continue his productive scientific research until he was in his late seventies. From the late 1870s until he returned the Royal Society's telescope in 1906, Huggins attacked several important problems. He continued his quest to photograph stellar and nebular spectra. He experimented with photographic techniques and published a major article on the photographic spectra of stars in the ultraviolet portion of the spectrum. He tried to make workable an idea he had to photograph the solar corona in the absence of an eclipse. Emerging victorious from a controversy with the solar physicist Lockyer, he proved correct his belief that the "chief nebular line," though extremely close to lines of both nitrogen and magnesium, did not, as he had earlier thought, correspond to the line of any known element. Most importantly, in 1882, he and Margaret were the first to photograph a nebular spectrum, that of the great nebula in Orion. The spectrum of the nebulae are more difficult to photograph than those of stars, because their light is generally fainter.

However, even with Margaret's help, Huggins could not stay ahead of the competition in the areas of investigation that most interested him. One of his great rivals was the New Yorker

Henry Draper, a wealthy and talented amateur whose influence on the study of the "construction of the heavens" reached far into the twentieth century.

Draper, 13 years younger than Huggins, had also been fascinated with daguerrotypes as a youth. He studied medicine at the University of the City of New York (now New York University) and incorporated daguerrotypes of microscopical images of blood cells in his thesis.

At 20, having completed his medical studies before the age at which he would be allowed to graduate, he made a trip to Europe with his older brother. In Ireland he visited William Parsons and his "Leviathan" reflecting telescope. During this visit he became interested in combining photography and astronomy. He explained in an article: "On returning home in 1858, I determined to construct a similar, though smaller instrument; which, however, should be larger than any in America, and be especially adapted for photography."[39] In 1861, after receiving some advice from John Herschel on the construction of reflecting telescopes, he erected a private observatory on his father's estate at Hastings-on-Hudson, New York, and began taking daguerrotypes of the Sun and Moon.

In the mid-to-late 1860s, when Huggins was publishing his work on stellar and nebular spectra and just beginning his work on the radial motion of stars, Draper was following a similar course, investigating the spectra of the elements and photographing stellar spectra. Draper married the former Anna Palmer during this period and, like Huggins in the late 1870s, found his wife to be an enthusiastic partner in the laboratory and at the telescope.

In August 1872 Draper succeeded, as mentioned earlier, in photographing the spectrum of a star, Vega, beating Huggins in this respect by four years.

In the spring of 1879, when Huggins was 55 and Draper 42, the rivals met face-to-face on the occasion of Draper's visit to the Tulse Hill Observatory. Draper learned of improvements in photographic emulsions from the Hugginses and returned to his own observatory with a renewed feeling of encouragement.

This was a period of great advances in stellar spectrum photography. Draper wrote in a scientific article, "It is only a short time since it was considered a feat to get the image of a ninth magnitude star, and now the light of the star of one magnitude less may be photographed even when dispersed into a spectrum."[40] Draper hoped to take long-exposure shots of the Orion nebula, a perennially interesting object, but he died before he could complete his plans, in November 1882. He suffered from double pleurisy contracted as a result of exposure to severe cold on a hunting trip in the Rocky Mountains.

Draper's influence on the course of astro-photography, and his challenge to Huggins, surprisingly, did not end with his death. Edward Pickering, the director of the Harvard College Observatory, persuaded Anna Draper to transfer her husband's telescopes to Harvard and to fund research on stellar spectra that he had hoped to continue. The resulting Henry Draper Memorial fund, established in 1886, paid for a photographic survey of bright stars and the detailed study of stellar spectra—just the kind of study Huggins contemplated. To help devise a classification scheme and to carry out the analysis of the spectra, Pickering hired a Scottish immigrant and single mother, Williamina ("Mina") Fleming. Pickering had first employed Fleming as his housekeeper, but gave her a more intellectually challenging job when he became aware that she had been a teacher and had a strong educational background. She was the first of a small army of women assistants at Harvard, many of them college-educated, who are still much spoken of today in the astronomical community.

Fleming and Pickering created a system ranking the stellar spectra with the letters A through M according to complexity (not to be confused with Fraunhofer's letter labels for individual lines in the solar spectrum). The simplest A-type spectra had a restricted set of lines later identified with helium, while other classes displayed hydrogen lines in varying degrees of intensity, and most red stars, classified as M, showed richer spectra with strong calcium lines.

In the late 1880s, Fleming was well on her way to classifying the more than 10 000 northern hemisphere stars that would be included in the first Draper catalog, which appeared in 1890. She was joined in 1886 by Annie Jump Cannon, who catalogued southern-hemisphere stars and made her own mark on the classification system. Other assistants in the "corps of women computers" made mathematical calculations—correcting star positions for the effects of precession, for example.

Pickering's "harem," as historians have termed it, posed a real threat to other astronomers working in the new field of spectrum analysis, for by employing women at a fraction of the pay male assistants would have received, Pickering could efficiently analyze the reams of data that were quickly being generated through photography.[41] An article in *Popular Astronomy* that appeared in 1898, shortly before Fleming published the first Draper catalog, publicized for a wide audience the opportunities Pickering had given to women.

"The application of photography to astronomy, whereby the determination of star-positions, spectra-type, variability, etc., [has] become laboratory work rather than Observatory work, has wonderfully increased the opportunities for woman in pursuit of the truths of nature," the author wrote. "But the most extended application of the aid of women in this specialty has been under the directorship of Professor Edward Pickering at Harvard Observatory, where a large force of women is constantly employed under the supervision of Mrs. Fleming."[42]

Whether or not Huggins read this article, he certainly knew of Pickering's corps of assistants and felt the pressure from across the Atlantic. He had been interested in the classification of stellar spectra, but it hardly seemed possible to make his mark in the field against the competition from Harvard. In 1887, shortly after the Draper fund became a reality, he wrote to the astronomer George Stokes for advice, noting particularly the "magnificent scale" of operations that the Draper endowment would allow.

"The question is, is it worth my while to continue working in this direction now that it is being done under circumstances with which no zeal and perseverance on my part will enable me to be in an equal position," he wrote. "It is scarcely worth while to do what will be done well, no doubt, elsewhere — I do not at this moment see clearly any entirely *new* direction of work."[43]

In 1888, the Hugginses received another jolt, this time in the area of research on nebulae. Isaac Roberts, an English amateur astro-photographer, had succeeded in taking a photograph of the faint Andromeda nebula, showing its oval form in detail. Huggins had previously obtained a faint spectrum of the same object, which appeared continuous and therefore coarsely resembled the absorption-line spectrum of a star or collection of stars, rather than rather than the bright-line spectrum of a gaseous nebula. Roberts's photograph, however, seemed to show a large nebulosity around a central bright core. Huggins quickly arrived at a view that synthesized elements of both the recent photographic view and the early attempt at a spectrum analysis: he saw in the Andromeda nebula a single star in the process of evolving from a nebula.

Huggins described the photograph in another letter to his friend Stokes. He wrote that the Andromeda nebula showed "for the first time to the eye of man its true nature. A solar system in the course of evolution from a nebulous mass! It might be a diagram to illustrate the Nebular Hypothesis! I never expected to see such a thing. There are some 6 or 7 rings of nebulous matter already thrown off, & in some of them we see the beginning of planetary condensation & one exterior planet fully condensed. The central mass is still larger, to compare it with the solar system, say as large as the orbit of Mercury. The rings are all in one plane & the position is such that we see it obliquely."[44]

In the observatory notebook Margaret noted the Huggins' somewhat competitive joint response to the photograph: "It would be of special interest we think to supplement this remarkable photograph with some photographs of the spectrum of this neb. Mr. Roberts' work gives the body: if we can get good spectra we should have the soul."[45]

Unfortunately for the Hugginses, and for all astronomers laboring to make sense of the nature of the nebulae, the light from the Andromeda nebula is so feeble that a photographic spectrum would not be obtained for years to come. In January 1899, the German astronomer Julius Scheiner, working at the Potsdam observatory, captured the exceedingly faint spectrum with a $7\frac{1}{2}$ hour exposure. As Huggins' earlier, non-photographic study had indicated, it proved to be a stellar spectrum, as would be expected for a distant cluster of stars. In the meantime, while waiting for the kind of strong evidence that the spectrum would provide, astronomers could only speculate, and the normally cautious Huggins clung to the Laplace "nebular hypothesis" as an explanation for the phenomenon he saw in Roberts's photograph.

Throughout this period of increasing competition with other astronomical spectroscopists and photographers, Huggins had never mentioned Margaret's assistance in his many scientific papers. But spurred, perhaps, by the threat of being trounced in one of his particular lines of investigation, he finally did so in 1899. Huggins was embroiled in a controversy over the distinction between the chief nebular line and a line arising from magnesium. To add weight to his argument, it was convenient for him to include Margaret's observations in his report, as those of an independent witness confirming his. Her name thus appeared with his as a co-author. He explained in the introduction to the paper, "I have added the name of Mrs. Huggins to the title of the paper, because she has not only assisted generally in the work, but has repeated independently the delicate observations made by eye."[46] In effect, he was able to claim a more authoritative result with her help, without entering into a collaboration with a peer or hiring a research assistant—and without acknowledging the full extent of her assistance.

Later years

In 1897, the year he became *Sir* William Huggins, Knight Commander of the Order of the Bath, Huggins wrote a lengthy article for the review magazine *The Nineteenth Century* entitled, "The New Astronomy: A Personal Retrospect." By "new astronomy" he meant astrophysics, the study of the physical properties of the stars and nebulae through spectroscopy, in contrast to the traditional focus on stellar positions and planetary dynamics.

In this article, Huggins described the early years of his scientific career, with considerable literary flair and a minimum of concern for historical accuracy as far as dates and scope of work are concerned.[47] His recollection of how he got his start in spectroscopy has proved particularly attractive to biographers, and is often quoted:

"I soon became a little dissatisfied with the routine character of ordinary astronomical work, and in a vague way sought about in my mind for the possibility of research upon the heavens in a new direction or by new methods. It was just at this time, when a vague longing after newer methods of observation for attacking many of the problems of the heavenly bodies filled my mind, that the news reached me of Kirchhoff's great discovery of the true nature and the chemical constitution of the Sun from his interpretation of the Fraunhofer lines.

"This news was to me like the coming upon a spring of water in a dry and thirsty land. Here at last presented itself the very order of work for which in an indefinite way I was looking— namely, to extend his novel methods of research upon the sun to the other heavenly bodies. A feeling of inspiration seized me: I felt as if I had it now in my power to lift a veil which had never before been lifted; as if a key had been put into my hands which would unlock a door which had been regarded as for ever closed to man—the veil and door behind which lay the unknown mystery of the true nature of the heavenly bodies. This was especially work for which I was to a great extent prepared, from being already familiar with the chief methods of chemical and physical research."[48]

This personal retrospective launched William and Margaret on an effort to begin collecting and editing the published papers that they felt most represented the legacy of the Tulse Hill Observatory. They continued to do original research—they collaborated, for example, on a series of papers between 1903 and 1905 on the spectrum of radiation from the radioactive element radium—but put much of their effort into a two-volume work, *Atlas of Representative Stellar Spectra* and *The Scientific Papers of Sir William Huggins*. These works included a history of the observatory, a description of the instruments and methods used, and reprints of papers, including those in both their names.

The Hugginses did not discuss William's unproductive efforts, such as those he made in the field of solar physics. Nor did they attempt to synthesize some new theory from his vast experience with the spectra of various kinds of nebulae. Perhaps they were right to hold back, because it was not until two years later that Scheiner obtained the photographic spectrum of the Andromeda nebula. That spectrum proclaimed, to those attuned to its message, the existence of "island universes" of stars, and contradicted Huggins' view of the Andromeda nebula as a Laplacian nebula in the process of forming a single star.

Perhaps the best synthesis of Huggins's views on the nature of our stellar system and the evolution of celestial bodies comes from an address he gave to a gathering of scientists in Cardiff, Wales, in 1891.[49] In words that to some extent echo William

202

Herschel's, describing "strata" of stars and nebulae winding across the sky, Huggins wrote:

"The heavens are richly but very irregularly inwrought with stars. The brighter stars cluster into well-known groups upon a background formed of an enlacement of streams and convoluted windings and intertwined spirals of fainter stars, which becomes richer and more intricate in the irregularly rifted zone of the Milky Way."

"We, who form part of the emblazonry, can only see the design distorted and confused; here crowded, there scattered, at another place superposed. The groupings due to our position are mixed up with those which are real."

Foreshadowing the words of Harlow Shapley, the subject of our chapter 8, he added that structures seen among the stars seemed to have been built up on a range of scales or levels. "We see a system of systems," he wrote, "for the broad features of clusters and streams and spiral windings which mark the general design are reproduced in every part."

Drawing on his pioneering studies of the motions of stars, Huggins emphasized that the components of the known universe were in motion, and that the motions of the stars might provide a clue to the universe's history. "Surely every star, from Sirius and Vega down to each grain of the light dust of the Milky Way, has its present place in the heavenly pattern from the slow evolving of the past," he wrote.

Thus, the picture he painted is of a stellar system slowly evolving before our eyes, with clusters in various stages of formation joining other clusters and sinking into the dense agglomeration that is the Milky Way. He concluded that "[t]he deciphering of this wonderfully intricate constitution of the heavens will be undoubtedly one of the chief astronomical works of the next century." Among the projects he predicted would provide important clues to the large-scale picture was the creation, by international cooperation, of a comprehensive photographic atlas of the stars. This project, known as the *Carte du Ciel* or Map of the Sky, would be an important part of the life of Jacobus Kapteyn, the subject of our next chapter.

An incident toward the end of Huggins' life puts his 30-year collaboration with his wife Margaret in a surprising perspective. In November, 1906, the Royal Society council met to vote on awarding its prestigious Hughes Medal to Hertha Ayrton, a woman physicist. Ayrton had worked on a number of phenomena, including electric arcs; she later became famous for inventing a type of fan to disperse the poisonous gases that threatened combatants during World War I.

Huggins had just completed his five-year term as president of the Royal Society. He was absent from the council meeting, having taken slightly ill. There is some speculation that Margaret persuaded him that he was not well enough to go out that day. In his absence, the council voted to award the medal to Ayrton, a move Huggins would certainly have argued against. On hearing the outcome of the vote, Huggins wrote in outrage to Joseph Larmor, the Royal Society's secretary:

"The papers will teem with publications from all the advanced women! I suppose the P [President] will invite her to the dinner, and ask her to make a speech. As the only lady — I should say woman — present, the P. will have to take her in, and seat her on his right hand! And all this comes from what appeared as the pure accident of my taking a chill on Wednesday. [...] Can we now refuse the Fellowship to a Medallist?"[50]

Huggins died in 1910, at the age of 86, following an operation. Since 1890, he had received a Civil List pension in recognition of his astronomical research, and after his death Margaret learned that she would continue to receive some of the money. She wrote to the same Joseph Larmor,

"No doubt you know about my Pension. £100 a year has been granted me, 'for my services to Science by collaborating with' my Dearest. This I *could* accept without *any* reflection on the memory of my Dearest — & with *honour* to myself as well as to *him*. I do regard the Pension as an honour *to him* although it is honourable also to me, & I humbly hope, — *really earned* for the 35 years of *very hard work*. None of you know how hard we worked here just our two unaided selves."[51]

Margaret's comment about the Hugginses working by themselves serves as a reminder that William Huggins was one of the last great amateur astronomers who have contributed so much to the field. After his time, astronomy became increasingly professionalized, so that it was necessary for aspiring astronomers to study astronomy and physics at university, and, usually, to pursue their research in connection with universities or observatories which supplied equipment and assistants and helped shape the research agenda. In part, this growth and professionalization of the field grew out of the application of photography to spectroscopy, which the Hugginses helped bring about. Photography yielded so much data so quickly, compared to previous methods of recording observations, that it became almost impossible for the amateur, working alone, to compete with the well-staffed professional observatory.

7

JACOBUS KAPTEYN: MASTERMIND WITHOUT A TELESCOPE

"Undoubtedly one of the greatest difficulties, if not the greatest of all, in the way of obtaining an understanding of the real distribution of stars in space, lies in our uncertainty about the amount of loss suffered by the light of the stars on its way to the observer."

J C Kapteyn, 1909[1]

Jacobus Kapteyn, a plainly dressed, thin man with a long neck and heavy eyelids, sat in animated conversation with his fellow travelers in the second-class compartment of a train. He spoke in the Low Saxon dialect of Groningen, a small provincial capital in the Netherlands, although he came from Barneveld, farther to the south. His seatmates, who were mostly traveling salesmen, had all but forgotten that he was not one of them but a highly educated professor of astronomy, a teacher, and a researcher who, like other university professors, had been appointed to his post by the Crown. University faculty members were rarely seen outside of the first class compartments.

The train clanked and screeched to a halt at Groningen, and Kapteyn hastily concluded his conversation. He threaded his way among the salesmen's large black bags and jumped off the train, quite forgetting his own luggage. This was not unusual; he routinely misplaced his wallet and forgot appointments, and his wife categorically refused to buy any more umbrellas because he promptly lost them.

A salesman ran after him, calling out, "Professor! You forgot your bag of samples!" Kapteyn stopped and acknowledged the man's help. "Thank you very much! But these are not samples, you know."

The salesman replied, "Well then, if you wish, your bag of stars!"[2]

The story, one of many fondly recalled by Kapteyn's friends, illustrates several of the personality traits that made him a much-loved figure among his neighbors and among astronomers all over the world. And the salesman's joke about the bag of stars is closer to the truth than he realized, for Kapteyn made it his specialty to study the known universe through representative samples of stars from around the celestial sphere. Throughout his career at Groningen, Kapteyn painstakingly added data to his "bag of stars" until, near the end of his life, he could pull out a carefully wrought model of the stellar system, the so-called "Kapteyn Universe."

Kapteyn (figure 7.1) was born on 19 January 1851 in Barneveld, the Netherlands, the ninth of 15 children. Such large families were

Figure 7.1 Jacobus Cornelius Kapteyn (1851–1922). (Credit: Yerkes Observatory.)

not unusual at the time, but Kapteyn had further cause to feel overlooked: his parents ran a boys' boarding school and treated both their own children and their pupils as one extended household of about 70 members. Kapteyn saw little of his parents or siblings, a fact which he later believed made him particularly appreciative of the warm friendships he developed as an adult.

Kapteyn inherited unusually long, heavy eyelids from his mother. This gave him an absent-minded look that his demeanor only reinforced. Henrietta, Kapteyn's daughter and biographer, wrote that his peers remembered him as a slim, pale youth, a sloppy dresser who was "constantly in deep thought."[3]

Kapteyn loved animals, especially birds, and bred canaries as a boy. An anecdote about an owl he caught illustrates his early interest in scientific experimentation. He had heard that owls could not see in daylight, but found this difficult to believe. To test the truth of the statement, he stretched a number of strings across a room in a web-like pattern and set the owl free in the enclosed space. The owl flew about without ever touching a string, contradicting the conventional wisdom and fueling Kapteyn's interest in checking scientific facts for himself.

Like many of his older brothers, Kapteyn showed a talent for mathematics and physical science. Even among these talented siblings—one of whom, Willem, became a professor of mathematics and a collaborator with Kapteyn on mathematical papers—he stood out as remarkably clever. One day when Kapteyn was about 10, his eldest brother Hubert, home from university for the holidays, offered to play chess with him. Kapteyn beat him three times in a row. He couldn't induce Hubert to play with him again.

At 14, Kapteyn came across a star chart that a younger sister had brought home, and began exploring the night sky with a small telescope in the garden. His father, noting his serious interest, bought him a larger telescope. Kapteyn must have been dissatisfied with the commercial star chart, because he set about making his own, with silver paper, paint and glue—foreshadowing his more sophisticated models of the stellar system in the 1920s.

Kapteyn passed the entrance examination for the university of Utrecht at the age of 16, but his father judged him too young to leave home and kept him at Barneveld another year. When

Kapteyn did enroll, in 1868, he found life as a student relatively carefree. His father's rigorous school had prepared him well, and he sailed through his studies of mathematics and physics.

During his last year at Utrecht, Kapteyn met his future wife, Elise Kalshoven. The Kalshoven family was, in daughter Henrietta's words, the "antipode" of the Kapteyn family. "There was no studying and working going on, everything was cheerful and cozy," Henrietta said of her mother's maiden household. Kapteyn admired Elise's confidence and cheerful nature, and "saved the picture of this radiant young girl in his heart until his time would come," Henrietta wrote.[4]

Kapteyn earned his PhD in 1875 with a dissertation on the vibration of membranes, a fundamental subject in physics. His father had hoped he would follow a family tradition of great longevity and become a teacher, but Kapteyn felt he did not have the personality for this career.

Kapteyn sought and obtained a position with a proposed observatory in Beijing, China. Since the 1840s, the Chinese had been studying Western science, industry, and diplomacy, and the peak of the so-called "Self-Strengthening Movement," which evidently included plans for a western-style observatory, occurred right about the time Kapteyn was looking for work. But plans for Kapteyn to join the observatory fell through. Instead, he accepted employment as a junior staff member or "observer" at the observatory at Leiden, a Dutch city known, like Utrecht, for its university. He stayed there two years, becoming adept at handling astronomical instrumentation and formulating the research questions he would pursue for the rest of his life.

An inauspicious start

In 1877, a royal decree established Kapteyn as professor of astronomy and theoretical mechanics at the University of Groningen, at the same time that his brother Willem obtained a mathematics professorship at the University of Utrecht. The University of Groningen, founded in 1614, had only 200 or 300 students, and some considered the town, in the far northeastern corner of the Netherlands, to be remote and provincial. However,

Kapteyn's professional success seemed assured. He moved to Groningen and then returned south for a visit to Utrecht, to ask Elise to marry him. A wedding date was set for the following year.

Kapteyn brought to his new post in Groningen high hopes of attacking and solving some of the fundamental problems confronting astronomers at the close of the nineteenth century. Upon joining the faculty, he was expected to present an inaugural address. Reflecting his interest in the scale of the universe and the distribution of stars, he chose to lecture on "The Parallax of the Fixed Stars." No text of his lecture survives, but it is likely that he explained to his non-specialist audience the importance of parallax measurements, and he may have hinted at his own plans to study the structure of the known universe by accumulating more information on stellar distances. Almost 40 years after Bessel, Struve, and Henderson first made convincing measurements of stellar parallax, data on stellar distances were still scarce and difficult to obtain.

Kapteyn joined the faculty shortly after the formation of the astronomy department, and naturally expected to avail himself of a telescope, photographic equipment, and other instruments. The university as yet had nothing, but Kapteyn set to work gathering the necessary resources. He located a suitable site outside of town where an observatory could be erected, and requested funds for a 6-inch aperture heliometer, an instrument like Bessel's, to collect parallax data. He sent proposal after proposal to the government, which controlled university spending. However, astronomers at Leiden and Utrecht, fearing that funds for astronomical research would be spread too thin, opposed his plans.

Kapteyn continued to petition the Dutch government for funds for 12 years after he arrived on campus. While he waited and hoped for a positive response, he collaborated on mathematical problems with his brother Willem and spent vacations at the observatory in Leiden. Use of the equipment in Leiden allowed him to publish a few papers on techniques for precise parallax measurements and to measure the parallaxes of 15 stars over the course of four years, but he was clearly still hungry for scientific data. At one point he even requested that the Dutch government help him ask the French government for the locations of 200-year old trees around a meteorological station in Paris, so he could study the growth of trees as a function of rainfall. He was stuck, and as the years slipped by he became rather depressed about his situation.

Kapteyn and his wife kept busy raising children during these professionally unrewarding years of the 1880s. Their eldest daughter Jacoba Cornelia was born in 1880, a year after their marriage. Henrietta was born a year and a half later, and son Gerrit in 1883. Henrietta wrote that her parents took an unusually rational approach to child-rearing; "[T]ogether they had made a thorough study of the principles of child-rearing," she noted.[5] They went so far as to challenge what they believed were unsound practices of doctors and midwives, whose authority was usually unquestioned.

Kapteyn took a more active role in the family life than most fathers of his day, according to Henrietta. "In contrast to the foolish etiquette of the time, mother was the first in her circle of friends who pushed her own baby carriage rather than have a nurse-maid do it. And if they went out together, the young professor used to push the baby carriage himself, despite the laughs of the street urchins, and the astonishment of his colleagues," she claimed.[6]

Henrietta's own favorite story about her father's philosophy centers on a luscious bit of fruit. "There was once a large bunch of grapes on the table, which itself was very rare in such a simple family," she wrote. "As a small child, I eyed them gleefully and sighed, 'Oh! If only I could eat that whole bunch alone, then I would be perfectly happy!'" Her father asked if she realized how unfair this would be—then relented and said, "Well now, then you shall enjoy perfect happiness this once, child, which is so seldom." And he gave her the whole bunch.[7]

Sky surveys

According to Henrietta, an article Kapteyn read during one Christmas vacation changed the course of his professional life and provided his *entrée* into a career of universe-charting in the tradition of William Herschel and Wilhelm Struve. The article, written by Kapteyn's acquaintance David Gill, described a new

211

project to photograph the southern sky, frame by frame, to produce a catalog that would complement existing catalogs of the northern sky.

David Gill, a Scotsman, had taken charge of Britain's Royal Astronomical Observatory at the Cape of Good Hope in what is now South Africa. The British Board of Longitude, which ran the Royal Greenwich Observatory, had set about establishing the Cape observatory as a southern-hemisphere counterpart in 1820; it was built and equipped by 1828. Gill, a jovial and out-spoken character who ran the observatory with an unprecedented degree of informality, took up his position there in 1879, when he was 36 years old.

Gill's self-imposed mission was to take advantage of recent improvements in photography. A large format photograph of a comet in 1882 showed a surprisingly vivid field of stars in the background, and Gill became convinced that photographic emulsions had become sensitive enough to record the positions of ordinary stars as well as capture the visual likenesses of bright or extended objects such as planets, comets, and nebulae. He proposed an international collaboration to photograph the stars of the entire sky, and began almost at once on his own photographic atlas and catalog of the southern sky.

In the article Kapteyn read in an astronomical journal, Gill described his project. He noted the immense volume of data that the project was generating—data that would require years of analysis, more data than Gill could hope to deal with on his own. The southern sky project, which came to be known as the *Cape Photographic Durchmusterung* or Cape sky survey, was meant to complement the *Bonner Durchmusterung* or Bonn sky survey. The Bonn survey was the most complete star-catalog to date, comprising 324 000 northern-hemisphere stars and indicating star brightness as well as position. But while Friedrich Argelander, compiler of the Bonn survey, had recorded star positions with a micrometer and had estimated the brightnesses by eye, Gill would record positions and brightnesses objectively and consistently on photographic plates. The hard part would be to measure the plates, that is, to extract the star coordinates and magnitudes from the photographs, and secondly to apply any necessary corrections to these raw data before assembling them into a catalog.

Kapteyn saw an opportunity. He dispatched a letter to Gill offering his help with the analysis. He had no illusions about the nature of the work; determining star positions and other information from the photographic plates would be tedious and repetitive, and he expected the work to take six or seven years. But the resulting data on the distribution of stars and their magnitudes would surely yield information on the architecture of the universe, a question of great importance to Kapteyn. From the beginning of his professorship, Kapteyn had envisioned studying the stellar system "from the ground up" by amassing large amounts of data and trying to infer the distribution and motions of the stars.

Kapteyn wrote to Gill, "If you will confide to me one or two negatives, I will try my hand at them and, if the result proves as I expect, I would gladly devote some years of my life to this work, which would unburden you a little, as I hope, and by which I would gain the honor of associating my name with one of the grandest undertakings of our time."[8]

Gill responded with warmth and enthusiasm. "It is not easy to tell you what I feel at receiving such a proposal," he replied. "I recognize in it the true brotherhood of science and in you a true brother."[9] And so, beginning in 1886 and continuing until 1892, Gill shipped negatives to Groningen and Kapteyn measured the star positions from them. Kapteyn, still strapped for resources, had no laboratory space of his own, but his friend professor Dirk Huizinga from the psychology department let him use two rooms in his building. At last Kapteyn was observing stars — if only on photographic plates.[10]

To obtain the star positions directly in right ascension and declination — rather than measuring them as x and y coordinates from a reference position on the photograph, and laboriously transforming those to coordinates on the celestial sphere — Kapteyn set the images on a stand and scrutinized them from a few feet away with a theodolite, a surveyor's instrument for the accurate measurement of small angles. This allowed him to progress faster than Gill had expected, although the project still took almost twice as long as Kapteyn had originally anticipated. Even when the right ascension and declination data were in hand, calculations were required, for example to correct for the effects of refraction and precession.

Kapteyn was equally creative when it came to finding an assistant, a "computer" like the women Pickering employed at Harvard. As Henrietta put it in her biography, "Intelligent and talented men who wanted to work hard for little money were difficult to find."[11] Kapteyn turned to the director of a vocational school in Groningen, who recommended a 19-year-old student, T W De Vries. De Vries proved adept and had a long career at the university.[12]

By the time Kapteyn and De Vries finished measuring the photographic plates, the international astronomical community had begun to recognize Kapteyn's ability and diligence. No new, comprehensive models for the distribution of stars had yet emerged, but the compilation of the data itself was seen as extremely valuable. The French government bestowed on him the Légion d'honneur, and the Royal Astronomical Society in England elected him a foreign member. These accolades lifted his spirits and helped turn the tide in his efforts to garner funds and equipment at Groningen, but did nothing to relieve the physical and intellectual burden of the Cape survey work. Some 450 000 stars were cataloged in total, and calculations on the stellar data continued through the publication of the catalog between 1896 and 1900. Kapteyn wore himself out, and developed eye strain that would bother him the rest of his life.

Kapteyn found some respite from the grueling calculations in attending international meetings. It was at these meetings that he glimpsed the future direction of astronomical research, and began to see how he might forge a niche for himself once the Cape survey work was completed. In 1887, the Paris observatory, acting on Gill's earlier idea, proposed a scheme to photograph both northern and southern hemispheres of the sky. This *Carte du Ciel*, as the survey came to be known, would employ a standard plate scale (1 arcminute on the sky per millimeter on the plate) and two standard exposure times, one to record stars for a basic catalog and one to record stars as faint as the 14th magnitude. The observatory convened a series of annual meetings to discuss the plan and to apportion the work to participating observatories.

Kapteyn proposed some ideas of his own at the first *Carte du Ciel* meeting. He wanted to take advantage of the experimental set-up of the survey to measure photographically a large number of stellar parallaxes, at the same time. His idea was not adopted, but his commitment to the study of the "sidereal problem," as he called it, of the distribution of stars in three-dimensional space, is reflected in the fact that he persuaded the director of the Helsingfors Observatory in Finland to carry out a limited version of the scheme. By 1900, already Kapteyn was able to add about 250 stars to the short list of those whose parallaxes had been determined.

Although his own parallax-measuring plan was not adopted generally, Kapteyn wholeheartedly supported the *Carte du Ciel* project. He and Gill attended the meetings and both played active roles in the project's planning and execution. Like the Cape survey, the *Carte du Ciel* took much longer to complete than the organizers had hoped.[13] However, Kapteyn remained impressed with the possibilities he had seen at these meetings for international cooperation. In the original plan, observatories at 18 sites each accepted responsibility for photographing one zone of the sky. Ten of the observatories were in Europe (Greenwich, the Vatican observatory, San Fernando in Spain, Catania, Helsingfors, Potsdam, Oxford, Paris, Bordeaux, and Toulouse), one in North Africa (Algiers), one in Central America (Tacubaya, Mexico), three in South America (Santiago, Rio de Janeiro, and La Plata in Argentina), one in southern Africa (the Cape of Good Hope), and two in Australia (Sydney and Melbourne). The *Carte du Ciel* project would come as close to being a worldwide effort as one could imagine in the late 1800s.

The absence of American observatories in this list is noteworthy, because a number of them had acquired significant equipment and staff by the time of the fourth *Carte du Ciel* meeting in 1890. Representatives from the United States Naval Observatory and the observatories at Yale University and Hamilton College in Clinton, New York attended the first congress in 1887. But the American observatory director who was in the best position to participate in the data-taking, Edward Pickering of the Harvard College Observatory, rejected the French proposal in favor of his own sky-survey plan.

About the time Kapteyn finished measuring Gill's plates and began calculations for the associated Cape survey catalog, University of Groningen authorities finally responded to his pleas for support. They approved his request for instruments to

215

measure data from photographic plates, and in 1896 found an entire building to put at his disposal.

In a speech at the building's dedication, Kapteyn explained the scope of the new Groningen Astronomical Laboratory (now called, in his honor, the Kapteyn Astronomical Laboratory) to his lay audience, who had never heard of an astronomical laboratory without a telescope. He foresaw that the work done there would eventually contribute not just to the Cape survey, the *Carte du Ciel*, and the analysis of parallax data from Helsingfors, but to subsequent research initiated by astronomers all over the world.

He explained, "At each observatory much more is produced than can be analyzed, because the work force available for measurements and all the other work, while adequate for the photographic recordings, is insufficient for data reduction." Borrowing a phrase from Charles Darwin, he described his laboratory's purpose as "the grinding of huge masses of fact into law."[14]

As Kapteyn predicted, raw data have continued to accumulate faster than astronomers can analyze them, so that institutes similar to his are now commonplace. His idea was a sound one. An historian of astronomy has also noted that the concept of a low-cost astronomical laboratory performing high-value calculations fit perfectly with his temperament and talents. "What satisfied him most about the project," the commentator wrote, was that "it enabled him to be both humble and extremely ambitious at the same time."[15]

Stellar distances and motions

Kapteyn found the years around the turn of the twentieth century, when he was in his fifties, to be among his busiest and most productive. In 1900, he began publishing regular updates on his laboratory's findings—often in English, which he expected would become astronomy's *lingua franca*. In 1901, he had the satisfaction of witnessing the graduation of his first doctoral student, Willem de Sitter. (De Sitter later became famous for applying Einstein's theory of general relativity to the question of the evolution of the universe—see chapter 9.)

216

Kapteyn's own children, meanwhile, were making their way in the world. He had sent both his daughters and his son to a boy's school, and his daughters helped set a precedent for women by studying medicine and law at university.

In his scientific work, Kapteyn stuck to "the grinding of huge masses of fact," while trying not to forget the larger questions that had drawn him into the effort in the first place. He doggedly pursued clues to the precise distances and distributions of stars, even as many of his contemporaries jumped ahead to the next step and advanced hypotheses about the nature of the stellar system based on scanty information or tentative analogies.

During this period, his peers argued both for and against the island universe hypothesis. The historian of astronomy Agnes Clerke represented a popular if not majority view when she wrote in 1886 that the Milky Way system was so vast, as astronomers had learned from parallax measurements, that the idea of "island universes" on a comparable scale to ours was untenable. But Julius Scheiner cast doubt on this view in 1898 when his long-exposure photograph of the Andromeda nebula's spectrum (the spectrum that eluded William and Margaret Huggins' attempt to photograph it) emerged as decidedly star-like, suggesting that this could be an "island universe." Shortly after the turn of the century, the Dutch science writer Cornelis Easton suggested our stellar system might look like a spiral system, similar to the spirals discovered by Lord Rosse, if we could see it from afar — but he based his prescient opinion on speculation rather than on observational clues. In short, the situation was confused, and called for more data.

Kapteyn needed to probe the greatest possible distances and the greatest number of distances to determine the structure of the stellar system. Initially, he naturally assumed he would proceed by accumulating the parallaxes of individual stars. The measurement of parallaxes was something of a growth industry in astronomy near the turn of the century: in 1882, only 34 parallaxes were in hand, while by 1914, thanks to the efforts of Kapteyn and other astronomers around the world, some 100 000 had been measured. But even this growth in the accumulation of parallax data was not enough for the man who would solve the "sidereal problem."

Kapteyn turned his attention to stars' proper motions. Astronomers had built up a database of many stars with measurable proper motion, i.e., a shift in position due to their real motion through space. Sometimes a given star had both its parallax and proper motion measured, as was the case for 61 Cygni, the fast-moving binary system Bessel had found success with. Kapteyn's main effort around the turn of the century was to try to use this relatively abundant proper motion data to lead him to more information about stellar distances. He attempted to correlate stars' proper motions with their parallaxes, so that, for stars for which only proper motion was known, he could derive some estimate of their parallax distances.

His reasoning was basically like that of Piazzi, the astronomer who first suggested 61 Cygni might be nearby, and a good candidate for parallax measurements, because of its high proper motion (see chapter 5). Assume that the stars, scattered about in space, have random motions in all directions. In that case, a sample of relatively nearby stars will have a higher average proper motion than will a group of distant stars. In the same way, a car moving past our window at 40 miles per hour on a street running by our house will have a higher proper motion than a car traveling 40 miles per hour on a distant street that we can just see on the horizon. The nearer car covers a greater angle in space in a given unit of time, so that, all other things being equal, the nearer car has a higher proper motion (see figure 7.2).

The argument does not apply to individual stars, because we have assumed that they might have any randomly given velocity. A distant star could, by chance, have a higher velocity and demonstrate the same proper motion as a nearer, slower star. However, the average proper motion of a sample of stars will tend to accurately reflect the distance of stars in the sample. Thus Kapteyn applied statistical methods—he used average values for subsets of the entire population of stars to gain insight into the overall pattern of their distribution and characteristics.

Working with his brother Willem at the University of Utrecht, Kapteyn compared the proper motions and parallaxes of stars for which both quantities were measured, and found a formula relating them. Then, for a star whose proper motion only was known, he could compute what he called the "mean parallax"—the parallax or distance he thought it should have, according to the formula.

218

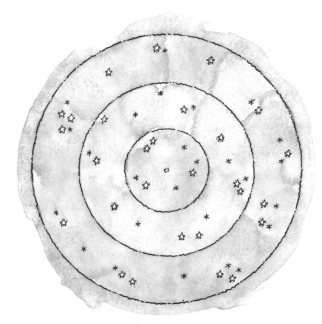

Figure 7.2 Mean parallax—a tool to estimate distances. The center of the diagram shows the position of an observer; encompassing circles divide the stars into three distance zones. If the velocities of the stars are random, an observer would measure a higher *average* velocity for stars in the inner circle, which are closer and which therefore appear to move faster than stars at greater distances. This establishes a connection between a star's typical distance (or mean parallax) and its average velocity; subsequently, if an observer acquires data about a given star's velocity, he can assign to it a distance, based on the distance–velocity relation. (Credit: Layne Lundström.)

This "mean parallax" tool was one of the ways by which Kapteyn studied the distribution of stars in space. Through its use he could in principle measure distances to about 1500–3000 light-years, while the technique of parallax (trigonometric parallax) was limited to about 300 light-years. Kapteyn no doubt expected that his mean parallax analysis would push back the limits of the known universe as it could be probed with any degree of accuracy, adding more distant stars to the sample of well-studied objects. To his surprise, however, his first results concerned not stellar distances but the organization of stars on a grand scale.

Kapteyn thought he was proceeding with all due caution in probing the parallax–proper motion relationship. In fact, he introduced refinements such as calculating separate formulas for stars of different colors, or spectral types. The key assumption he made, that stars' velocities are random, would have passed muster with any of his peers. And yet in 1902, he came to the conclusion that something was wrong with his study, or that the universe was not what he and his fellow astronomers had imagined. His results suggested that, contrary to all expectations, the motions of the stars are not random. Kapteyn was in the habit of visualizing his data by plotting points or drawing vectors with white chalk on globes covered with blackboard-material, and when he did this with the velocities of the stars in his study, he saw a distinct pattern emerging on the celestial sphere.

Today we understand that the phenomenon he uncovered is explained (in a rather complicated way) by the rotation of our galaxy. To Kapteyn, the results simply suggested that there was some large-scale systematic motion of the stars. It was a bizarre but possibly very significant conclusion, he knew.

Kapteyn waited until 1904 to present and explain his findings to the astronomical community and to the public. He was waiting for the publication of more extensive data that he might use to verify his results — data that he believed had already been acquired at some American observatories, but was not yet made available. But in 1904, he decided to go ahead and present his results as they stood. That year he received an invitation to lecture at an astronomers' convention that his American friend Simon Newcomb planned for September, in conjunction with the World Exhibition or World's Fair in St. Louis, Missouri.

Kapteyn was about to make a big splash in a country that had lagged behind in astronomical clout, but was rapidly catching up to Europe. American observatories made a rather belated emergence on the international scene. In 1825, President John Quincy Adams bemoaned the lack of observatories in his country. "It is with no feeling of pride, as an American, that the remark may be made, that, on the comparatively small territorial surface of Europe, there are existing upward of one hundred and thirty of these *light houses of the skies*; while throughout the whole

American hemisphere, there is not one," he told Congress. "If we reflect a moment upon the discoveries, which, in the last four centuries, have been made in the physical constitution of the universe, by means of these buildings, and of observers stationed in them, shall we doubt of their usefulness to every nation?"[16]

By the time Kapteyn made his first visit to the United States in 1904, however, stops at several astronomical observatories were *de rigeur* for someone of his stature. The first to achieve some measure of prominence, as mentioned earlier, was the Harvard College Observatory. Pickering assumed the director-ship in 1876, and made it his life's mission to expand on the promise of spectroscopy. With funds from Henry Draper's estate—the endowment that so frightened William Huggins, because it threatened to put him out of business—Pickering established Harvard as a center for the analysis of photographic stellar spectra.

Two other major American observatories made their mark before the turn of the twentieth century. In 1888, Lick Observatory unveiled the largest lens-based or refracting telescope of its time, 36 inches in diameter, atop Mount Hamilton near San Jose, California. Lick observatory was and is operated by the University of California. Yerkes Observatory, on the shores of Lake Michigan, operated by the University of Chicago, opened in 1897. Its 40-inch diameter refracting telescope improved on Lick's.

Star streaming and the plan of selected areas

In their characteristically modest way, Kapteyn and his wife Elise set off by bicycle from their home near Groningen. (One must presume that they sent their luggage on ahead!) They caught a train, then boarded a ship in Rotterdam for their first trip to the United States. A few weeks later they arrived at the scene of unparalleled festivities.

The focal point of the 1200-acre St. Louis Exhibition was Festival Hall, a temporary but extremely ornate building set behind an artificial lagoon and flanked by sculptures and fountains. Its gold-leaf dome glinted in the Sun; inside, tourists found an auditorium seating 3500 and the largest pipe organ in the world. At night, half a million electric light bulbs and colored

221

searchlights illuminated the buildings and fountains around the lagoon.

The International Scientific Congress formed but a small part of the World Exhibition, and most visitors probably overlooked Kapteyn's lecture, which he contributed to the astronomy section. However, his lecture was arguably one of the best examples of the official theme of the Exhibition, "the progress of man since the Louisiana Purchase" about 100 years earlier.

To the astonishment and admiration of his peers, Kapteyn revealed during his scheduled presentation his finding that the motion of stars was not random. He had plotted the stars' motions as vectors on a diagram of the sky. The pattern he pointed to reflected, first, the well-known effect of the Sun's motion in space, which causes the stars to appear to sweep away from the direction in which the Sun is headed. This motion was not expected to be random. It is simply analogous to the apparent backward motion of trees and buildings when we see them from a train moving forward.

Secondly, superimposed on that well-known effect of solar motion, Kapteyn had found that stars tended to drift in one of only two directions. They did not, after all, move randomly in all directions. One set of stars appeared to drift in the direction of the constellation Orion, while the other set moved in the direction about 140° away, toward Sagittarius. All stars appeared to belong to one or other of these two streams.

Kapteyn's careful analysis, involving thousands of stars, convinced his associates that the "two stream" effect was real — although it would take astronomers until 1927 to understand it correctly as a reflection of the rotation of stars, with various kinds of circular and elliptical orbits, around the center of the galaxy. The great English astronomer Arthur Eddington wrote later that Kapteyn's study "revealed for the first time a kind of organization in the system of the stars and started a new era in the study of the relationships of these widely separated individuals."[17]

For Kapteyn, the two-stream effect suggested that our stellar system consisted of two giant clouds of stars moving through one another. Stars from the first mingled in space with stars from the second, but each star retained its original speed and direction. Thus his study of the structure of the stellar system led him to ponder also the history of the system, by extrapolating backward to a time when the two star clouds may have first drawn closer together.

Kapteyn's discovery of the star-streaming pattern left his audience with plenty of food for thought—and that wasn't all he had to say. Before closing his lecture he issued a proposal to astronomers to attack the problem of the structure of the universe in a coordinated way. The unexpected discovery of star streaming, he felt, called for a renewed and systematic investigation into the stellar system. This was his famous *Plan of Selected Areas* proposal.

The "selected areas" were representative patches of the sky, much like those William Herschel delineated when he sought to map the heavens. Some 200 000 stars would be observed, about twice the number Herschel had counted. Kapteyn was particularly eager to extend the collection of data to faint stars, which of course are more numerous than bright ones, to better test any theoretical models that might be constructed. He was aware that in selecting bright, high proper-motion stars for parallax determinations, astronomers were optimizing their chances of finding nearby stars with measurable parallax—but Kapteyn worried that the very act of selecting stars this way introduced some bias or skew in the conclusions drawn from the data. Collecting information on fainter stars impartially from all areas of the sky, he believed, would help assure the reliability of astronomers' view of the construction of the heavens.

Kapteyn's proposal called for not just counting the stars in the selected areas, but for measuring all their relevant attributes: position, magnitude, proper motion, spectrum, radial velocity (i.e., the velocity in the line of sight between the observer and star, as Huggins had first measured), and parallax. The proposed method of attack would make efficient use of astronomers' observing time, for it would ensure that a complete set of useful measurements would be taken in widely scattered parts of the sky, without duplication or gaps in coverage. Laboratories like Kapteyn's in Groningen would carry out the analysis of data collected at observatories all over the world.

Kapteyn's listeners received his *Plan of Selected Areas* proposal well, too, although it was some years before the astronomical community agreed on the specifics of the plan. His more immediate

Figure 7.3   George Ellery Hale (1868–1938). (Credit: Courtesy of the Observatories of the Carnegie Institute of Washington.)

effect was on the grand master of American telescope-building, George Ellery Hale, who sat in his audience (figure 7.3). A young, highly entrepreneurial astronomer, Hale listened to these expansive ideas and felt himself drawn to Kapteyn as to a kindred spirit. Hale was then director of the Yerkes Observatory, the world's largest refracting or lens-based telescope. His main interest was the Sun, and he had organized a meeting of solar astronomers in St. Louis, alongside the International Scientific Congress, specifically to urge international cooperation in this field.

In the same year as the St. Louis conference, Hale obtained an endowment from the Carnegie Institute of Washington for a new solar observatory on Mount Wilson. The main telescope of this solar observatory, which became operational in 1907, was in turn the world's largest — until 1917, when Hale again surpassed his own previous accomplishments and garnered funds for a 100-inch diameter telescope to be erected near the 60-inch telescope atop Mount Wilson.

Hale later wrote about his first meeting with Kapteyn as follows: "My plans for the then nascent Mount Wilson Observatory were chiefly confined to an attack upon the physical problems involved in the study of stellar evolution, based upon a thorough investigation of the Sun as a typical star. Researches on the distribution of stars in space did not then enter into the scheme. However, as I listened to Kapteyn's masterly paper and realized the wide scope of his plans and the skill with which he availed himself of international cooperation in assuring their execution, I was deeply impressed by his appeal. Could we not help him to secure the data needed for the fainter stars and at the same time broaden and strengthen the attack in our own problem of stellar evolution? [...] The genius of Kapteyn and the personal charm which brought him the unqualified support of astronomers the world over, convinced me at once that the Mount Wilson Observatory ought to profit by his cooperation as soon as circumstances might permit."[18]

As we shall see, Hale followed through on his inclination to help Kapteyn, and at the St. Louis meeting took steps that would eventually tie not only Kapteyn, but generations of Dutch astronomers, to Mount Wilson Observatory.

Kapteyn presented his Plan of Selected Areas proposal again a year later, at a gathering of the British Association for the Advancement of Science in South Africa. By 1907, when he was back home in Groningen, Kapteyn had secured all the cooperation and input he could hope for. Pickering, at Harvard, had not only agreed to help but had suggested 46 additional areas that he thought showed peculiarities in the distribution of stars that should be investigated.

Kapteyn's Astronomical Laboratory had come into its own. One of his associates noted, "Kapteyn presented the unique figure of an astronomer without a telescope. More accurately, all the telescopes of the world were his."[19]

Mount Wilson and the problem of interstellar absorption

Kapteyn's presentations on star-streaming not only aroused general interest but also provided new grist to the mills of several of his peers, notably, as mentioned above, the great mathematician and astrophysicist at Cambridge University, Eddington. Eddington saw the elucidation and mathematical description of the star-streaming phenomenon as a critical problem for twentieth-century astronomy, and essentially took the problem over from its discoverer. Many astronomers found his writings on star-streaming more understandable than Kapteyn's.

Kapteyn, meanwhile, returned to the problem of determining the fundamental characteristics of the stellar system, but now without making the assumption that the stars' motions are random. He also allowed himself to take a step back from the actual data analysis and spent more time in discussion with observatory directors and their staff members, suggesting observations that he believed would contribute to uncovering the architecture of the universe. Kapteyn's three-month summer vacations, which he, and later, his wife, began spending at Mount Wilson at Hale's invitation, proved to be the ideal opportunity for him to step back and take the "long view." Kapteyn made his first visit there in 1908.

The Mount Wilson Observatory's main offices stood then, as they do now, in the city of Pasadena, near the foothills of southern California's San Gabriel mountains. From the outskirts of town a rugged road wound up Mount Wilson, whose highest point is almost 6000 feet (1800 meters) above sea level. The road led to the dome sheltering the 60-inch diameter telescope and to a wooden dormitory building for observers.

There were no accommodations for families, and the first time Elise accompanied her husband, they camped in a tent. The next year, however, they found, to their surprise and delight, that Hale had ordered a small wooden house constructed for them on the mountaintop. The Kapteyn Cottage, as it came to be known, still stands, although it was enlarged and modernized in 1995.

The Kapteyns' visits to Mount Wilson allowed them to learn more of the personal history of their host, the director. Hale was 17 years younger than Kapteyn, and more mechanically adept, but both men had similar charm, modesty, and restless energy.

Hale was born to a wealthy family in Chicago and never suffered the lack of resources or institutional support that plagued Kapteyn in his early professional career. At an early age he showed a talent for building instruments: he made a spectroscope at 13. He studied physics at the Massachusetts Institute of Technology, and while in Boston made the acquaintance of Edward Pickering at the Harvard College Observatory. Upon graduation, he returned home to Chicago and completed the construction of his own observatory, the Kenwood Observatory, in his back yard. Hale used it mainly to study the Sun. It was equipped with a 12-inch aperture telescope, spectroscope, and camera.

Within a year of Kenwood's dedication, the journal *Science* was comparing it to Henry Draper's former observatory at Hastings-on-Hudson, New York.[20] Hale's reputation quickly grew. When he traveled to Europe in the summer of 1891, accompanied by his wife Evelina, whom he had married in 1890, the venerable William Huggins invited the young couple to visit the Tulse Hill Observatory. The visit went so well that Huggins and Hale remained on friendly terms the rest of Huggins' life.

Hale's exceptional talents — and resources — did not go unnoticed by the president of the newly founded University of Chicago, William Harper. In 1892, he succeeded in convincing both Hale and his observatory to migrate to the university. But the victory was not only Harper's; Hale agreed to becoming a faculty member on the condition that the university undertake to build a large telescope that would enable him to carry out his plans for studying the Sun. The result was the Yerkes Observatory, paid for, through Hale's fund-raising efforts, by the Chicago streetcar magnate Charles Yerkes.

In 1908, when the Kapteyns arrived for the summer at Mount Wilson, Hale was at the crest of an extraordinarily active period in his life. The founding of Mount Wilson was just one of his exploits; he had made the momentous discovery of magnetic fields in sunspots, and he had been engaged in transforming or building up three institutions: the National Academy of Sciences, the California Institute of Technology, and the International Union for Cooperation in Solar Research. He would shortly suffer one of several major nervous breakdowns as a result of his relentless work. However, Hale never lost sight of the scientific goals of his observatory, and he saw Kapteyn's presence on the mountain as a valuable aid to keeping the members of his professional staff focused on the most important problems in astronomy.

One of the questions that Kapteyn urged Mount Wilson astronomers to look into was the presence or absence of absorbing material in space that would affect the amount of light detected from stars. Since about 1904, Kapteyn had worried about this possible light-absorption effect, which would make stars appear dimmer and hence more distant than they actually were.

In 1847, when Wilhelm Struve speculated on the existence of absorbing material in his *Etudes d'Astronomie Stellaire*, astronomers largely ignored his comments, or scorned them. By the time Kapteyn turned his attention to the question, the accurate determination of stellar brightness had become a significant area of research, and astronomers no longer dismissed the possibility of interstellar absorption out of hand. Still, little progress had been made to understand the nature of any obscuring material or the amount of absorption occurring. John Herschel had discovered dark patches of the southern hemisphere skies that seemed to indicate the presence of thick interstellar clouds and, in the first decade of the twentieth century, the American astronomer Edward Barnard was photographing similar "dark nebulae" in the northern hemisphere. (See also chapter 2, figure 2.9, the "horsehead" nebula.) The "Coal Sack nebula" near the Southern Cross constellation, visible through binoculars or small telescopes, is an example of one such dark cloud. However, the presence of obscuring clouds, whose existence was not in dispute, did not necessarily imply that absorbing or nebulous material was scattered throughout space, veiling all the stars to some degree. In other words, the existence of some obscuration was not questioned, but astronomers disagreed on how widespread the problem was.

But how could one go about searching for diffuse clouds? Kapteyn approached the problem from different angles and changed his mind more than once about the severity of the interstellar absorption. One line of reasoning suggested to him that interstellar absorption might affect not only star brightness, but also color. The more distant a star, the more material would lie in front of it and the redder one would expect its color to appear, due to refraction. For the same reason, the setting Sun appears red because atmospheric refraction is greater on the horizon than overhead.

After Kapteyn urged them to study the matter, Mount Wilson astronomers found such a reddening effect in 1914. But just at that time, detailed investigations into the spectra of stars seemed to suggest that the apparent reddening of distant stars might be a true difference in color. Unfortunately for the fate of his models of the universe, Kapteyn began to doubt his previous conviction that there was significant interstellar absorption.

Hertzsprung

Back in the Netherlands after his summers in the United States, Kapteyn discovered that his ties to Mount Wilson naturally made him the contact person of choice for northern European astronomers who dreamed of access to the world's largest telescope. Among these was Adriaan van Maanen, who later played a prominent, if not infamous, role in twentieth-century debates concerning island universes. Between 1908 and 1910, van Maanen, although enrolled as a student at the University of Utrecht, resided at Groningen so he could use the plate-measuring equipment in Kapteyn's Astronomical Laboratory. He was among the first European astronomers to join the staff at Mount Wilson, thanks to Kapteyn's recommendation to Hale.

In 1911, Kapteyn received a visit from another young hopeful, the Danish astronomer Ejnar Hertzsprung, who worked at the Potsdam Observatory in Germany. Hertzsprung, 38 years old, happily got more than he bargained for from the visit. He gained Kapteyn's support for a letter of recommendation to Hale, and he met Henrietta Kapteyn, his future wife.

Hertzsprung had studied chemistry and worked in that field in St. Petersburg and Leipzig until family circumstances dictated that he return to his mother's home in Copenhagen. There, as a private scientist, he pursued his interests in photography, spectroscopy, and astronomy.

In 1905, he came out with a remarkable paper on the classification of stars according to their spectra. He showed that stars that are as red or redder than the Sun can be divided into two unequal groups: those of approximately the same brightness as the Sun, and those, less numerous, that far outstrip the Sun in luminosity, or intrinsic brightness. The physics of the relationship between color, size, and luminosity implied that the more luminous stars must be physically larger than their counterparts

229

of the same color, so Hertzsprung designated them as "giant" stars. Solar-like stars, in contrast, came to be called "dwarf" stars, although "normal size" might have been a better term, since there is nothing unusual about the size of a solar-like star.

The existence of these two classes of stars, while not yet universally accepted, had implications, if true, for the kind of star-gauging that Kapteyn was perpetually engaged in, and for the study of the evolution of the stars in time, which Hale was interested in. Hertzsprung brought his work to the attention of established astronomers, and within a few years was recognized as an expert on stellar spectroscopy and had made the transition to professional astronomer. He remained particularly interested in establishing the possible connections between the spectrum and luminosity of stars.

Hertzsprung found support for his "giant and dwarf" theory in research conducted at Harvard, carried out by Antonia Maury. Maury was a niece of Henry Draper, and was one of the first women Pickering hired to help analyze spectra. She had attended a women's college, Vassar, and like Hertzsprung himself had studied chemistry.

Pickering assigned Maury to the task of analyzing in detail the spectra of the brightest stars in the northern sky. Maury compared the stellar spectra to a spectrum of the Sun represented on the same scale—several inches wide, and showing thousands of the lines across the wavelength range of visible light. She classified the spectra, as usual, according to the relative strength of certain lines. Maury went further than her colleague Williamina Fleming, however, in dividing the classes more finely and in adding the designations "a," "b," "c," or a combination thereof as a shorthand way to represent the sharpness and width of the lines.

Pickering dismissed Maury's system as too complex and time-consuming to implement. He argued that the lines might appear sharp or blurry according to the conditions under which the photographer had captured them. Hertzsprung, however, had discovered a real difference between the c-type stars and the rest: c-type spectra pertained to giant stars. In 1908, he chided Pickering for abandoning Maury's a-b-c distinction, writing in a letter, "To neglect the c-properties in classifying stellar spectra, I think, is nearly the same thing as if the zoologist, who has detected the deciding differences between a whale and a fish, would continue classifying them together."[21] Pickering stood firm — but in 1922, the astronomical community adopted Maury's "c" notation, thanks to Hertzsprung's attention to her work.

When Hertzsprung appealed to Kapteyn, he had yet to convince the astronomical community of the giant and dwarf distinction, and some of his most important work was yet to come. But he showed promise, and Kapteyn approved of Hertzsprung's short-term goal, which was to use Mount Wilson resources to study the spectra of the faintest stars. He wrote to Hale, and Hale in turn offered Hertzsprung observing time on the 60-inch telescope and a stipend to support his trip. Hertzsprung arranged to accompany the Kapteyn couple on their annual voyage to the United States the following summer.

In June 1912, Hertzsprung made his way to Groningen for the start of the voyage to the United States. On the 8th of that month, he and Henrietta surprised their families by announcing their engagement; on the 9th he left for six months overseas, accompanied by her parents. Henrietta spent her summer at home studying Danish.

Hertzsprung and the Kapteyns stopped in London on their way to Liverpool, where they would embark for New York. Eddington, who had adopted the star-streaming problem, was always glad to see Kapteyn. He later wrote, "We rejoiced to hear again the familiar gutteral exclamations and quaint expressions, as with youthful spirit and enthusiasm he unfolded his latest ideas."[22] Hertzsprung and the Kapteyns also visited Kapteyn's old friend David Gill, who had retired to London from his duties at the Cape Observatory.

Although their ultimate destination for the summer was Mount Wilson, both Kapteyn and Hertzsprung had business with the Harvard College Observatory on the way. Kapteyn was counting on Pickering's help on the *Plan of Selected Areas*, and in return performed some data analysis for Pickering. Hertzsprung was even more keenly interested in the work going on at Harvard than Kapteyn, for much of it related directly to his research on stellar spectra. As it turned out, Hertzsprung's research and some of the work going on at Harvard contributed to the development of new distance measurements and directly affected discussions of the scale of the galaxy, the validity of the "Kapteyn Universe," and the question of the island universe hypothesis later in the century.

New distance measures

Two new distance measures for stars were developed in the early years of the twentieth century, complementing and extending the traditional measures of trigonometric parallax and what Kapteyn called "mean parallax." The first is based on the connection between a star's spectrum and its intrinsic luminosity, and the second on a type of star known as a Cepheid variable. These tools were developed during the latter years of Kapteyn's career and he did not incorporate into his models of the universe all the new results that astronomers ultimately obtained with them.

At the time of the visit by Hertzsprung and Kapteyn, Pickering's astrophysics industry at Harvard had ramped up production to the point of classifying 5000 stellar spectra per month. The data came primarily from telescopes in Cambridge (Massachussetts) and Arequipa, Peru, where the observatory had established a station in 1891. Annie Jump Cannon, the assistant Pickering had hired to classify the stars of the southern hemisphere, was the most prolific classifier. Her ability to recognize spectra by simple inspection is legendary. She could consistently and accurately classify 300 spectra per hour, without reference to comparison spectra. When she needed to concentrate for such a task, she would turn off the hearing aid she had needed since suffering a bout of scarlet fever as a young woman, and ignore the hubbub around her.

In the course of her work, Cannon revised the system of classification that Pickering and Williamina Fleming had devised before her to arrive at the classification system that is still used today. She again used the letters between A and O in the alphabet, but she arranged the classes in temperature order — something that was not possible in the early stages of stellar spectrum analysis, before the connection between temperature and the appearance of the spectrum was confirmed. Arranging the classes in temperature order destroyed the alphabetical order. From hottest to coolest, the divisions are: O, B, A, F, G, K, M, which generations of astronomy students have memorized with "Oh, Be A Fine Girl (or Guy), Kiss Me." The white, hot helium-line stars correspond to the O, B, or A types, the cooler, red, calcium-line stars are those of K or M type, and those in the middle, yellow in color and of types F, G, or K, have the hydrogen lines in varying degrees of intensity.

Hertzsprung and the American astronomer Henry Norris Russell, working independently, developed an extremely useful chart that embodies the temperature sequence and almost every-thing else astronomers knew about the stars in the early part of the twentieth century. The chart, known as the Hertzsprung–Russell or "HR" diagram (figure 7.4), led indirectly to a technique for estimating distances.

As Hertzsprung and Russell plotted the absolute or intrinsic luminosity of stars as a function of their spectral class or tempera-ture, they saw that most of the data points cluster in a broad line running from the upper left to the lower right of the diagram. The line is called the main sequence; it encompasses the dwarf stars of all spectral types, A through M.

The clustering of data points on the main sequence shows that the temperature (and hence color) of a star and its luminosity are related. A star of a given color — for example, one which has its most intense emission at a wavelength of 5000 Ångströms, or 5×10^{-7} m, in the yellow part of the spectrum — is most likely to be a dwarf star. In this example, the star would have the same luminosity as the Sun because it has the same color.

Points that lie above the main sequence generally represent the rarer giant stars. For example, Betelgeuse in Orion has a similar color to dwarf stars of the M class, but Betelgeuse is more luminous and has the c-type spectrum with sharp lines. The data point corresponding to it, at the upper right of the HR diagram, helps define the "giants and supergiants" area of the graph. Some points lie below the main sequence. White dwarf stars, for example, have a relatively high temperature like B or A stars but are very small and have low luminosity, so their data points lie in the lower left-corner of the HR diagram.

Astronomers sometimes use the relationship between a star's spectrum and intrinsic luminosity to obtain an estimate of stellar

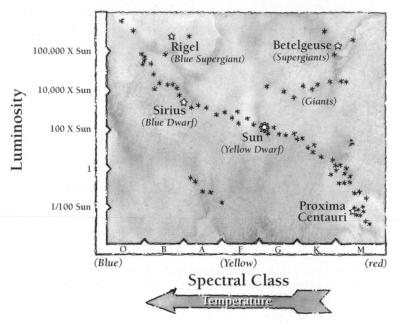

Figure 7.4 The Hertzsprung–Russell or HR diagram, showing the distribution of stars in color and luminosity. The x axis shows color from blue to red, or temperature decreasing to the right. The y axis shows the luminosity in units of the solar luminosity. Most stars have a color (or, equivalently, a temperature) and a luminosity that put them in the so-called main sequence, running from the upper-left to the lower-right corner of the diagram. The blue-white star Sirius, our yellow Sun, and red Proxima Centauri (our nearest neighbor star) all lie on the main sequence, and are therefore also known as dwarf stars, to distinguish them from the rarer giant stars.

A smaller number of stars are found with different combinations of color and luminosity. The red supergiants such as Betelgeuse (in the upper-right corner) are unusually luminous red stars, while the blue supergiants such as Rigel (upper-left corner) are more luminous than similarly colored stars on the main sequence. Stars in the lower-left corner of the diagram are hot but dim.

At one time it was thought that stars evolved along the main sequence—that is, that a star like Sirius might end up as a Proxima-Centauri look-alike. In fact, a star's position on the HR diagram is mainly determined by the mass it has at the outset, and hot, high-mass

234

stars like Sirius do not evolve into cool, low-mass stars like Proxima Centauri. Stars spend most of their lives with the characteristics they have on the main sequence. Stellar evolution also takes them briefly off the main sequence. For example, our Sun will eventually become a giant (evolving more or less vertically on the diagram) and will end its life as a white dwarf, in the lower left corner. (Credit: Layne Lundström.)

distances. A star with a G-type spectrum identical to the Sun's, for example, must have an energy output like that of the Sun and would appear as a 5th magnitude star if placed at a distance of about 30 light-years from the observer. Similarly a red star with the characteristics of a giant, like Aldebaran, the eye of the bull in the constellation Taurus, has an intrinsic energy output about 100 times that of the Sun, and we could compare its apparent magnitude with its true energy output to gauge its distance. Astronomers refer to distances estimated this way as spectroscopic parallaxes, using the word "parallax" as a stand-in for "distance estimate."

The second new method of estimating distances, which emerged directly from work done at Harvard, was that based on Cepheid variables. Henrietta Leavitt, one of Pickering's assistants, found the first clue to the usefulness of this category of star. Her specialty was not the classification of stars but the search for variable stars, which fluctuate both in luminosity and spectral type.

Leavitt had graduated from the women's college later known as Radcliffe, and joined the observatory initially as a student and unpaid research assistant. She took charge of a number of projects at Harvard College Observatory, including Harvard's contribution to Kapteyn's *Plan of Selected Areas* study.

Between 1904 and 1908, by a laborious comparison of before-and-after photographic plates, Leavitt identified more than 1700 variable stars in the Large and Small Magellanic Clouds of the southern hemisphere. The stars had similar patterns of variability, repeatedly brightening and dimming (see figure 7.5). They were known as Cepheid variables, so called because a bright star of the constellation Cepheus exhibits this same pattern.

In 1908, Leavitt noted that the brighter variables had longer periods. The observation was significant because, if period and

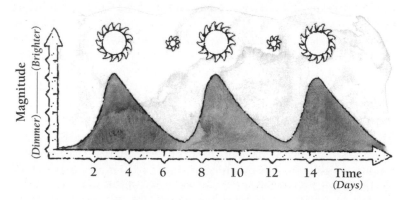

Figure 7.5   Cepheid variables. Cepheid (SEF-ee-id) variable stars, named after the prototype star exhibiting the behavior, Delta Cephei, brighten and dim in a regular pattern as they swell and shrink. At their maximum size, they have their peak brightness. The fact that the maximum brightness of a Cepheid variable is related to the period of time over which the pattern repeats has led to the use of Cepheid variables as distance indicators, as explained in the text. (Credit: Layne Lundström.)

luminosity really were related in these special stars, then one could determine a Cepheid variable's absolute luminosity simply by measuring its period of variability.

The astronomical community failed to pick up on her remark, so in 1912 Leavitt restated it, this time in eye-catching graphical form. The plot showed a one-to-one correspondence between a Cepheid variable's maximum intrinsic brightness and the period of its variability.

Hertzsprung realized the significance of Leavitt's finding, namely, that Cepheid variables could be used as standards of luminosity, and hence as indicators of distance. It is fairly easy to measure the period of variability of a star. The Cepheids Leavitt studied had periods between about 1 day and 270 days. Leavitt's plot allowed one to read off the intrinsic luminosity corresponding to a given period. Once the intrinsic luminosity was known, it was theoretically possible to derive the distance of any Cepheid variable by comparing its apparent luminosity and the luminosity it should have according to the period–luminosity law. All that remained was to calibrate the formula,

236

by determining the specific energy output for a Cepheid variable at a known distance.

Hertzsprung calibrated the Cepheid formula in 1913, shortly after his visit to Harvard, and used it to derive the distance to the Small Magellanic Cloud. His result—some thousands of light-years—contained a computational error, and was off by a large factor, but it placed the Magellanic Clouds at a great distance compared to the stars of known trigonometric parallax, which lie within about 100 light-years. Even Kapteyn's estimates of star distances using "mean parallax" correlated with proper motion could reach stars at distances only up to about 3000 light-years. Hertzsprung's result was momentous; he had taken the first step that would lead astronomers to fathom the stars beyond our own Milky Way system.

Searching for a synthesis

While Hertzsprung thrived on the connections he made in the United States, Kapteyn, during the years he spent summers at Mount Wilson, felt somewhat ill at ease. On the one hand, he worried that he was not earning his keep at Mount Wilson, and Hale had to reassure him that his role there of advisor and motivator to the observers was a useful one. At the same time, he could not envision a successful conclusion to his years of research. He wrote to Henrietta, "My life is nearly consumed with problems of knowledge—and seldom do I bring them to some kind of satisfactory resolution, and then it's off again to search for something better. [...] Sometimes I have a yearning for my retirement so that I can put the arithmetic marker down and enjoy all that is human. Will I still be able to do that?"[23]

Nevertheless, he continued trying to elucidate the structure of the Milky Way. In a paper published in 1913 and reprinted for a less specialized audience in 1914, he synthesized much of what he had learned about the universe from the distribution of stars according to their positions, velocities in space, and spectral type, and came up with some remarkably accurate conclusions.

Kapteyn noted that the two star streams he had discovered earlier consisted of unequal distributions of stars. The first stream was rich in what he called helium stars, that is, stars

237

with strong helium lines, of spectral type O or B. Astronomers generally believed these stars to be young, or recently formed from some primordial matter. He noted that these stars tended to have low velocities. The second stream was rich in older stars, and contained hardly any young stars. These older stars, for reasons he could not yet understand, had acquired higher velocities.

When he compared the velocities and directions of the nebulae with those of the young and old stars, a pattern emerged. Although very few measurements had been made of the velocities of nebulae, it seemed clear to Kapteyn that the velocities of the so-called planetary nebulae, like the Cat's Eye nebula Huggins had studied, put them in a separate category from the "irregular" nebulae like the great nebula in Orion. As Kapteyn put it, "There are nebulae and nebulae."[24] The planetary nebulae had rather high velocities, and should, he thought, be associated with the endpoint of stellar evolution rather than the beginning. The Orion nebula's motion through space, on the other hand, resembled that of the young stars. This, together with the fact that young stars were often found in association with irregular nebulae, suggested that irregular nebulae might be the birthplace of stars.

Kapteyn envisaged a stellar evolution scenario in which stars were born from irregular nebulae as helium stars, and evolved from O and B stars into the K and M stars, then into planetary nebulae. Some parts of this scenario are correct. We now understand that the Orion nebula lies at the edge of the Orion spiral arm of our galaxy, and that it is one of the nearest and largest stellar nurseries. Planetary nebulae are indeed associated with the opposite endpoint of stellar evolution: the nebulous matter there consists of remnants of a star's atmosphere, thrown off after a star has gone through a nova phase.

Kapteyn's only significant mistake was to suppose that the stellar classes form a series in time. It is true that stars are born in irregular nebulae, and that many end up as planetary nebulae, but they do not evolve from one spectral class to another. A young O star evolves from the main sequence or dwarf state into a red giant star, and a class K dwarf star similarly evolves into a red giant star, albeit of a different luminosity.

By correlating stars' velocities and ages, Kapteyn had learned something about the history of our galaxy. His *magnum opus* on the structure of the entire Galaxy was yet to come.

World War I

Early in 1914, the Kapteyns were saddened by the loss of their good friend Gill, who died in London after a short illness. It was, as Henrietta Kapteyn-Hertzsprung remarked, the sad beginning to a disastrous year. When World War I broke out in the summer, Elise and Kapteyn were at Mount Wilson, and were too fearful of crossing the Atlantic to go home as planned. They stayed on the American continent until January 1915, worrying about the situation at home. Meanwhile, Kapteyn, maintaining a neutral Dutch outlook on the politics of the war, incurred the enmity of some of his colleagues by accepting a prestigious scientific award from the German Kaiser.

Kapteyn found distraction from the hardships of the war years in work and family. Results came in from the *Plan of Selected Areas* collaboration. His student Peter van Rhijn, who was to collaborate with him on his model of the universe, obtained his PhD in 1915. In 1916, Henrietta and Ejnar Hertzsprung produced a granddaughter. Her parents named her Rigel after the bright blue-white star marking one of the feet in the constellation of Orion.

Although Kapteyn never returned to the United States after his departure in 1915, he kept in contact with some of his American colleagues. Harlow Shapley, the subject of our next chapter, sent some of the most interesting mail. Hale recruited Shapley to Mount Wilson in 1914, and there he began to raise questions about Kapteyn's earlier estimates of the amount of absorbing material between the stars. Shapley did not see the amount of reddening that Kapteyn thought should exist for distant stars. Kapteyn was not convinced that Shapley was correct, but he could find no flaw in the young man's arguments and maintained a polite correspondence with him.

The question of interstellar absorption became even more controversial around 1917, manifested in the problem of the so-called "zone of avoidance." Spiral nebulae which are tilted so that we view them edge-on show a dark lane running through the disk, like a spread of vegemite between two pieces of bread (see figure 10.7 in Chapter 10). The dark lane indicates a layer of thick dust in the middle plane of a spiral system's disk. Heber Curtis, an astronomer at Lick Observatory who became familiar with the various aspects of spiral nebulae from photographing them, pointed out that a similar layer of obscuring material in the plane of the Milky Way might be expected if our stellar system were a spiral nebula—that is, if our system were an island universe like countless others. Furthermore, Curtis said, the existence of such a layer of dust would explain why he had to aim the telescope and camera out of the plane of the Milky Way to find spiral nebulae. As others had already observed, few spirals could be seen in the plane of the Milky Way; they seemed to "avoid" this zone of our stellar system. Curtis concluded, correctly, that a layer of obscuring material in the central plane of the Milky Way would explain this effect. The spirals lie in all directions far outside our own system, but when astronomers try to look through the plane of the Milky Way, their view is blocked.

Kapteyn took an interest in the absorption controversy, but did not actively pursue a resolution. Similarly he stayed out of the fray as Albert Einstein introduced his theory of general relativity in 1915, with implications for the nature and evolution of the universe. Kapteyn attended lectures on Einstein's theory, but confessed that he couldn't comprehend them. As chance would have it, however, one of his first students, Willem de Sitter, became very interested in the observational consequences of Einstein's theories. In any case, Kapteyn did not live long enough to see observational confirmation of some important aspects of Einstein and de Sitter's models of the universe.

At the close of the war, and as his retirement approached, Kapteyn became embroiled in a different sort of to-do. Allied governments had decided to withdraw from existing international scientific organizations, which had grown up around an association of German academies. The Allies proposed to form a new organization, the International Research Council, excluding Germany. Neutral countries, such as the Netherlands, would be invited to join.

Kapteyn, the idealist, could not countenance the exclusion of German scientists, who, he felt, were not responsible for their government's policies. He wrote to Eddington, a pacifist Quaker and the first Englishman to attend postwar astronomical conferences in Germany, "If the men of science give an example of hate and narrow mindedness, who is going to lead the way?"[25]

Kapteyn and one of his colleagues petitioned the International Research Council to reconsider their statute against German participation and pleaded with the Dutch Academy of Sciences to resist joining, but to no avail. Not until 1926 did the Council move to open membership to all nations.

Later years: the Kapteyn universe

After Kapteyn's retirement from the University of Groningen in 1920, he took a part-time position at the University of Leiden, where he had carried out some research during the dismal early days of his career at Groningen, and where his former student de Sitter was director of the observatory. As a favor to de Sitter, Kapteyn filled a post in positional astronomy until university officials could find a permanent replacement. His bimonthly trips to Leiden allowed him to visit with Henrietta, Hertzsprung, and granddaughter Rigel, as Hertzsprung became a professor there in 1921. Henrietta's marriage to Hertzsprung dissolved in 1922, however.

Kapteyn devoted a large part of his energy at this time to his own final big project, a synthesis of all his research into the distribution of stars. He collaborated with Pieter van Rhijn at Groningen, his former student. Together they assembled the data from the *Plan of Selected Areas* on star counts and luminosities, and analyzed them with the assumption that space is free of any absorbing material. Their first paper on the subject appeared in 1920.

Kapteyn and van Rhijn were able to derive the density of stars—the number of stars per volume of space—as a function of distance from the middle of the plane. Within the plane of the Milky Way, they found that the density of stars decreased with distance. Around the Sun, the density is about one star per 300 cubic light-years. Their data implied that, at 2000 light-years from the Sun, the density decreased to 60 percent that of the solar neighborhood, or one star per 500 cubic light-years. Beyond that, their results were less certain, but an extrapolation of the density function indicated that the density fell to 1 percent of the solar neighborhood density, or about 1 star per 30 000 cubic light-years, at a distance of 30 000 light-years. They concluded

241

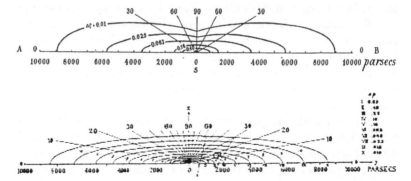

Figure 7.6 The "Kapteyn Universe." Top panel: 1920 model of the distribution of stars in our galaxy, derived by Kapteyn and his student van Rhijn from an analysis of star-counts. The system is assumed to be symmetric, so only the "top half" of the galaxy is shown. The line AB represents the plane of the Milky Way. The Sun is at S, the center of the system. Distances are given along the x axis in parsecs; 1 parsec is equivalent to 3.26 light-years. The lines represent contours of equal density. Bottom panel: 1922 version of the "Kapteyn Universe." Kapteyn no longer placed the Sun at the center of the system, but at a distance of 650 parsecs or about 2000 light-years. (In fact, the Sun is some 26 000 light-years from the center of the galaxy; Kapteyn was misled by assuming that interstellar absorption of light was not significant.) Kapteyn's universe is about five times as wide as it is thick, and includes about 50 billion stars. (Credit: Adapted from the originals in *Astrophysical Journal* 1920, **52**, 23 and 1922, **55**, 302.)

that the boundary of the disk of the Milky Way lay at about 60 000 light-years; perpendicular to this plane, the Milky Way extended only about 8000 light-years. The system as a whole, they estimated, included some 50 billion stars.

Kapteyn and van Rhijn illustrated the density function graphically using a contour plot (see figure 7.6, top panel). The Sun occupied the central position. Ellipse-shaped curves around the Sun indicated lines of constant density, rather like lines of constant altitude on a topographic map.

This 1920 model was static—it did not incorporate the information Kapteyn himself had gleaned on star-streaming. In 1921, then 70 years old and in failing health, he summoned the last of his strength to revise his "Kapteyn Universe" to remedy this flaw. His academic duties were finally at an end, and he looked forward to a restful retirement. Still, Henrietta wrote, "An important new theory based on the results of his work of the last few years so enthralled him and kept him so busy that his books and papers accompanied him on his vacation to Switzerland."[26]

Kapteyn called his last effort, "a first attempt at the theory of the arrangement and motion" of the stellar system. He considered that the star-streaming phenomenon might be due to stars in concentric rings orbiting the center of the system. The Sun, according to this theory, must be in one such ring, orbiting the center of the stellar system at a distance of about 2000 light-years.

Thus, he removed the Sun from the center of the stellar system in accordance with his theory for the cause of star-streaming, although he displaced it only about 3 percent of the way to the edge of the system (see figure 7.6, bottom panel). He knew that the central position of the Sun in the 1920 model had caused some concern among his peers, who looked with suspicion on any model that placed the Sun and its accompanying solar system in a special or unique point in the universe. At the same time, he did not take into account new data from Shapley, the subject of our next chapter, which correctly indicated that the Sun might be even further removed from the center of the stellar system. By this time, the Kapteyn Universe and Shapley's "big galaxy" model stood in opposition to each other, and astronomers were lining up with one or the other. Kapteyn, always seeking harmony, hoped that the difference between the two models might eventually be attributable to their entirely different approaches, and that further refinements to both models might bring them closer together. He wrote to Shapley in 1919, "You are building from above, while we are up from below. You start from the general system in its greatest extension, we try to struggle laboriously up from our nearest surroundings. When will the time come that we thoroughly mesh?"[27] In 1920, Kapteyn added, "I hope to live in order to see your and my studies meet."[28]

Kapteyn presented his results before publication to a gathering of astronomers in Edinburgh in 1921, and again at an informal gathering in Leiden that included Einstein. The paper appeared

243

in print in 1922. Shapley visited Leiden in May 1922, but Kapteyn was too ill to take advantage of the opportunity to meet his principal challenger face-to-face.

Kapteyn died in June 1922 at the age of 71. A lifetime of slow, methodical work sampling the stars and "grinding huge masses of fact into law" had come to an end. Perhaps he still wished, as he had remarked in earlier times, to be reincarnated as a graceful and fast-flying swallow, his favorite bird. He lived to see his "Kapteyn Universe" model come to light, but not long enough to see the perplexing differences between his model and Shapley's unravelled by new discoveries.

8

HARLOW SHAPLEY: CHAMPION OF THE BIG GALAXY

"We are on the brink of a big discovery – or maybe a big paradox, until someone gets the right clue."

Henry Norris Russell (1920),
in a letter to Harlow Shapley, on the difficulty of reconciling
evidence for and against spiral nebulae as island universes[1]

In 1917, a self-assured but hardworking young astronomer named Harlow Shapley penned a note about his work at Mount Wilson to Kapteyn, whom he had met on the elder astronomer's last visit to Pasadena. Shapley wrote that "the work on clusters goes on monotonously – monotonous so far as labor is concerned, but the results are continual pleasure. Give me time enough and I shall get something out of the problem yet."[2]

Shapley's confident prediction that he would obtain interesting results from his study of globular clusters was not just correct – it was, in retrospect, a considerable understatement. Later that year, synthesizing his work on globular cluster distances and mapping their distribution in space, he asserted that the Sun was not at the center of the stellar system, but belonged in an undistinguished corner of a newly re-conceptualized Galaxy. In effect, as Shapley himself saw it, he had fomented a Copernican revolution, once again moving man's place in the cosmos from the center to the periphery. But he would need all of his characteristic aplomb to see him through the overthrow of the "Kapteyn Universe."

The story of Shapley's insight and the debate that ensued in the astronomical community is particularly satisfying because it

245

ties together two threads that have run through this book: the inquiries into the distribution of stars in our stellar system and the elucidation of the nature of the nebulae.

Our account began with a connection between the two problems: Thomas Wright depicted various possibilities for the three-dimensional shape of the stellar system, and Immanuel Kant related the concept of a disk-shaped stellar system — which he attributed to Wright — to the nebulae, imagining our system, and the nebulae, as "island universes." William Herschel attempted to trace the contours of the known universe with his star-gauges, and, separately, considered the possible evolution of nebulae into individual stars and stellar systems.

In the nineteenth century, astronomers generally approached these problems without intertwining them. Struve and Kapteyn — each in his own way — attempted to put an absolute, numerical scale on maps of the star system, while Huggins applied spectroscopy to the study of the nebulae. The fact that the nebulae turned out to come in at least two varieties, gaseous and star-like, left open the question of whether some of the nebulae might be island universes.

Shapley's work on the size of the Galaxy and the location of the Sun within it brought to the fore the question of the nature of the nebulae. The stupendous size he inferred for the Galaxy made it easy to believe that our stellar system, if not identical with the universe, at least filled a large part of it; the nebulae must then be much smaller systems, of a very different nature. But those who made nebulae their specialty had good reason to think they were distant galaxies of millions or billions of stars. These questions, and others, demanded an answer in the early twentieth century.

A Missouri childhood

Harlow Shapley (figure 8.1) and his fraternal twin brother Horace were born on 2 November 1885, on a farm near the small town of Nashville, Missouri. The family included an older sister, Lillian, and, later, a younger brother, John. Their father grew hay and occasionally taught school in Nashville. Their mother, a descendant of New England abolitionists, read to the children and encouraged them to "amount to something, get somewhere, go

Figure 8.1 Harlow Shapley (1885–1972). (Credit: AIP Emilio Segrè Visual Archives, Shapley Collection.)

to school," as Shapley recalled.[3] Half a century later, Shapley fondly remembered her introducing them to the English Victorian comic adventure *Three Men in a Boat* by Jerome K Jerome.

Shapley described his own life as something like a series of adventures in his informal autobiography, *Through Rugged Ways to the Stars*. He painted a picture of his youth that emphasized his self-reliance, hard work, and unusual academic inclinations. He pitched hay from a wagon and took care of livestock, but he also recited poems of Tennyson as he milked cows to "keep the rhythm going" and he developed an early botanical interest in wildflowers.[4]

Although the elder Shapley taught school in Nashville, Harlow and his brother attended a one-room schoolhouse at the edge of the farm, taught by their sister. Lillian nurtured

Shapley's talents and prodded the twins to seek more education than was locally available. However, Shapley's early education followed no traditional structure and was interrupted by his work as a newspaper reporter.

Inspired in part by Lillian's advocating a career in writing, Shapley spent his mid teens as a journalist for the Chanute, Kansas *Daily Sun* and the Joplin, Missouri *Times*. The boldness that later characterized his professional dealings manifested itself in his pursuit of sensational stories. He prided himself, for example, for his role in the downfall of a local politician. After being kicked out of the politician's office for representing an unfriendly editor, Shapley sat outside the door as he talked to reporters from rival newspapers, and took notes in shorthand. The politician's language was foul, but Shapley's newspaper was able to report on it explicitly by printing a picture of the shorthand notes. The cursing cost the politician the election. In another coup for Shapley, he unmasked the secret of a circus horse who supposedly solved equations by pawing out the answer with his hoof: Shapley asked the horse what the square root of four was. Not all of this lively journalism was fun, however. At 16, Shapley covered a shooting "duel" in which one of the contestants died.

Eventually, Shapley and his twin brother decided to save their money for college—but that meant they would have to finish high school first. They presented themselves at an "elegant" high school about 20 miles from home, but were turned away for lack of preparation. They enrolled instead at a small Presbyterian school in the same town, the Carthage Collegiate Institute. Shapley took special examinations and studied Latin and geometry on trips home to work on the farm. His efforts paid off. He graduated first in his class of three, delivering an address on Romantic values in Elizabethan poetry. In 1907, he was admitted to the University of Missouri in Columbia. His brother went back to farming.

Undergraduate years in Columbia

Shapley had intended to study journalism, but found that the school of journalism was not yet open. As he told the story, he

248

decided to take classes in astronomy on a whim, simply because he searched for an alternative course of study by looking through the course catalog alphabetically, and couldn't pronounce "archaeology." Once exposed to astronomy, he discovered an outlet for his talent in science, and had the good fortune to be mentored by excellent teachers.

His astronomy teacher, Frederick Seares, was to become a lifelong colleague. Seares had toured the great observatories of Europe—he had studied in Paris and Berlin. When Shapley arrived Seares was doing a remarkable job of carrying out a research program at Laws Observatory on campus, which was equipped with an old $7\frac{1}{2}$ inch refractor and a micrometer. Seares managed to make a significant contribution to the determination of comet positions and to the cataloging of the changes in light output of variable stars.

Shapley cut his observing teeth at the Laws Observatory as one of two assistants. They got some of their education by finding faults with the old instruments, Shapley noted. He also helped Seares analyze his lightcurves of variable stars. Seares was delighted to have the help of such a talented student, although even he was awed by Shapley's industry. He later wrote in a letter of recommendation that Shapley worked "incessantly—much too continuously for his own good."[5]

Shapley claimed to have learned very little science in the one-room schoolhouse or at the Carthage Collegiate Institute. As an undergraduate, he retained an interest in the humanities, even as he applied himself to courses in physics, mathematics, and astronomy. "It may seem strange that, with no experience or particular interest in astronomy, I went on to make a career of it, at a time when prospects for a degree were not very promising," he wrote in his autobiography. "The explanation is, I think, that when I got to the University I found—and I know it was a genuine finding—that all fields of learning are exciting. I came very close to accepting a classics scholarship that would have given me a chance to be a classicist. Perhaps I shouldn't have got excited about physics," he said, recalling his difficulties with the subject, "but about astronomy I could and I did."[6]

Shapley indulged his fascination for the classics by writing a paper on "Astronomy in Horace," which was published in *Popular Astronomy* magazine in 1909.[7] However, as he spent time doing astronomy he found he was less attracted to other fields and no longer interested in returning to journalism. He had been at the university two years when the editor of the *Daily Sun* offered him an interest in the paper and the job of managing editor. By this time, Shapley knew he would not go back. Astronomy had him.

During his third year, Shapley met his future wife, Martha Betz. She was a gifted mathematician and linguist who would receive a bachelor's degree in education in 1910, a bachelor of arts degree in 1911, and a master's degree in 1913, all from the University of Missouri.[8] They met in a mathematics class, he related: "She sat in the front row and knew all the answers."[9] Martha was brilliant intellectually and "retiring" in nature, according to friends. She did not develop an interest in astronomy until she met Shapley, but, she pointed out to friends, she had a grandfather from Hanover who had several times, as a youth, seen the aging Caroline Herschel being driven around the city in a coach.[10]

Shapley obtained his bachelor's degree in 1910. Seares had left the previous year to become a staff astronomer at Mount Wilson, but stayed in touch with his protégé in Columbia, trying to get him a position where he would have access to better observing facilities. Shapley decided to stay another year in Missouri for a master's degree.

During his final year, another one of Shapley's influential teachers, the mathematician Oliver Kellogg, helped him secure a prestigious fellowship to Princeton. Kellogg had studied at Princeton himself, as an undergraduate. Kellogg urged the deans of the faculty and graduate school at Princeton to attract Shapley before other schools could seize on his great talent and industry. The deans took note, and arranged for Shapley to begin his doctoral studies in September, 1911.

Eclipsing binaries at Princeton

At Princeton, Shapley once again had the good fortune to associate with distinguished scholars. Outstanding among these was his thesis supervisor, Henry Norris Russell.

Russell, only eight years older than Shapley, already claimed the status of an important personality on the Princeton campus.

Shapley described him as "a high-class Long Island clergyman's son and very high hat."[11] He found Russell aloof and "shy" at first, although Russell's excitable, driven personality must have quickly overriden any reserve he exhibited initially around his new graduate student.

Shapley recalled that Russell became friendlier after Shapley used one of his methods to solve a problem in orbital mechanics. Thereafter they treated each other more like colleagues than student and teacher. Wrote Shapley, "Students were much interested when Shapley, the Missourian, and Russell, swinging a cane, would stroll across the campus. If students got in the way, Russell would just brush them off with the cane. We got along well, and we both learned a great deal."[12]

As a graduate student, Shapley took courses related to his thesis work, including theoretical astronomy and classes in the astronomical uses of the photographic camera, spectroscope, and photometer. As he had been at Missouri, he was attracted to other fields too. His natural curiosity always made other subjects very enticing. At Princeton he found time to audit classes in physiology and paleontology.

Russell kept abreast of developments in many branches of astronomy, but his steadfast quest, beginning with his postgraduate studies at Cambridge University, was to apply the principles of physics to the problem of stellar evolution—how stars form and evolve. Shapley, as Russell's graduate student, naturally contributed to this area of research, but his interests took him in a different direction and eventually led him to the quite distinct "sidereal problem" of the arrangement and extent of stars in space.

At the time of Shapley's move to Princeton, the basic idea Russell worked with was that stars condensed out of nebular gas and dust and gradually became smaller and denser as gravity pulled the stellar material together. The denser a star, the older it must be. What happened to the temperature of a star as it aged was unsettled; Russell was inclined to think that the temperature of the visible, outer layers of a star would rise as the star contracted. Double stars, too, were thought to form together from a common cloud. The leading exponent of the theory of double star formation from a nebular cloud was George Darwin, Plumian Professor of Astronomy and Experimental Philosophy at Cambridge and son of the famous naturalist.

Russell was mainly a theorist, but the resources of Princeton's Halsted Observatory, manned by his junior colleague Raymond Dugan, were at his disposal. To put his ideas about stellar evolution on a quantitative basis, Russell examined data on so-called eclipsing variables that Dugan and, later, Shapley collected. These objects are binary systems, consisting of two stars in orbit around each other. Their name comes from the fact that they have variable light output. The plane of their orbits is such that, from our vantage point, each star periodically passes in front of the other, causing an "eclipse" of the star on the far side.

The advantage of studying binary systems is that detailed information about the component stars can be obtained — more information than is available for isolated stars. Kepler's laws of motion allow one to deduce the combined mass of the two stars from the orbital size and period. Knowing the combined mass of the system allows one to put limits on the possible mass of each star. Furthermore, spectroscopic measurements of the stars' velocities as they orbit each other give the size of the orbit. Finally, the information on mass and orbit, combined with an analysis of the lightcurve — the variation of light as each star is eclipsed in turn — yields the dimensions of the stars themselves. Russell was interested in deriving from these quantities the density of stars, to see how the density related to age.

Shapley's arrival gave Russell's program a boost. During the two and a half years that it took him to complete his doctoral dissertation, Shapley made thousands of observations of variable stars at the Halsted Observatory and computed the orbits of 90 eclipsing binaries. The calculations involved were laborious; before Shapley published his thesis, fewer than a dozen orbits had been computed although some 50 000 eclipsing variables had been discovered. Shapley and Russell used slide rules and "little calculating machines," as Shapley called them — electric multipliers and adders.[13]

Shapley apparently benefited from his fiancée's mathematical expertise in reducing all these data for his dissertation. In 1913, Martha was affiliated with Bryn Mawr, a women's college near Philadelphia, as a "scholar in German."[14] She intended to pursue a PhD in Teutonic Philosophy.[15] According to an astronomer who collaborated with her in the 1940s and

1950s, Shapley would take a train from Princeton to meet Martha at the Broad Street Station in Philadelphia, where he would collect finished calculations from her and give her new ones.[16] Certainly it is true that she became interested in the mathematical challenges of eclipsing variables, and published a number of papers on eclipsing systems under her own name or as co-author in the *Astrophysical Journal*, beginning in 1916.[17]

Among the first exciting results to emerge from the study of eclipsing binaries were estimates of the sizes and densities of the stars. Shapley and Russell found hundreds of stars that far surpassed the Sun in radius — giant stars like those Hertzsprung described. Because their masses were not too different from the Sun's, the densities of the stars had to be very low. Shapley called them "enormous gas bags."

The importance of Shapley's thesis work for our story is that the eclipsing variables got him interested in another type of variable star, the Cepheid variable (see chapter 7). In the early part of the twentieth century, astronomers assumed that Cepheid variables were a type of eclipsing binary system. But the Cepheid variables did not seem to fit the pattern of the other variable stars Russell and Shapley studied, and during Shapley's tenure as a graduate student, he and Russell discussed the possibility that the Cepheids might be single stars whose brightness varied intrinsically. Shapley later elaborated on this theory, and showed that Cepheids could be used, along with other distance indicators, to study the distances of globular clusters and the structure of our stellar system.

Shapley completed his dissertation in 1912 and turned to the question of his future. He wrote to his former Missouri teacher, Seares, at Mount Wilson. The 60-inch reflector there had been in operation since 1908, and the five-year task of grinding the glass mirror base for the 100-inch telescope was already under-way. Seares arranged for George Ellery Hale, the founder and director of the observatory, to meet Shapley in New York. The meeting went well, and in due course Shapley received an offer to go to California. However, Shapley elected to defer the start of his job there for a year.

Shapley had some loose ends to tie up in the binary work, and Seares was also recommending that he travel. In 1912, Shapley attended his first meeting of the American Astronomical

Society, in Pittsburgh. In 1913, Shapley and his younger brother John traveled to Europe.

For John, the Atlantic crossing was the beginning of doctoral studies in Vienna in linguistics, archaeology, and the history of art. For Shapley, it was an opportunity to visit observatories and meet astronomers. He got as far east as Hungary, as far north as Sweden, and as far south as Algiers. He attended a meeting of the Royal Astronomical Society in London, where he met "almost everybody of stature," and a meeting of the international astronomical society the *Astronomische Gesellschaft* in Bonn, where he met Ejnar Hertzsprung.[18] As we saw in chapter 7, Hertzsprung was working along lines very similar to Russell's, and correlating stars' spectral types with luminosities. Shapley's enjoyment of the trip came to an abrupt end in Paris, where he found a telegram informing him that his father Willis had been killed by lightning. He wandered the streets of Paris in shock.

Back in the United States in 1914, Shapley made a shorter trip up and down the East Coast, visiting American observatories. At Harvard, director Edward Pickering invited Shapley to his house for dinner, and welcomed Shapley to search the Harvard archives for data that interested him. Shapley ate a second dinner in Boston with "the famous and jolly Miss Annie J Cannon," the classifier of stellar photographic spectra.[19] But the most important meeting he had on this visit was that with Solon Bailey.

Bailey had come to Harvard in 1887 as a college-educated but unpaid assistant. At the time of Shapley's visit he was a full professor and, informally, assistant director. He had proven his worth by establishing Harvard's observing station at Arequipa, Peru, where he had photographed the entire southern sky with a succession of telescopes and, almost single-handedly, had measured the magnitudes of about 8000 stars.

It was during one of his multi-year stays in Arequipa that Bailey had developed an interest in variable stars. He was specifically interested in variables in globular clusters, which at the time were thought to be stellar systems comparable to our own, and in the Magellanic Clouds, which, though irregularly shaped and very different from globular clusters, were also assumed to be "island universes" of some sort. When Bailey began his study, only a few of the 400 known variable stars were in clusters. Within five years he discovered 300 variable stars in clusters. His favorite objects were the globular clusters 47 Tucanae and Omega Centauri; he knew them so well that on at least one occasion he identified a variable star simply by noticing visually a change in a photograph. The usual method that Bailey employed, and that Henrietta Leavitt used on the plates he sent back to Harvard, was to overlay a negative taken on one date with a positive image taken on a different date. Variable stars stood out because they left a slightly larger photographic impression when they brightened, while the stars that shone steadily matched exactly in the overlays.

Shapley sought Bailey out in his office during his 1914 visit, and found a quiet but warm reception. Shapley reported that Bailey said, "I have been wanting to ask you something. We hear that you are going to Mount Wilson. When you get there, why don't you use the big telescope to make measures of stars in globular clusters?"[20] In later years, Shapley always credited Bailey for leading him to the rich field of variables in globular clusters.

Arrival at Mount Wilson

In April 1914, on his way from Princeton, New Jersey to Pasadena, California and the Mount Wilson Observatory, Shapley stopped in Kansas City, Missouri. His fiancée Martha Betz had arrived earlier from Bryn Mawr, and was waiting for him at her parents' home. They were married at her home by Dr. George Hamilton Combs, a well-known pastor in the Disciples of Christ church in Kansas.[21] After the wedding they boarded a train for California. Shapley recollected, "It was a long trip, but I had some nice observations with me, and we worked on the orbits of eclipsing binaries on the honeymoon. Mrs. Shapley was very quick at computing, so we enjoyed ourselves for a couple of days."[22]

The city of Pasadena, where the Mount Wilson Observatory offices and workshops were located, had recently evolved from a prosperous agricultural community, known in the 1880s for its citrus groves and vineyards, to an even more prosperous city that attracted visitors as a winter resort town. Hale had embarked on a quest to turn the Throop Polytechnic Institute into the famous California Institute of Technology. The Shapleys settled into a house very close to the observatory offices. For the first few years of his tenure at Mount Wilson, Shapley would spend most of his time at the Pasadena offices; he traveled the nine miles up the mountain to observe at the telescope only three or four nights a month.

Shapley's job, initially, was to help his former teacher Seares with observations of the colors and magnitudes of stars. Some of these observations contributed to the catalog of Kapteyn's "Selected Areas." After he learned to operate the 60-inch telescope, Shapley was granted occasional access to it for his own observing projects.

Shapley was among the first of several post-doctoral researchers to join the Mount Wilson observatory staff. He soon made friends with Adriaan van Maanen, a Dutch astronomer his age who had arrived in Pasadena a few years earlier. Van Maanen, a lively storyteller with a reputation as a playboy, had studied astronomy in Utrecht and Groningen and had found a mentor in Kapteyn. His task at Mount Wilson was a painstaking one—to measure the proper motions and parallaxes of stars, from photographic plates taken at intervals of months or years. Van Maanen's research apparently did not interest Shapley very much, for he appears not to have paid much attention to it initially, but within a few years van Maanen was to play a crucial role in the development of Shapley's thinking about the scale of the universe.

Gregarious and outgoing by nature, Shapley maintained good relations with almost everyone at the observatory. He enjoyed observing with Seth Nicholson, a solar astronomer who occasionally used the 60-inch telescope at night to study the satellites of Jupiter. He and Nicholson took one of the observatory's caretakers, Milton Humason, under their wings, recognizing his talent for observing. With their help, Humason rose to a staff position in the photography department; he later sharpened his skills working under Hubble, and finally forged a career in astronomy as an assistant astronomer.

Shapley developed a cordial rapport with Kapteyn, despite their difference in age. As we saw in chapter 7, Kapteyn had arrived at Mount Wilson for his annual summer sabbatical in

July 1914, and remained there through the beginning of 1915 because of the difficulty of securing a safe passage back to the Netherlands when World War I broke out.

Perhaps the only Mount Wilson astronomer whom Shapley had trouble getting along with was Walter Adams, who ran the observatory in Hale's absence. Hale by 1916 was spending most of his time in Washington, DC, preoccupied with establishing the National Research Council, a vehicle for advancing scientific research in the service of America's national security and welfare. Adams, a rather severe character who strove for excellence in equal measure whether he was shooting billiards or recording photographic spectra, may have found it difficult to mesh gears with his more impulsive younger colleague. Adams and Shapley also felt differently about the war in Europe. Shapley (who was exempted from military service) reported that Adams was "pretty rough on the Germans," while he and van Maanen "were pretty well suspect" because they "thought there might be another side to some issues."[23]

Shortly after joining the observatory staff, and before he had even begun his own research at Mount Wilson, Shapley published a seminal paper on Cepheid variables. Building on ideas he had discussed at Princeton with Russell, he disproved the widely-held idea that Cepheids were eclipsing binaries. He suggested that the variation in their light output was due instead to their physical change in size. According to Shapley's theory, which is at the root of our current understanding of Cepheids, the outer atmosphere of this type of star quickly balloons in response to increasing temperature. The star appears larger and brighter in this phase. Eventually the pressure in this outer "envelope" of the star drops, and the star slowly dims as it shrinks back to its original size. The cycle repeats as the atmosphere heats up and expands again.

Understanding the cause of light variation in Cepheids gave Shapley and other astronomers the confidence to develop Cepheid variable observations as a tool to determine interstellar distances. If the Cepheids were eclipsing systems, there would be no reason to expect their period of variation to be correlated with their peak luminosity; the frequency of eclipses would depend on the masses and the size of the orbits of the two stars, and it would be difficult to explain how these factors could be tied to the combined brightness of the stars. The period–luminosity relation would seem fortuitous — and not to be relied on to predict the luminosity from the period. Shapley's idea, which the theoretician Arthur Eddington elaborated on, provided a physical basis for the period–luminosity relationship. The maximum brightness of the Cepheid depended on the size to which it expanded, and physicists readily understood how physical conditions in the star could lead to a regular oscillation in the star's outer atmosphere.

Shapley settled in quickly: his first official day of work was the day he arrived. As he carried out his duties for Seares, he also devised a research program of his own. Solon Bailey at Harvard had suggested he study variable stars in globular clusters, and indeed, Shapley soon found many new variables there. His thesis work with Russell had shown him how fruitful the study of variables could be: he had obtained many physical parameters of variable star systems, and he had determined rough distances too. Not surprisingly, then, the research program he developed at Mount Wilson incorporated elements from these past experiences. His aim, he announced in one of his early papers, was to extract as much information as possible from globular clusters — to learn their sizes, distances, and compositions. When his turn came to use the 60-inch telescope every month, he suffered through the mountaintop's long, cold nights to collect data on the colors and magnitudes of cluster stars, and to search for variables among them. At the offices in Pasadena, he analyzed his data.

Martha continued to assist her husband with the data analysis and computations. She was expecting their first child in the early spring of 1915, and it appears that she found in mathematical analysis an enjoyable way to pass the time before delivery. The first paper she published as a co-author with Shapley appeared in 1915, about the time their daughter Mildred was born. Most of her subsequent papers appear to have been written during times when she would have been expected to rest.[24]

From 1914 to late in 1917, as Shapley gathered and analyzed his data, he looked for clues to the relationship of the globular clusters to the Milky Way and pondered the nature and origin of the stellar system.[25] The universe was his jigsaw puzzle, and the theory of island universes his framework. Soon he noticed that the pieces he carved from his detailed studies of clusters and the pieces of evidence from other established facts did not fit together. Toward the end of this period, late in 1917 or early in 1918, he saw a way to make the pieces fit, but the solution he found called for drastic action that he was at first unwilling to take: discard the theory of island universes.

Globular cluster distances

Omega Centauri, the Hercules cluster, M3: together with 47 Tucanae, visible from the southern hemisphere, these are three of the biggest, brightest, and most beautiful of the globular clusters. On dark, moonless nights, they are visible to the naked eye — for northern hemisphere observers, M3 and the Hercules cluster are high in the sky on summer nights, while Omega Centauri, at a declination of about −47°, is visible to an observer at the latitude of Mount Wilson, in the spring, above the southern horizon. Even small telescopes or binoculars reveal their blazing centers and sparkling outer edges.

Shapley chose these objects and a few others for his intensive studies of the colors and magnitudes of stars in globular clusters. When he began his studies, he thought at least some of the globular clusters comparable to the Milky Way system in size, about 20 000 light-years in diameter. Indeed, in the mid-1910s the belief that globular clusters were island universes, and evolutionarily related — somehow — to spiral nebulae, was not uncommon. But clearly, the globulars differed in shape from the flattened Milky Way system, and Shapley noted other differences. For one, the brightest stars in the neighborhood of the Sun are the massive blue stars (the O or B type stars), while in the globular clusters, the brightest stars are red. This difference in the color of the brightest stars suggested that the mix of star types was different in globular clusters and in the Milky Way system.

One of Shapley's most significant papers of this early period, a publication from 1915, bore the title "Thirteen Hundred Stars in the Hercules Cluster (Messier 13)."[26] As the name implies, it was an intensive study of that cluster. In it he also announced the distances he had derived to Omega Centauri, the Hercules cluster,

M3, and the Small Magellanic Cloud. For this early study he used a combination of techniques to arrive at the distances, including a version of the "faintness mean farness" principle (see chapter 4) and the distances as indicated by the Cepheid period–luminosity relationship.

As we saw in chapter 7, Henrietta Leavitt had noticed that the brighter a variable appeared at its maximum, the longer the time it took to cycle from maximum to minimum and back to maximum. This meant that Cepheid-type variables could, in principle, be used as "standard candles." If Cepheid A were brighter than Cepheid B of the same period, then Cepheid A must be closer. If just one Cepheid "candle" could be found in a cluster and compared with another Cepheid, the distance of the cluster relative to the comparison Cepheid could be determined. The Cepheids' distinctive light variation and high brightness at the maximum of their cycle made them easy to pick out of a crowd of stars. The hitch was that astronomers could use them only to determine relative distances – until Hertzsprung established a scale according to which the period of a Cepheid variable could be used to predict its absolute luminosity.

Using these measures, and Hertzsprung's calibration, Shapley obtained what he felt were rather large distances: 10 000 light-years for Omega Centauri, 30 000 for M3, 50 000 for the Small Magellanic Cloud, and a staggering 100 000 light-years for the Hercules cluster. All but the first of these values put the globular clusters well outside what Shapley then believed were the limits of the Milky Way system, and seemed to confirm their status as "island universes."

In another of his early papers Shapley drew attention to the fact that the globular clusters are asymmetrically distributed in the sky, with about one-third of them lying in the direction of the constellation Sagittarius. The phenomenon baffled him, and he worried that it might be significant. He was not the first to comment on this skewed distribution; John Herschel, son of William Herschel, had noted it in the course of his search for nebulae from the southern hemisphere. A Swedish astronomer, Karl Bohlin, had put forward the little-appreciated suggestion that the swarm of globular clusters centered on the center of the Milky Way system, like bees around a hive. The center of the system, in his view, was not coincident with the Sun, giving

260

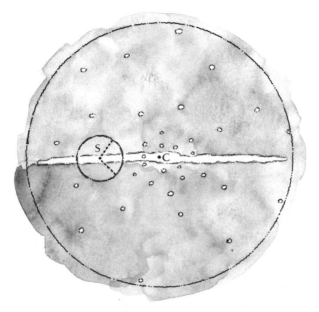

Figure 8.2 Distribution of globular clusters with respect to the Galaxy. Globular clusters fill the spherical space around the disk of our galaxy; in other words, the center of the globular cluster distribution coincides with the center of the galaxy, "C." This is the situation imagined by the Swedish astronomer Karl Bohlin, and later confirmed by an initially skeptical Harlow Shapley. As viewed from the position of the Sun ("S") — in the disk, but far from the galactic center — the globular clusters thus appear concentrated in one part of the sky. An observer at or near the Sun sees more globular clusters in the area of sky spanned by the dashed lines than in other areas of the celestial sphere. (Credit: Layne Lundström.)

rise to the appearance of asymmetry from our location (figure 8.2). Shapley recalled Bohlin's suggestion for the sake of argument, but dismissed it because Bohlin's proposal was incompatible with the distances Shapley had derived for the globular clusters. However, Shapley would later return to the problem of the asymmetric distribution, having concluded that it provided an important clue to how his puzzle pieces fit together.

Before Kapteyn departed Mount Wilson from what was to be his last visit to the United States, Shapley stopped to see him and

261

show him the newly calculated globular cluster distances. For Kapteyn, who worked with distances for individual stars derived from trigonometric parallaxes and the "mean parallax" method, the distances of tens of thousands of light-years that Shapley derived were unthinkable, but he admired Shapley's initiative and wanted to encourage him. Shapley reported Kapteyn's reaction to the distance to M3 in his memoir: "He looked and suggested that I check my observations again. In other words, he would not accept the result. But he was kind about it, because I was a nice young man and he was a nice old man."[27]

No doubt Kapteyn urged Shapley to check into the possibility that interstellar absorption was affecting his results. Shapley might have attributed the dimness of globular cluster stars to distance and not to the fact that they were partially obscured, in other words. Kapteyn had a life-long fear—well justified, as it turned out—that his star-counts would be vitiated by obscuration. Edward E Barnard, an astronomer at Yerkes Observatory near Chicago, had shown that obscuring matter definitely existed in clouds. The only lingering question that the astronomical community faced was whether obscuring gas and dust was also more thinly distributed throughout space.

Shapley did his homework. To detect dust and gas particles he looked for their effect on the colors of stars. As in the Earth's atmosphere, particles in space scatter blue light and allow red light, of a longer wavelength, to continue unimpeded its journey between the source and the observer. Like Kapteyn before him, he therefore looked for reddening of stars. But how to tell if the stars appeared redder than they would have in the absence of dust? Shapley had to rely on the assumption that in general, samples of stars nearby have the same mix of colors as stars in more distant globular clusters. The greater distance of the globular cluster stars, in this analysis, should allow for more reddening from interstellar dust, and a mix of stars that appeared to include more red ones. The assumption was a risky one, but one he had to make in order to proceed at all.

In 1916, and again in 1917, he reported that he had found no such reddening effect. In the plane of the Galaxy, he acknowledged, space might not be perfectly transparent, but in clusters from near the galactic pole down to a few degrees above or below the plane of the Galaxy, he had found faint, very blue stars. It didn't seem possible that these faint blue stars would be visible if interstellar dust acted as expected. He stood by his large distances to the globulars.

The "big galaxy" hypothesis

Sometime before November 1916, Shapley came up with a distance measurement that surprised him even more than the globular cluster distances had. Looking in the plane of the Milky Way — or what Shapley called the galactic system — in the direction of Scutum, near Sagittarius, he examined the stars of the open or galactic cluster known as M11. He found faint blue stars there, similar to the ones in the globular clusters. Applying the same techniques he had used on the globular clusters, he was forced to the conclusion that M11, firmly a member of the Milky Way system, was some 50 000 light-years distant.[28] This was as far as some of the globular clusters, and larger than the diameter of about 20 000 light-years that he had previously ascribed to the Milky Way system. Here was a clue that the Galaxy might be larger than anybody expected — though still an "island" in the universe. Perhaps a new, larger estimate of the size of the Galaxy was what Shapley had in mind when he predicted to Kapteyn in February 1917 that he would "get something out of the problem yet."

At about the same time that he was considering revising his estimate of the size of the Galaxy, Shapley's thoughts about spiral nebulae were thrown into confusion, also. In 1917, several novae had appeared in a number of spirals, and these inspired a search through old photographic plates at Mount Wilson and Lick Observatories for more "new" stars. Shapley joined the fray and in October 1917 announced that the novae he had examined in the Andromeda nebula — with the single exception of one that had flared in 1885, S Andromedae, whose brightness had been exceptional — indicated that the nebula was about 1 million light-years away.[29] This result from the novae posed no challenge to Shapley's belief in the "island universe" theory.

But contradicting such a large distance for a spiral nebula was the work of Shapley's friend Adriaan van Maanen. Shapley had paid scant attention to it at first.

Van Maanen had been studying pairs of images of the spiral nebula M101 with a blink stereocomparator. This instrument allowed him to view an old photograph and a new one of the same subject in rapid alternation through the same optics. The resolution of the images was not fine enough to permit him to see stars, but distinct nebulous points that he took for star clusters appeared in the nebula. Van Maanen searched for movement of those nebulous points. If any had shifted position within the nebula between the times when the two photographs were taken, he would detect the shift by eye. To obtain the amount of shift, van Maanen selected comparison stars on the image, which he assumed were foreground stars not associated with the spiral nebula, and measured the shifted points with respect to the comparison stars using a kind of micrometer.

To his own surprise, van Maanen noted shifts indicating what he called internal proper motions or rotation of the spiral. Not that rotation, per se, was unanticipated; ever since Lord Rosse had drawn the spiral M51, which he had seen through his enormous telescope, astronomers had speculated that the whirlpool structure of the spirals indicated their rotation. Heber D Curtis voiced a widespread opinion in 1915 when he said, "The spirals are undoubtedly in revolution since any other explanation of the spiral form seems impossible."[30] But if the spirals were vast assemblages of stars so distant that individual stars could not be seen telescopically, then their undoubtedly stately pace of rotation should not be so easily discerned in photographs taken only a few years apart.

Van Maanen published the first of his results in the spring and summer of 1916, and began searching for similar motion in other spiral nebulae. Shapley, when he finally took note of van Maanen's work, was perturbed. If the spirals rotated at the rate van Maanen indicated, they must be much closer than anticipated. In September 1917, Shapley wrote to his former teacher, Russell: "Do you sometimes suspect the internal motions in M101? V. M. [van Maanen] does a little, Hale more, and I much."[31] About two months later he sent off another letter to Russell reporting on the distances of about 30 globular clusters, the nearest about 20 000 light-years away, and the farthest about 200 000 light-years. "This is a peculiar universe," he wrote.[32] Evidently Shapley still clung to the framework of island universes, was inclined to think the Galaxy might be exceptionally large, and wished he could dismiss van Maanen's results.

Russell replied in a letter dated 8 November 1917. "I am at present inclined to believe in the reality of the internal proper motions, and hence to doubt the island universe theory."[33] He puzzled over what the spiral nebulae might be, if not distant agglomerations of stars.

Russell's opinion may have swayed Shapley and persuaded him to try to fit the puzzle together with van Maanen's piece. Shapley held Russell in very high esteem; one of Shapley's graduate students in the 1920s said many years later, "Russell had been Shapley's teacher and mentor, and his word was law. If a piece of work received his imprimatur, it could be published; if not, it must be set aside and its author had a hard row to hoe."[34] In any case, Shapley reconsidered his distances, the novae in spirals, van Maanen's rotations, and the island universe theory itself.

Elucidating the sidereal structure

In the fall of 1917, Shapley sought to put his results on the distances and arrangement of the globular clusters on a firmer footing, and for this he needed to re-calibrate the period–luminosity relationship of Cepheid variables that he had borrowed from Hertzsprung. In other words, the Cepheids' periods of variability were proportional to their absolute luminosities, and he sought the constant of proportionality.

Unfortunately, no Cepheids are close enough that their distances can be determined directly by trigonometric parallax. If he could get an independent distance measurement to a few local Cepheids, the general constant of proportionality would be known. With characteristic panache, Shapley made a series of assumptions that would allow him to obtain an answer. First he borrowed from Kapteyn's technique of statistical parallaxes. He collected data on the proper motions and radial velocities of 11 nearby Cepheid type variables — those not in globular clusters, but in our own system — to find their three-dimensional motion through space. He determined the average speed of these

Cepheids as a function of their luminosity, and, like Kapteyn, assumed that the average speed was related to a star's distance. In this way, he related the luminosity and absolute distance of nearby Cepheids, and established the proportionality between the Cepheid period of variability and distance.

Next, Shapley applied the period–luminosity relationship to calculate the distances of Cepheids in globular clusters. He had to assume, as astronomers frequently must, that the same physical principles governed the nearby and more distant objects, so that the nearby variable stars he had used to calibrate the relationship behaved the same way as the variables in globular clusters. This assumption turned out to be ill-advised, but was not shown to be somewhat in error until the 1950s. Shapley published his new calibration in a paper he submitted in November 1917.

The Cepheid variables could be used in about a dozen of the 69 known globular clusters. How to proceed with those clusters so far away that even the bright Cepheid variables could not be discerned? Again, Shapley made some bold assumptions to pierce deeper into space than anyone had before. The next link in the chain of distances was provided by the highly luminous stars. Shapley contended that the most luminous stars were intrinsically similar from one cluster to another, and used the principle of "faintness means farness" to gauge the distances to the clusters. But then, some clusters were so distant that he could not even isolate their most luminous stars; in this case, Shapley assumed that globular clusters come in a single size. The apparent diameter of the cluster then yielded an estimate of the distance.

Shapley revealed the distances and locations of the globular clusters he had mapped out in this way in a paper he submitted for publication in November 1917. He plotted the positions of the globular clusters on a set of concentric circles centered on the location of the Sun (see figure 8.3). The plot clearly showed that the globulars are clustered in the direction of Sagittarius, at galactic longitude 325°. Shapley also drew arrows on his diagram (not shown here) to indicate the distances of the globular clusters above or below the plane of the Galaxy. The arrows showed (though not very legibly) that the globular clusters tended to avoid the plane of the galactic system, and hovered at distances of thousands of light-years from the plane.

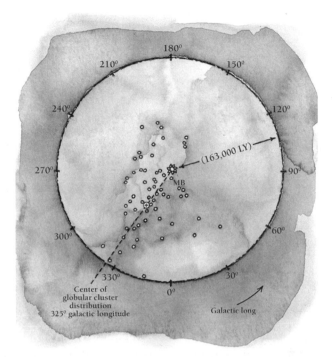

Figure 8.3 Distribution of globular clusters mapped by Shapley. The diagram is a kind of bird's-eye-view of the Galaxy, showing the globular clusters asymmetrically located in galactic longitude with respect to the sun. Shapley found that the center of the globular cluster distribution (marked by a + symbol), which he correctly assumed matched the center of the galaxy, lay tens of thousands of light-years from the Sun. (Note that the longitude system Shapley used is no longer the standard.) (Credit: Layne Lundström.)

Late in 1917 or very early in 1918, Shapley had an epiphany of sorts. In a letter to Eddington dated 8 January 1918 he wrote, "I have had in mind from the first that results more important to the problem of the galactic system than to any other question might be contributed by the cluster studies. Now, with startling suddenness and definiteness, they seem to have elucidated the whole sidereal structure."[35] Indeed, the elucidation of the structure of our system, which Shapley accomplished at the age of 32, proved to be a high point of his career and the most significant contribution he made to our understanding of the Milky Way galaxy. But it would be decades before his insights could be fully appreciated, for his hypothesis was not without difficulties.

Shapley had plotted the distances and positions of all 69 known globular clusters. He had known they were asymmetrically arranged in the sky; what leaped out at him from his plot was the picture of a swarm of globular clusters centered on a point in the plane of the Galaxy in the direction of Sagittarius, which must be the center of the whole system (see figure 8.2). In effect, he resurrected Bohlin's hypothesis, which he had explicitly dismissed in 1915, and placed it on a firm foundation. The Sun could no longer be considered the center of the sidereal system. It lay about 60 000 light-years from the center.

Shapley had found a way to fit the pieces together: abandon the island universe theory. The Milky Way system was very big — a "continent" among islands, and the Sun occupied a position about one-fifth of the way from the center to the edge. The Milky Way galaxy included the globular clusters, as shown by Shapley's distances. It included the spirals as well, as shown by van Maanen's data. The novae in spirals had always yielded ambiguous results; one must either equate the fainter, more numerous novae in spirals with their galactic counterparts and infer large distances, or one must assume that the bright flaring of S Andromedae in 1885 was an outburst comparable to that of a galactic nova. Once he had concluded that the spirals were near, Shapley placed his faith in the latter interpretation.

Within a week or two of writing to Eddington, Shapley reported with equal ardor to his chief, Hale. He described the Milky Way as no one had envisioned it before, an "enormous, all-comprehending galactic system." The plane of the Galaxy extended to a diameter of some 300 000 light-years — at least ten times farther than most of Shapley's fellow astronomers would have it. Strikingly, to those accustomed to thinking of the Milky Way as a flattened or lens-shaped system, Shapley asserted that the galactic system's boundaries reached far above and below the plane as well: Its diameter "may be much the same at right angles [to the plane], but except for the first few thousand light-years from the plane there is little evidence of isolated stars — only clusters, spiral nebulae, the Magellanic Clouds." After stating the facts as he knew them, he allowed himself to hypothesize:

268

"There is no plurality of universes of which we have evidence at present."[36]

Characteristically, Hale responded with both encouragement and a note of caution. He liked Shapley's "daring hypotheses," but he urged Shapley to "substitute new hypotheses for old ones as rapidly as the evidence may demand."[37]

In the meantime, Shapley was buffeted by feelings of satisfaction at having sorted out the sidereal system and vexation over some petty issues with Adams, the acting director of the observatory. Late in 1917, Adams had written to Hale, complaining that Shapley had been rather cavalier with the evidence for island universes, and, more seriously, that Shapley had "never given the credit where it belongs" for some aspects of his distance methods. Shapley in turn wrote to a correspondent, "If I did not take great joy in the actual learning of things, I would feel that scientific labors are after all quite futile, for the body suffers through the necessary privations and the spirit through clashes of professional jealousies."[38]

"Formicid" and other adventures

Shapley's seven-year tenure at Mount Wilson Observatory saw the publication of about 80 papers in his name as author or co-author. Most of these, of course, concerned variable stars, his cluster studies, or topics relating to the theory of island universes. But even in this most intense period of research in his career, Shapley could not confine his intellectual activity to his primary topic of star clusters. He lent his advice and support to other projects at the observatory, such as an effort to measure the radius of the star Betelgeuse in Orion using the technique of interferometry. He discovered an asteroid, which he named after his daughter—it is number 878 Mildred. Perhaps the only activity on the mountaintop that Shapley stayed away from was the readying of the 100-inch telescope, which saw "first light" in 1917. He was, however, one of the first to use the telescope when it was finally ready.

And Shapley had energy to spare. At Missouri he had delved into classics and poetry, and at Princeton he had audited courses in physiology and paleontology. At Mount Wilson he made a hobby of myrmecology, the study of ants. Shapley liked to refer to his "formicid episodes," using the Latin for "ant."

While resting from a climb up a canyon near the observatory one day he noticed a line of ants running busily to and fro along a concrete wall. "I had not been interested scientifically in ants up to that time," he wrote in his memoir, "but I noticed that when the ants went into the shade of the manzanita bushes, they slowed down—just as I would have done. It was cool and nice, and I supposed that they slowed down for comfort. I began to wonder about this, however, and soon I got a thermometer and a barometer and a hydrometer and all those 'ometers' and a stop watch. I set up a sort of observing station while resting and getting ready for another night's tussle with the globular clusters. With a flashlight I followed those ants in the dark, I found it great fun to watch them."[39]

Shapley set up "speed traps" for the ants and found that they ran quicker when the temperatures were higher. He had discovered the "thermokinetics" of ants. His quantitative experiments resulted in several myrmecology papers that the publishing arm of the National Academy of Sciences and other journals saw fit to publish, beginning in 1920. Shapley maintained a lifelong interest in ants and collected them around the world on his travels.

In 1920, Hale, who was in California between trips to the East Coast, invited Shapley to participate in a venture that would draw on his experience both as a journalist and as a scientist. The millionaire newspaper publisher Edward W Scripps, who had founded the independent news gathering company United Press (later United Press International), had a plan to form a service that would supply science news and features to the nation's daily newspapers. Hale, Shapley, and other prominent scientists representing the American Association for the Advancement of Science, the National Research Council, the American Academy of Science, and other professional organizations assembled at Scripps's coastal ranch, Miramar, near San Diego. The outcome of the meeting was the formation of Science News Service, later called Science Service. Science Service published what is now known as *Science News*, a weekly news magazine, sponsored science fairs, and conducted a science talent search among high school students. Shapley served on the board of Science Service and later also as president.

Shapley's outside interests, his liberal politics, and his bold speculations on astronomical matters probably contributed to the friction that developed between him and an astronomer of an entirely different disposition who arrived in Pasadena in 1919. This was Edwin Hubble—"Major" Hubble, freshly returned from military service. Shapley found him unapproachable right from the start. "He was born in Missouri not far from where I was born and he knew the Missourian tongue," Shapley wrote in his memoir. "But he spoke 'Oxford.' He would use such phrases as 'to come a cropper.' "[40] While at Mount Wilson, the two doubtless avoided each other as much as possible. The days when their astronomical fortunes would intersect were still some years in the future.

Shapley marshals his arguments

A pair of lectures that Shapley and Heber D Curtis delivered in Washington in 1920 is known in the astronomical community as the "Great Debate," implying perhaps that controversy over the island universe hypothesis had reached some sort of crisis point, and that astronomers felt a pressing need to hear proponents from each side articulate their arguments in a face-to-face meeting. In fact, the island universe hypothesis that Shapley came to cast doubt on was well accepted, and the idea of bringing matters to a head came mostly from Hale, who was in a position to notice the challenge that Shapley's findings posed for the existence of stellar systems comparable to our own.

In late 1919, Hale, as an influential member of the National Academy of Sciences, proposed that the program for the April 1920 meeting include a debate or discussion about either the island universe hypothesis or about general relativity. In suggesting island universes as a topic, he may have been influenced by a March 1919 lecture that Curtis, a Lick Observatory astronomer, gave at the Washington Academy of Sciences. Curtis's title was, "Modern Theories of the Spiral Nebulae." General relativity would also have been in the news in 1919. That summer, Einstein's theory was put to the test when astrophysicists such as Eddington attempted to measure its effects during a solar eclipse.

271

The Home Secretary of the National Academy of Sciences feared that island universes, "notwithstanding their vast extent," would bore the audience, and squashed the idea of a lecture on general relativity completely; no one, he felt, would be able to understand more than a few words, and he himself hoped "that the progress of science will send relativity to some region of space beyond the fourth dimension, from whence it may never return to plague us."[41] In the end, they settled on "the scale of the universe" as a topic. Hale originally thought of inviting Wallace W Campbell, director of Lick Observatory, to present one of the lectures, but Curtis was clearly at the top of the field when it came to arguments concerning the spiral nebulae. Telegrams went out to Shapley and Curtis.

The invitation caused Shapley some concern, as well as, undoubtedly, satisfaction. Much more would be at stake for him than swaying an audience of scientists over to his view of the Galaxy. In 1919, the venerable director of the Harvard College Observatory, Edward Pickering, had suddenly passed away. Within days, a flurry of letters began circulating among observatory directors discussing his replacement, and Shapley's name had come up more than once. Kapteyn suggested him, among others. Hale initially thought Shapley might be right for the job; later he wondered if Shapley might lack the maturity for it. Some favored Russell, even though his style was more that of a "lone wolf" than a leader of the pack. The debate in Washington, DC, would put Shapley's poise and maturity on display to Harvard overseers and patrons of the observatory.

Accordingly, when Shapley marshaled his arguments for the debate and for a later published version of his lecture, he concentrated on the evidence he believed he had for a large galaxy — a "continent" among islands — and did not confront the evidence on the other side directly. He wrote to Russell that he would only "touch lightly" on the spiral nebulae, for he had "neither time nor data nor very good arguments."[42]

Naturally, one of Shapley's main points would be the distances to the supposed center of the Milky Way system as indicated by the Cepheid variables. However, Shapley's confidence wavered as he considered that his calibration of the Cepheid period–luminosity relationship rested on only 11 stars. He therefore planned to emphasize other distance indicators.

The gist of Shapley's presentation would be that the Galaxy is far larger than previously thought, and that the Sun-centered Kapteyn universe was but a local cluster of stars within the Galaxy. The spiral nebulae he shrugged off as probably truly nebulous, not starry, entities, distant but still part of the larger Milky Way system.

To back up his point of view, Shapley summoned the following lines of evidence.

1. B-type stars in the solar neighborhood have an average luminosity about 200 times that of the Sun, and their individual luminosities do not vary from the average very much. In the Hercules cluster, which Shapley used as a convenient example of the application of his methods, stars with the very same spectrum, that is, B-type stars, appeared very dim. A comparison of their apparent brightness and true luminosity — the "faintness means farness" principle again — showed that they must be about 35 000 light-years away.

2. The dimness of the B-type stars in globular clusters such as the Hercules cluster could not be due to the absorption of light by small interstellar dust particles. Dark nebulosity is, indeed, present in the plane of the Milky Way, but studies of the colors of stars outside the plane of the Milky Way do not show any reddening effects as would be expected if the nebulosity were widespread.

3. The globular clusters are arranged symmetrically about a point which must be the center of the Milky Way system. The distances to the globular clusters give the scale of the system, and the results show that the Sun is about 60 000 light-years from the center of the Galaxy.

    In the published version of his lecture, which differed considerably from the oral one, Shapley elaborated on the usefulness of Cepheids as distance indicators and added other arguments. These included the following.

4. If the spiral nebulae are comparable to the Milky Way system in size, then they must be very far away, to judge from their small apparent size; and if they are so far away, then some of the new stars or "novae" that have been seen in them, such as S Andromedae, are impossibly bright.

273

5. The rotations of spirals recently measured by van Maanen are "fatal" to the interpretation of spirals as island universes, for if the spirals were very distant, the rotations measured would correspond to orbital speeds of the stars in the system comparable to or greater than the speed of light.

Shapley did not consider data on spiral nebulae that, in fact, threatened his arguments. His thinking about the spirals was rather vague, and he satisfied himself that the spirals were truly nebulous objects, within the boundaries of the galactic system.

Heber Doust Curtis

On the way to the debate in Washington, DC, Shapley and Curtis discovered they were riding the same Southern Pacific train. They had met and corresponded before; Shapley had visited Curtis at Lick Observatory, and the two may have crossed paths at astronomical congresses. The long train ride gave them the opportunity to get to know each other better, but they refrained from divulging their plans for the lectures in Washington. Shapley wrote in his memoir, "When the train broke down in Alabama, we walked back and forth and talked about flowers and classical subjects. It was quite pleasant. But we deliberately kept away from the controversial subject—the Great Debate."[43]

Curtis, 13 years older than Shapley, had a talent both for mathematics and languages, and, as Shapley certainly learned on the train, he had studied classics at the University of Michigan. After graduating he took a job teaching Latin and Greek at a small college in Napa, California. He became interested in astronomy, however, and resumed his schooling in 1900, when he and his wife and two children moved across the country so that he could pursue graduate studies in astronomy at the University of Virginia.

After earning his PhD, Curtis accepted a staff position at Lick Observatory near San Jose, California and began assisting director W W Campbell in spectroscopic studies of stars. In 1905, Campbell sent him to Santiago, Chile, to man the observatory's southern hemisphere station. In 1909, Campbell called him back again, this time to continue an observing program on the nebulae that former director James Keeler had begun some years before. This was the project that made Curtis an acknowledged expert on both spiral and planetary nebulae.

The first world war—which broke out while Curtis was on a solar eclipse expedition in Russia—slowed Curtis's progress, but did not curtail his research on the nebulae. He taught navigation for a while, then moved to Washington, DC, to develop cameras for the Bureau of Standards.

In Washington, DC, Curtis had the opportunity to discuss his astronomical work with a wide audience. By 1917, he had become convinced that the spiral nebulae, which he estimated numbered more than 700 000, were island universes. He presented his views in March 1919 at a joint meeting of the Washington Academy of Sciences and the Philosophical Society of Washington. He may have discussed them, too, at a dinner with Hale within a week of his lectures. Shortly thereafter he returned to Lick Observatory. When he met Shapley on the train to Washington, he was on the verge of accepting the directorship of the University of Pittsburgh's Allegheny Observatory.

The Great Debate

On the evening of 26 April 1920, members of the National Academy of Sciences and their guests, and interested members of the public, made their way to the Smithsonian's Natural History Museum and through its cavernous rotunda to hear Shapley and Curtis speak on the scale of the universe. The debate began at 8:15 p.m. in the museum's auditorium. Each speaker was allotted 40 minutes.

Shapley spoke first. Historians are not aware of any newspaper or eyewitness account of the debate, but it appears that Shapley read his lecture and illustrated it with lantern slides. His notes indicate that he aimed his talk at the general public, tarrying through a rather lengthy introduction, showing photographs of open and globular clusters and defining for his listeners the term "light-year." He was careful to build up an image of the Milky Way system as a collection of clusters: he spoke of a "clustering motive" or motif that ranged from the loose open clusters in the plane of the Milky Way to the rich, dense globular clusters farther away from the plane.[44] Implicit in his argument was the idea that the "Kapteyn universe" constituted a local cluster within the Milky Way system.

Having prepared his audience to accept that various types of clusters pertained to a single, large system, Shapley outlined the methods of determining distances to the globular clusters. Without going into "the dreary technicalities," as he called them, he discussed the use of B stars and Cepheid variables as distance indicators, emphasizing, too, the lack of absorption by dust and gas in space, except in the plane of the Galaxy. He then summarized the results: the distances of the globular clusters, and their arrangement, implied that the Milky Way system is 300 000 light-years in diameter. A consequence of this "cluster theory of the galactic system," he noted, is that the Sun is very distant from the center of the Galaxy — 60 000 light-years, according to his reckoning. In conclusion, he offered the opinion that the spiral nebulae "can hardly be comparable galactic systems," if the Milky Way system was as large as he maintained. Only if his distances were off by a factor of about 10, he suggested, was it reasonable to suppose that the Milky Way was a spiral system and an "island" in the universe.[45]

Curtis, a more experienced teacher and speaker, presented a completely different sort of lecture. He had been expecting a spirited but technical give-and-take on the subject, and had summarized his arguments on typewritten slides. Curtis was taken aback by Shapley's general, and hence mostly uncontroversial, presentation. He later wrote to Shapley that five minutes before Shapley's time was up, he thought about changing the character of his presentation to match Shapley's, but decided to go ahead with what he had planned.

In a sense, Curtis and Shapley talked past each other. Shapley followed the established theme of the debate — the scale of the universe — more closely, and rather neglected to consider the implications for the island universe hypothesis. Curtis's agenda, on the other hand, was to demonstrate the strength of the arguments for spirals as systems comparable to ours. He attacked Shapley's distance determinations and presented cogent arguments for the spiral nebulae as distant stellar systems, not associated with the Milky Way.

276

Curtis's main arguments, some of which he emphasized more in the published version of his lecture, were as follows.

1. The blue (B-type) globular cluster stars that Shapley used as distance indicators were not as intrinsically bright as Shapley believed. Shapley was comparing local giant stars—very bright—with distant dwarf stars that are intrinsically dimmer, so the distances he derived were too great.

2. Spirals demonstrate a stellar-type spectrum. Curtis may not have dwelled on the long history behind this assertion, but he could have made the case that the stellar spectrum of the spiral nebulae had been known since Julius Scheiner captured that of the Andromeda nebula in 1899 (see chapter 6).

3. The distribution of spiral nebulae in the sky seems to delineate a "zone of avoidance," which makes sense if the spirals are island universes like the Milky Way. Many spiral nebulae seen edge-on have dark bands running along the center line, indicating a layer of obscuring matter. If the Milky Way system is a spiral galaxy and has such an obscuring layer, we would—as is in fact the case—see the spirals arrayed around the north and south galactic poles, and apparently avoiding the plane of the Galaxy.

4. About 25 novae are known in the spiral nebulae; 16 of these are in the Andromeda nebula. We can compare them with about 30 known novae in the Milky Way system. Most of the novae seen in the spiral nebulae are relatively dim and indicate a distance of about one million light-years for the Andromeda nebula. The nova known as S Andromedae is much brighter, and gives a conflicting result, but it may be anomalously bright. Perhaps there are two types of nova.

Curtis added a fifth argument based on data taken at Lowell Observatory in Arizona by Vesto M Slipher. In the 1910s, Slipher recorded the faint spectra of stellar nebulae with very long exposures and found, to everyone's surprise, that the spectral lines were shifted to the red end of the spectrum. A notable exception to this rule was the spectrum of the Andromeda nebula, which showed a blue shift. No one was sure what to make of these results, but they clearly implied, based on the kind of analysis that Huggins had first made on the radial motions of individual stars, that these stellar or spiral nebulae were almost all receding

277

from the Milky Way as though repelled by some force, with velocities up to several hundred kilometers per second. While this repulsion was not easy to explain, Slipher's data, Curtis believed, were solid. These space velocities were unlike those of any galactic objects; the spirals must be extra-galactic. Curtis was willing to take Slipher's data at face value; years later, Hubble confirmed the data and investigated the phenomenon more closely, as we shall see.

Curtis did not trust van Maanen's measurements of the apparent rotations in spirals, which, if true, implied that the spirals were small and nearby. He was aware of other measurements that reported rotations in the opposite sense—van Maanen claimed the spirals were rotating such as to "wind up" the arms, while Slipher, at Lowell Observatory, showed they were "unwinding." Unfortunately, as Curtis knew, there would be no quick way to resolve the issue, because the rotations were in any case slow and would require a long time interval between comparisons on photographic plates. In the published version of his talk Curtis said diplomatically that "should the results of the next quarter-century show close agreement among different observers" indicating a certain minimum speed of rotation, then the island universe theory "must be definitely abandoned."[46]

Curtis came away from the evening feeling that he had won the debate—or rather, since it had not been a debate, that he had made the best impression with his arguments. He reported to his family back in California that his friends told him he "came out considerably in front." Shapley must not have felt as confident; in later years he acknowledged that Curtis had been more articulate. Russell felt the same way, and wrote to Hale that Shapley should teach a lecture course to cultivate a "gift of the gab."[47]

The debate was not reported in the general press. It did, however, find its way onto the pages of science-oriented semi-popular publications, at least in Europe. Bart Bok, a Dutch astronomer who began working with Shapley in the 1930s, claimed that "the impact of the Shapley and Curtis presentations upon astronomical thought was terrific."[48] He read about the debate in a Dutch magazine and immediately petitioned the public library in The Hague to obtain a copy of the published versions of the talks. Walter Baade, an astronomer about seven years younger than Shapley who worked at Mount Wilson from the 1930s through the 1950s, did not comment on the debate specifically but emphasized that Shapley's ideas contributed to stirring up the astronomy community on the scale of the universe. He said, "It was a very exciting time, for these distances seemed to be fantastically large, and the 'old boys' did not take them sitting down. But Shapley's determination of the distances of the globular clusters simply demanded these larger dimensions."[49]

With the benefit of hindsight, we can arbitrate this debate-that-was-not-a-debate. But it is not easy to call a winner. Both contestants were right in some respects. Shapley was correct to draw attention to the asymmetrical distribution of globular clusters, and to suggest that the center of the globular cluster distribution coincides with the center of the Galaxy.

Shapley's distance determinations were riddled with errors and invalid assumptions, as we shall see. Most importantly, he assumed, incorrectly, that interstellar absorption could be discounted. However, his errors did not mask the fundamental truth that the Galaxy is, in fact, much larger than his contemporaries assumed, and that the globular clusters constitute a sub-system of the Milky Way. (See chapter 10 for more on globular clusters and the structure of our galaxy.)

Curtis's defense of the popular island universe hypothesis was very strong; his instinct about what evidence to consider and what evidence to view as suspect was excellent. His remark that there might be two categories of nova, for example, was prescient. S Andromedae turned out to be a supernova, 10 000 times brighter than an ordinary nova. He was, of course, correct to point out that spiral nebulae have exactly the type of spectra one would expect if they are made up of stars. And Curtis was right to suspect van Maanen's measurements of the rotation of spirals.

Years passed before the dust settled on the Great Debate; astronomers did not resolve all the issues raised by discordant data until the 1950s, and most astronomers who attended the debate or read about it did not change their views as a result. However, the debate did help sway the astronomical community on one important point that Shapley argued: the Sun is located far from the center of the Milky Way galaxy. The center lies in the

279

direction of Sagittarius, behind vast star clouds and obscured by interstellar dust and gas.

Shapley to Harvard

The immediate outcome of the debate for Shapley was that he almost lost the Harvard job to his former teacher Russell.

Russell himself was not eager to take on the administrative responsibilities that came with the directorship of Harvard College Observatory. His position at Princeton suited him very well, as he could leave the running of its small observatory to Dugan, and his teaching duties left him time to pursue his research. He warned Shapley that the Harvard job might cost him the freedom to do astronomy.

However, in discussions with Harvard faculty members and patrons of the observatory after the debate, Russell had articulated Shapley's arguments on the scale of the universe so well that in July 1920, Harvard president Abbott Lawrence Lowell, at the urging of his advisors, offered Russell the job. And for a while, Russell considered taking it. He envisioned himself as director with Shapley as second in command. Only when Russell made up his mind not to accept the job, and lobbied for Shapley, did Lowell seriously consider Shapley for the job.

Shapley finally got an offer from Harvard in January 1921. The offer came in the form of a temporary appointment, which he could take while on a year's leave from Mount Wilson Observatory. In the end, Shapley garnered the approval of those in charge and won the full position in October 1921, before his trial period ended.

In April 1921, Shapley moved his family, which then included sons Willis and Alan as well as daughter Mildred, to Cambridge, Massachusetts, and settled into the observatory director's residence. This large, rambling house stood behind the domed building that housed Harvard's 15-inch aperture "Great Refractor," the twin of Pulkovo Observatory's "Great Refractor." The director's residence also stood adjacent to a building known simply as "the brick building," which housed Harvard's massive collection of photographic plates.

The observatory at that time included a number of small telescopes on the main campus, a small observatory in Jamaica run by Edward Pickering's brother William, and the station in Arequipa, Peru, that Solon Bailey had directed in the 1890s.

Bailey, though he still maintained an office under the dome of the 15-inch telescope, departed for Peru after Shapley's arrival, so as to leave the new director a free hand. Annie Jump Cannon, whom Shapley had met on his 1914 visit to Harvard, was still classifying spectra, as she continued to do until her retirement in 1940. She received an honorary doctorate from Groningen University, at Kapteyn's behest, the year Shapley took over at Harvard. Henrietta Leavitt, the discoverer of the period–luminosity relationship in Cepheid variables, was still there, but died in December 1921. Antonia Maury, who had developed the "c" classification for giant stars, was a frequent visitor; she ran the Henry Draper Memorial Museum in Hastings-on-Hudson, New York, but returned to Harvard occasionally in connection with her independent research on stellar spectra. About a dozen other women rounded out the complement of female staff members who analyzed the photographic data and maintained the plate collection.

Shapley's administrative assistant also served as his research assistant and co-author. Adelaide Ames graduated from Vassar College north of New York City, which was then a women's college with a tradition of observational astronomy, and in 1922 enrolled in a graduate astronomy program at Radcliffe College, Harvard's sister institution. She finished her master's degree in 1924, at the age of 24. Despite her youth she proved to be a very capable assistant, organizing meetings on the one hand and examining photographic plates to identify galaxies on the other.

Shapley's desk at Harvard, which elicited numerous comments, also allowed him to work effectively. His friend Hudson Hoagland, a prominent physiologist, called it "symbolic of his way of life," and described it this way: "It was a desk in the form of a great wheel mounted on a vertical axle—a kind of rotating galaxy for ideas. Near the hub of the wheel were radially arranged compartments, cubby holes, and drawers. The disk of the wheel extending beyond the radius of these containers gave ample writing space for any position of the wheel. Thus sitting in one place, by turning the wheel, Shapley could bring before him any one of his divergent fields of interest. This marvelous desk thus allowed him, from his chair, to marshal the contents of half a dozen desks and files on as many topics by merely a twist of the wrist."[50]

One of Shapley's first endeavors—and a highly successful one—was to establish a graduate program in astronomy at Harvard. Previously, the observatory had, curiously, no connection to the teaching of astronomy at the university; its mission was purely research. At first, members of the scientific staff cobbled together a lecture series based on their individual areas of expertise. Shapley then recruited Harry H Plaskett, a Canadian based at Oxford, to pull together a graduate program and head the new department.

Shapley unwittingly recruited one of the first graduate students while on a trip to London shortly after he settled in at Harvard. She was Cecilia Payne, later Cecilia Payne Gaposchkin, who was to make a remarkable contribution to the field of astrophysics in her doctoral dissertation and later became the first woman to chair Harvard's astronomy department.

Payne wrote in her autobiography, *The Dyer's Hand*, that she heard Shapley speak just at the time when she began to despair of ever being more than a teacher, even though, as an undergraduate at Cambridge University, she was a protégée of the great astrophysicist Eddington. Shapley impressed her with his familiarity with the stars and his masterful, direct style. After the lecture she got a friend to introduce her to Shapley, and she told him forthrightly that she would like to come and work under him. He reportedly answered, charmingly, that she could succeed Miss Cannon when she retired. Payne wrote, "Knowing him as I did later, I doubt whether he took me seriously, or gave me a second thought. But I took *him* seriously."[51] Payne secured enough fellowships and grants to get her started with a year at Harvard, and in 1923, set sail for the United States.

Payne marveled at the freedom and intellectual stimulation she found at Harvard. She found the brick building "a hive of industry."[52] Shapley gave her an office there—the one Leavitt had used. Although she never warmed to Shapley personally, she found him a wonderful scientific guide and an effective manager. She vividly described his style in her memoir:

"The young director was everywhere, running upstairs two steps at a time, pushing his soft sandy hair off his forehead, greeting everyone with the same casual cheerfulness. He knew exactly what each member of the staff was doing. He made a regular stop at each desk, and with a few well-chosen words (I use the overworked expression advisedly), made each of us feel important. If by chance one of us was not at work when he paid his daily visit, a little note would soon appear there, calling attention to the fact that we were expected to put in regular hours."[53]

Payne and Shapley's assistant Adelaide Ames became fast friends. Payne said they were known as the "Heavenly Twins." They shared a telling joke about Shapley and his ambitions for Harvard: they chuckled that he had "found a Dear Little Observatory, and intended to leave it a Great Institution."[54]

Shapley's hopes for a Great Institution got a boost from the private Rockefeller Foundation. In the late 1920s, the observatory was in the process of moving the 24-inch aperture telescope at the Arequipa station (known as the "Bruce" telescope after a donor) from Peru to a less isolated and more convenient southern hemisphere location near Bloemfontein, South Africa. Shapley convinced the Rockefeller Foundation to provide the funds for a 60-inch aperture telescope to join the Bruce telescope there.

The late 1920s and early 1930s saw tremendous growth in the graduate program. Payne obtained the first PhD in 1925.[55] Nearly a dozen students followed her in quick succession. Helen Sawyer Hogg, a graduate student who began working for Shapley in 1926, remarked on the excitement he generated at the observatory: "His exuberant personality, his flair for ideas and his equally great flair for words, his phenomenal memory, his enormous interest in his students and associates as individuals, his amazing capacity for work made a profound impression on all of us."[56]

Shapley did not neglect building up the post-graduate scientific staff, either. In 1929, he hired Bart Bok, the Dutch astronomer mentioned earlier who insisted that his home town library acquire a copy of the Great Debate proceedings. Bok was educated at Leiden and Groningen by Kapteyn's students. The first of several staff members and students Shapley attracted who later became very influential, Bok introduced radio astronomy to Harvard, helped develop radio astronomy in Australia as director of the Mount Stromlo Observatory, then directed the Steward Observatory at the University of Arizona during a major phase of expansion in facilities and staff.

Lingering questions from the Great Debate

When Shapley and Curtis debated the scale of the universe in 1921, both knew that the continued accumulation of observations with ever-improving telescopes and instruments would eventually make sense of the conflicting evidence they argued over. In 1921, when Shapley was still settling in at Harvard, he got news of an apparent clarification of this kind. Van Maanen wrote to say that measurements on the spiral nebula M81 showed rotation similar to that of M101. Shapley responded gleefully: "Congratulations on the nebulous results! Between us we have put a crimp in the island universes, it seems — you by bringing the spirals in and I by pushing the galaxy out. We are indeed clever, we are. It is certainly nice of those nebulae to have measurable motions."[57]

The surprise must have been all the greater, therefore, when Shapley found some pieces of his carefully constructed puzzle rearranged as early as 1924. In that year, Hubble announced in a letter to Shapley that he had found Cepheid variables in the Andromeda nebula, and they implied a large distance for it and other spirals. The news made it clear that van Maanen was wrong about the internal motions of the spirals; the spirals could not possibly rotate as fast as van Maanen said they did, given their large true size. Shapley could and did still believe in a big Galaxy — a continent among islands — and the news had no impact on his contention that the Sun lay far from the galactic center. He had been dealt a blow by a powerful rival, however, and had to admit the status of spirals as independent stellar systems.

Cecilia Payne witnessed the fateful letter and its impact. She recalled, "I was in his office when Hubble's letter came, and he held it out to me: 'Here is the letter that destroyed my universe,' he said."[58] She also heard him say, ruefully, that he had believed van Maanen's result, in part because van Maanen was his friend.

More unraveling of the Great Debate problems was in store. In 1930, Lick Observatory astronomer Robert Trumpler showed convincingly that interstellar dust and gas dimmed and reddened starlight, and did so in all directions. His argument was quite elegant. He had estimated the distances to open or galactic clusters using two methods, and compared the results. First he assumed that clusters had similar linear diameters, therefore those that appeared smaller were more distant. Secondly, he used spectroscopic parallax: a star of a given spectral type should have a certain intrinsic luminosity, and a comparison of its apparent and intrinsic luminosity yielded an estimate of the distance. The two methods gave discordant results and pointed to the culprit: interstellar absorption dimmed the light from the stars, but did not affect measures of cluster diameters.

Shapley was, surely, disappointed over the blunders he had made in assessing the merits of the island universe hypothesis. He had been cautioned about the potential effects of interstellar absorption but had satisfied himself, through detailed observations, that it was not a problem. But even if he felt some lingering frustration over the turn of events, he did not allow his past errors to get in the way of further research. He accepted his mistakes and forged ahead with studies of external galaxies and a new favorite subject, the Magellanic Clouds.

The Magellanic Clouds, and other metagalactic subjects

Shapley harbored a fascination for the Magellanic Clouds that his position as observatory director at Harvard allowed him to satisfy. Indeed, his colleague Bart Bok asserted that "For thirty years, from 1922 to 1952, Harlow Shapley was 'Mr. Magellanic Clouds.'"[59]

Shapley viewed the Large and Small Magellanic Clouds together as a "gateway" to the universe, or at least to the group of galaxies to which the Milky Way belongs.[60] Their significance for research on galactic systems first became apparent in 1913 when Hertzsprung, seizing on the fact that the Clouds contain large numbers of Cepheid variables, calculated their distance from the Milky Way system using the Cepheid period–luminosity relation. By the mid-1950s, when Shapley wrote a review article on the Magellanic Clouds, it was clear that they constitute our nearest galactic neighbors, and so provide a close-up view of important features that are difficult to study in more remote systems.

Although the Clouds are less massive than our own galaxy, they are by no means sidereal lightweights, and are richly endowed with interesting and unusual stars, clusters, and nebulae. Shapley and his collaborators accumulated thousands of photographic plates and spectra in the course of charting these two objects alone. They found "supergiant" stars hundreds of thousands of times more luminous than the Sun, indicating the enormous possible range of stellar characteristics. They saw regions they took to be stellar nurseries, where the processes of stellar evolution might be studied, and they marveled at the extreme phenomenon of the Tarantula Nebula in the Large Magellanic Cloud. This gargantuan emission-line nebula is intrinsically more luminous than many small galaxies. These and other attractions make the Magellanic Clouds useful "gateways" in galactic astronomy even today.

Nor did Shapley ignore the more distant island universes whose existence he had at one time doubted. One of his best-known legacies is the so-called Shapley–Ames catalog of 1932, now much revised and expanded upon. Shapley and Ames prepared a survey of all of what he called "external galaxies" — Shapley consistently used the term "galaxies," while Hubble insisted on calling them "nebulae" — down to a limiting brightness of 13th magnitude. Other extensive studies followed; some of his publications have titles such as, "A Study of 7900 External Galaxies" (1935) and "A Survey of Thirty-Six Thousand Southern Galaxies." These surveys permitted a statistical analyses of the properties of galaxies, complementing Shapley's in-depth studies of the Magellanic Clouds. Sadly, Ames died in the summer of 1932, the year the first survey was published. She drowned in a canoeing accident.

In 1938, Shapley and his collaborators discovered two peculiar objects that, like the Magellanic Clouds, are cohorts of the Milky Way system. The objects resemble extremely faint and extended globular clusters — "phantom" systems that are so rarefied, they showed up on photographic plates as mere smudges. We know them now as "dwarf galaxies," an important class of object. The two that Shapley found, in the southern constellations Sculptor and Fornax, are more distant than the Magellanic Clouds, at about 270 000 light-years and 630 000 light-years, respectively. Still, they form part of the so-called "Local Group" of galaxies, dominated by the Milky Way and Andromeda systems. (chapter 10 discusses dwarf galaxies and their hypothesized role in galaxy evolution.)

World War II and its aftermath

Even before formal hostilities began in the Second World War, Shapley focused some of his boundless energy on helping European scholars who lost their university posts because they were Jewish or held unpopular political opinions. As part of his agitation on behalf of refugees, Shapley sat on the executive committee of the Emergency Committee in Aid of Displaced Foreign Scholars. This organization formed in May 1933 in New York City with the aim of finding new positions in the United States for professors at German universities, and later expanded its scope to include all Western Europe. The Committee met the needs of 335 displaced scholars, out of about 6000 who asked for assistance.[61] Shapley noted in his memoir that about 100 "rescues" went through his office.[62]

After the United States joined the war in 1941, some members of Shapley's staff took "war jobs," although the routine work of the observatory continued. Martha Shapley composed mathematical firing tables for the Navy; the five Shapley children by then were of high school age or older. Some staff members taught navigation to aviators, and Shapley and Bok wrote a book for service men and women, *Astronomy from Shipboard*. The Red Cross later reprinted the book as *What to do Aboard a Transport*.

After the war, Shapley lost some of his momentum in astronomy and devoted more of his time to national and international affairs. Bok wrote later, "Looking back at it all, it seems a pity that about 1946 he did not resign his post as Director of Harvard Observatory to assume an important administrative post in the national or international realm."[63] Bok may have had the National Science Foundation in mind.

In July 1945, Vannevar Bush issued a report calling for the US government to capitalize on the scientific research conducted

287

during the war and to continue to support research activities by public and private organizations. Bush had been Franklin D. Roosevelt's trusted advisor on scientific matters when the president died in office in April; Bush delivered the report instead to Harry Truman. *Science: the Endless Frontier* broke new ground by insisting that the government had the responsibility to support basic research inside and outside government laboratories, even in peacetime. In fact, the report argued that the nation's prosperity and security rested on scientific and technological progress. Bush suggested that grants should be dispensed through an independent agency, which he called the National Research Fund.

US House of Representatives members favorable to Bush's proposal promptly introduced a bill to implement his plan for a National Research Foundation, but the bill ran into trouble for the simple reason that the president could not agree to an agency that would control public funds without being in any way "part of the machinery of government."[64] Furthermore, some scientists opposed the plan, fearing a limitation of scientific independence. Responding to this turmoil, the American Association for the Advancement of Science in March 1946 appointed Shapley and two other prominent scientists to solicit the views of professional scientific organizations on how a national research fund might operate.

Proponents of a national research fund — now referred to as the National Science Foundation — were still divided and at an impasse more than two years later. Both chambers of Congress passed legislation for a fund to be managed by a presidentially-appointed board, but Truman vetoed the National Science Foundation Act of 1947 because the agency it would have created would still have been isolated from normal political processes. Work continued behind the scenes, however. A Truman advisor, John Steelman, issued his own report three weeks after the veto, making an even stronger case for a science agency than Bush had made, and calling for presidential oversight.

Shapley secured his place in the annals of National Science Foundation history as a member of a committee that agreed on a compromise for its governing structure — a board and director appointed by the president with the advice and consent of the US Senate. Shapley by then had been elected president of the

American Association for the Advancement of Science, and reportedly had "good personal contacts" with fellow-Missourian Harry Truman.[65] His fellow committee members were Cornell University president Arthur L Day, Dael Wolfle of the American Association for the Advancement of Science, and William Carey and Elmer Staats from the government's Bureau of the Budget. Congress established the National Science Foundation through legislation passed in 1950. Its budget today is over $4 billion.

An unpopular internationalist stance

The immediate post-war era provided Shapley with ample opportunity to cultivate support for science both at home and internationally. He was particularly proud of his role in the formation of the United Nations Educational, Scientific and Cultural Organization (Unesco). In the spring of 1945, delegates from 50 countries met in San Francisco to deliberate on the United Nations Charter. Shapley had been involved in getting a resolution through Congress supporting the basic idea of Unesco, and he stayed in close contact with the assistant secretary of state for cultural and public affairs, the poet Archibald MacLeish, who represented the United States at the San Francisco meeting.

The story, as told by Shapley, is that the delegates wanted to settle on a United Nations agency for education and culture, eliminating science in name at least. When Shapley heard this he placed a rare cross-country telephone call to MacLeish. He threatened "action from scientists' groups" and wheedled with his friend.[66] MacLeish was probably not a hard sell on science; it was he who composed a famous poem on the occasion of the first moon landing in 1969, which the *New York Times* published on its front page. At any rate, the "S" in Unesco was saved, thanks in part to Shapley's efforts. The historical record shows that, in fact, Shapley gave up his fight for the "S" in August 1945, knowing that the English scientists Joseph Needham and Julian Huxley would insist on it at the London conference of November 1945, when representatives from 37 countries met to sign Unesco's constitution.[67]

Shapley's efforts to bring the National Science Foundation and Unesco into being continued even as he encountered personal difficulties stemming from his support for international cooperation among scientists. In 1945, when American distrust of communism and internationalism ran very high, Shapley traveled to Moscow as Harvard's representative at the celebration of the 220th anniversary of the Academy of Sciences. The next year the House Committee on Un-American Activities sub-poenaed him. Bok vividly recalled that Shapley returned to Harvard from his interrogation with Senator John Rankin and said, with his voice breaking, "That man had the nerve to tell me that I am un-American."[68] Some Havard alumni, alarmed at the flap, called for Shapley's dismissal.

In his memoir, Shapley recounted the episode with charac-teristic bravado. He recalled that he started taking notes of the Senate hearing in shorthand, which made Senator Rankin even more antagonistic. "He came crawling over the intervening table and grabbed the notes out of my hand," Shapley wrote. "I rose in great dignity and said, "This is a case of assault."[69] After the hearing, Shapley added, Rankin blustered to the press that Shapley had treated the Committee on House-Un-American activities with contempt.

Shapley admitted that the situation was "rather tough" on his nerves.[70] He refused to be intimidated, however, and maintained his involvement with left-wing organizations. In March 1950, Senator Joseph McCarthy identified him as a Communist sympathizer connected with the State Department, and pursued interrogations with his wife Martha and one of his graduate students, despite the fact that Shapley did not have a formal connection with the State Department. Later that year the Senate Foreign Relations Committee exonerated him, but it was not until after he retired from Harvard in 1952 and retreated from politics that his family life became more serene.

Later years

In 1952, Mount Wilson astronomer Walter Baade resolved one of the last lingering questions from the era of the "Great Debate:" the difficulty of reconciling intra- and inter-galactic distances using Cepheids and other indicators. The difficulties had allowed Shapley to maintain, on the basis of his distances to globular clusters, that the Milky Way galaxy was larger than typical spirals.

Baade discovered that stars in globular clusters have a different chemical composition from otherwise similar stars in the disk of the Galaxy. In particular, he discovered that galactic Cepheids of a given period are intrinsically brighter than their counterparts of the same period in globular clusters. In fact, these globular cluster "Cepheids" deserve a different name — the W Virginis type variables.

The discovery undermined the commonly made assumption that stars everywhere in the Galaxy are comparable. Shapley's calibration of the Cepheid period–luminosity relationship had been correct for the distances to globular clusters, but incorrect for the distances to disk-type Cepheids, such as those Hubble saw in the Andromeda nebula. Hubble's distances had to be revised upward, and so also did the dimensions of the spirals. Finally, the pieces settled into place: the Milky Way was a spiral, far larger than astronomers had assumed in the 1910s, but comparable in scale to the spiral nebulae.

In retirement, Shapley and his wife moved to a country house in southwestern New Hampshire, within easy reach of Harvard and Boston. Shapley accepted invitations to lecture on astronomy to college students and public audiences around the country. He kept up with the scientific literature on ants and continued to look for wildflowers; he was proud of the fact that he had identified 121 species of flowers on his New Hampshire farm.

Among Shapley's achievements outside astronomy, he was quick to point out, was the fact that he had "collaborated in producing a rather remarkable family."[71] His daughter Mildred Shapley Matthews enjoyed a long career at the University of Arizona's Lunar and Planetary Laboratory and is known in the field of planetary science as the editor or co-editor of numerous widely-read standard references, including *Comets*, *Asteroids II*, and, more recently, *Mars* and *Resources of Near Earth Space*.[72] Son Willis held high administrative positions in the Bureau of the Budget — the government agency that Shapley dealt with in organizing the National Science Foundation — and at NASA. Alan became a physicist and administrator at the National Institute of Standards and Technology in Boulder, Colorado, directing the internationally famous National Geophysical and Solar Terrestrial Data Center. Lloyd became a prominent mathematician in industry and academia. Carl founded a private school in New England and later opened another in Italy. Some of Shapley's grandchildren pursued careers in science or science journalism.

In 1972, Shapley, 86, suffered a heart attack and died while visiting his son Alan in Colorado. Martha Betz Shapley passed away in 1981.

At the dawn of the twentieth century, when Shapley was attending a one-room school on the edge of the family farm in Missouri, most astronomers subscribed to the island universe theory and believed that other stellar systems, probably comparable to our own, lay scattered in an infinite and inaccessible space. A small number of astronomers, such as Kapteyn and the German astronomer Hugo von Seeliger at the Munich Observatory, were engaged in laborious star counts and statistical studies to ascertain the shape and extent of the sidereal system. These star counts resembled William Herschel's star gauges of the late 1700s, except that they incorporated quantitative measures of the star density and the dimensions of the system. The stellar system appeared to extend some 10 000 light-years from a centrally-placed Sun.

Around the time Shapley retired as director of the Harvard College Observatory, and thanks in no small part to his studies of globular clusters from Mount Wilson, the astronomical community had developed the necessary tools to span intergalactic distances. Astronomers understood the obscuring effect of interstellar dust on their measurements. The Milky Way galaxy had come into sharp focus as a disk of stars some 100 000 light-years in diameter, accompanied by a "halo" of hundreds of globular clusters. The Sun was seen to occupy a position in one of the spiral arms of the disk. External galaxies — formerly the spiral nebulae — had become a distinct field of study.

Man's understanding of the universe of stars had grown by leaps in the span of only half a century. Yet as always, new findings raised new questions. Hubble, as we shall see in the next chapter, took the study of the Milky Way one step further. His research opened up the question of the galaxies' place in space and time.

9

EDWIN HUBBLE: REDEEMER OF ISLAND UNIVERSES

"Whether true or false, the hypothesis of external galaxies is certainly a sublime and magnificent one. Instead of a single star system it presents us with thousands of them....Our conclusions in Science must be based on evidence, and not on sentiment. But we may express the hope that this sublime conception may stand the test of further examination."

A C D Crommelin, 1917[1]

On the evening of his eighth birthday, Saturday 20 November 1897, Edwin Powell Hubble looked forward eagerly to a special gift: he would be allowed to stay up late to look through his grandfather's telescope. Unlike Harlow Shapley, who, some 120 miles to the west in Nashville, Missouri, had just celebrated his 12th birthday, Hubble developed a fascination for astronomy as a young boy. His maternal grandfather, William James, introduced him to stargazing. James, a medical doctor and drugstore owner, had built his telescope himself.

The night of 20 November was evidently a clear one in Hubble's hometown of Marshfield, Missouri, for his sister later reported to his biographer that he had had a great time.[2] We have no record of what objects grandfather James pointed the telescope to that night, but some at least are fairly good bets. At that time of year, and from that location, the Andromeda nebula would have been high overhead shortly after Hubble and his family finished dinner. Amid the stars of those constellations visible all night—including Andromeda, Ursa Major, Ursa Minor, Cepheus, and Taurus—James might have located

293

double stars or individual stars famous for their particular characteristics. Mizar, the middle star of the three in the "handle" of the Big Dipper asterism of Ursa Major, for example, appears to be double upon close inspection. As seen through a small telescope, Mizar turns out to be not just a double, but a triple star system. In the constellation Cepheus, near Andromeda, James might have pointed out one of the reddest stars of the sky, a star Herschel had dubbed the "garnet star." The Pleiades asterism would certainly have been on James's observing list, as his telescope, no matter how small, would have made the six naked-eye stars of the group multiply spectacularly to dozens.

Neptune rose in the east at about 7 p.m., although this dim planet would have been a challenge for James to find, unless he had detailed information about its location. At about 10 p.m., Orion rose, and no amateur astronomer would fail to put the famous nebula in Orion's sword on display. Then, if young Edwin was allowed to stay up really late, his grandfather may have shown him Jupiter and its four principal moons, first seen through a small telescope by Galileo. Jupiter rose above the horizon at about 3 a.m., followed the next hour by the waning Moon. Certainly Hubble and his grandfather saw many meteors, for Hubble's birthday fell near the peak of the Leonid meteor storm. Mars, a much-discussed planet at the time, was not visible that evening, having set before dark, along with Saturn and Venus.[3]

Early interest in astronomy

Hubble (figure 9.1) was the second son and third of eight children born to John and Virginia Hubble. His father had attended law school, although without earning a degree, and had run a legal practice for a few years. During Hubble's childhood he worked for a number of insurance companies and was often away from Marshfield on business travel for weeks at a time. His children feared his return as much as they looked forward to it; he was stern and religious, although if they behaved themselves, he might amuse them by blowing smoke rings from his pipe or cigar. Virginia, who had completed two years at a women's college before being married, had a more patient, humorous

Figure 9.1   Edwin Powell Hubble (1889–1953). (Credit: Palomar Observatory, courtesy AIP Emilio Segrè Visual Archives.)

disposition. Both parents expected much of their children and supervised their schoolwork closely.

Hubble had two brothers: Henry, three years older and somewhat withdrawn, and Bill, two years younger. The family as a whole was quite musical. John played the violin, Hubble's older sister Lucy played the piano, and Bill played the mandolin. Hubble contented himself with singing at the family musical soirées, but taught himself to play the mandolin as an adult. He and Bill both did well in school and excelled in athletics, but the similarities did not run deep, for while Bill was friendly and outgoing, Edwin acted aloof even toward his few friends.

Hubble's friends described him as a nature-lover. He liked to roam the fields and woods looking for wildlife, while keeping an eye out for flint arrowheads left by Osage Indians and their ancient predecessors. The countryside around Marshfield is now, and was in Hubble's time, known for its apple orchards, which in late summer yield an abundance of red fruit. Beyond the agricultural lands near town, Hubble liked to explore the rolling hills and well-timbered rivers. In winter, temperatures sometimes dipped below freezing overnight. Then he liked to show off his ice-skating prowess on a lake close to home.

Hubble learned to read early and evidently found his school work tedious, for his marks in deportment were never exemplary. But outside the classroom he found intellectual stimulation through his relationship with his maternal grandfather, the amateur astronomer William James, and his paternal grandfather, Martin Hubble. Martin Hubble, a notably tall and broad-shouldered man, divided his time between homes in Marshfield and Springfield, 20 miles away. He had fought on the Union side of the Civil War, as a captain and quartermaster in the Enrolled Missouri Militia, one of 89 small regiments funded by the state government to defend cities, towns and railroads. "Captain" Hubble, as his friends and neighbors knew him, believed strongly in education—he helped to endow Drury College, a four-year Christian school in Springfield—and engaged his grandson in discussions about astronomy and American history. He lived longer than Hubble's maternal grandfather, and figured prominently in the boy's early life.

Hubble's late night at the telescope on his eighth birthday may have been the spark that lit a life-long fire for astronomy. A little less than two years later, he told his friend Sam Shelton about an upcoming total lunar eclipse, when the alignment of the Moon, Earth, and Sun would cast the Earth's shadow across the illuminated face of the Moon. Hubble persuaded his parents to let him and Sam stay up all night together to witness every moment. Although any lunar eclipse, on a clear night such as the boys enjoyed, is an intriguing spectacle, it also required unusual patience and imagination on nine-year-old Edwin's part to grapple with the abstraction behind the phenomenon and to stay focused on the event for many hours.

In 1899, shortly after Hubble's lunar eclipse viewing, the family moved to Illinois. They lived briefly in Evanston before settling in Wheaton, about 25 miles west of John's office in downtown Chicago. They lived comfortably there in a succession of large houses. Though not well-to-do, they could afford the help of cooks and housekeepers. The family by then included Henry, Lucy, Edwin, Bill, and baby Helen. Two more daughters, Janie and Betsy, were born in the 1900s; another daughter, Virginia, had died in Marshfield at the age of 14 months.

Around the time of the family's move to Wheaton, probably before his 12th birthday, Hubble corresponded with his grandfather Martin Hubble, in Springfield, about the planet Mars. The young Hubble had been caught up in a kind of "Mars fever" that spread through Europe and the United States in the late 1800s. The excitement culminated in 1910, about a decade after Hubble's correspondence, with the publication of Percival Lowell's book, *Mars as the Abode of Life*.

In 1877, as the orbital motions of Mars and Earth brought the two planets as close together as they ever get, the Italian astronomer Giovanni Schiaparelli had taken advantage of the opportunity to study the surface of the red planet telescopically. Schiaparelli reported seeing dark streaks on the surface, which he called "canali," or channels. In English translation, the word became "canals" — calling to mind the great engineered waterways of the nineteenth century, the Erie canal (completed in 1825) and the Suez canal (completed in 1869). Newspapers editorialized on the possibility of life on Mars, and the French astronomer Nicolas Flammarion published an entire encyclopedia of Mars and its "living conditions." In 1897, H G Wells' *The War of the Worlds*, the story of an invasion of Earth by Martians, appeared in serial form, in Great Britain in *Pearson's* magazine and in the United States in *Cosmopolitan*.

Thus Mars was a very popular topic the year of Hubble's eighth birthday, and it is not surprising that he maintained an interest in the planet. We don't know the content of his letter to his grandfather, but given his style as a researcher in adulthood, it would not be surprising if the youth echoed the opinion of most professional astronomers, that the existence of canals was still speculative. In any event, Hubble family history has it that a proud Martin Hubble had Edwin's letter printed in a Springfield newspaper.

At school in Wheaton, Hubble had some of the same problems he had encountered in Marshfield. He loved to read, but was an atrocious speller. He antagonized his teachers by challenging them in class. Among his peers, he sought respect rather than friendship: one of his closest associates recalled that he was a schemer and a dreamer who "always seemed to be looking for an audience to which he could expound some theory or other."[4] And yet, Hubble could easily have been popular with his classmates. By the time he entered high school he was developing into a tall and handsome young man. Throughout his high school years, furthermore, and especially in his senior year, he made a name for himself in athletics. As a member of the basketball team, he helped lead the Wheatonians to victory in the state championships of 1905. The next year, as a senior, he distinguished himself as a member of the football team, playing tackle, and in high jump, pole vault, hammer throw, and discus, as a member of the track team. At a meet at Northwestern University in May 1906, he even set a state record with a high jump of 5 feet $8\frac{1}{2}$ inches.

Hubble's lack of interest in his school's academic program did not mean he avoided exposure to science and the use of scientific instruments. At home, he read everything he could get his hands on pertaining to astronomy. With his brothers, he also attended occasional public lectures and scientific demonstrations at nearby Wheaton College. If any of those lectures dealt with current topics in astronomy, he would have been exposed to the technology of photography and spectroscopy. He may also have heard of recent observations made with what was then the world's largest telescope, the 40-inch refractor at Yerkes Observatory, some 70 miles north of Wheaton in the Wisconsin town of Williams Bay.

The high point of Hubble's early life, according to his sister Helen, was the summer he spent assisting a team of surveyors in the wooded plains of northern Wisconsin. Hubble had just finished his junior year of high school and was probably glad to escape the company of his classmates, who found him so difficult to get close to. In Wisconsin he relished the outdoor life, as he had enjoyed exploring the area around Marshfield and as he was to enjoy hiking, fishing and camping in later life. And as an avid student of astronomy, he must have been aware of the connection between surveying and techniques of determining distances in astronomy through parallax. Whether he knew it or not, Hubble was following in the footsteps of Thomas Wright, Friedrich Bessel, Wilhelm Struve, and other explorers of the cosmos who had practiced their skills in geometry by measuring angles and baselines on the ground.

A reluctant student of law

To his own surprise, Hubble learned on the day he graduated from Wheaton High School in June 1906 that the University of Chicago had awarded him a scholarship, one of dozens given to promising high school students around the country. The University no doubt viewed the so-called "Entrance Scholarship," which covered a student's first year's tuition, as a kind of recruiting tool, for it had opened its doors and admitted its first class only 14 years earlier. Although the stature of its first faculty members ensured its greatness as an institution of higher learning and research, the university, with its Gothic style buildings modeled on those of Oxford University, still faced a struggle to establish itself as a rival of its older counterparts in the east.

Hubble lived on campus, so as to participate fully in the social life of the university. But membership in one of the university's sports-oriented fraternities did not remove Hubble from his father's stern influence. Hubble's fervent hope was to become an astronomer, but to satisfy his father, he had to prepare for admission to law school, and to take the science classes relevant to astronomy as time permitted. In his free time, he wanted to play football, but John forbade that too, fearing his son would suffer debilitating injuries. The injunctions against football and astronomy rankled: Hubble, who once told one of his sisters that their father had "blighted his life," remained bitter about the sacrifices he had made to please his father decades later.[5]

Hubble did join the basketball and track teams, and boxed in off-campus venues. With the basketball team, he traveled all over the Midwest and to Philadelphia and Washington, DC, sharing the glory of the team's national championship in the spring of 1908. However, he never again achieved the success in sports that he had in high school. It was rather in the scholastic sense that Hubble shone, and this success nurtured his dreams of earning a prestigious Rhodes Scholarship to study at Oxford.

During his second year at the University of Chicago, Hubble, hoping for some change of heart in his father, began in earnest to acquire the background in physics that he would need later if he were to take graduate courses in astronomy. His teachers and guides in the physics department could hardly have been better qualified. Albert A Michelson, originator of a famous set of experiments to measure the speed of light and winner of the 1907 Nobel Prize in physics, headed the department. Associate Professor Robert A Millikan, from whom Hubble took courses in "Mechanics, Molecular Physics and Heat" and "Electricity and Light," would later win the Nobel prize himself, for work he did during Hubble's student years at Chicago. Hubble did well enough to earn the University's Junior College Scholarship in physics, and, as he had hoped, Millikan chose him as his laboratory assistant for the 1909/1910 academic year.

Hubble's favorite science courses were those taught by Forest Ray Moulton, an associate professor in the astronomy department. With Moulton, as with Millikan, Hubble observed a talented scientist and teacher at work during his most creative years. Moulton was the co-author, together with geology department chairman Thomas Chamberlin, of a new theory of the origin of the solar system. For some time, this theory was thought to point to the spiral nebulae as progenitors of solar systems.

Moulton had obtained a PhD in astronomy and mathematics at the University of Chicago in 1899. George Ellery Hale, who at that time was deeply involved in building up the staff of Yerkes Observatory and carrying out his own solar observations, had served on Moulton's thesis committee and continued to correspond with him and to help publicize his work.

Moulton and Chamberlin began attacking Laplace's eighteenth-century nebular hypothesis while Moulton was still a graduate student, and gradually drew together elements of a new theory to replace it. Among their objections to the nebular hypothesis were two problems. First, they said, the Earth could not have retained its atmosphere if it had cooled from a molten state, as Laplace stated. The temperature of the molten Earth would have had to exceed 3000°C and, under such conditions, the atmosphere would have escaped like steam from a kettle, unrestrained by the planet's gravity. Second, a close examination of Laplace's notion of planets forming out of rings of nebular material revealed a host of dynamical problems. A single planet would not simply or neatly collapse out of the ring material, Chamberlin and Moulton argued.

The Chicago theorists also pointed out that no annular nebulae of the kind described by Laplace were known, and that the appearance of the spiral nebulae could not easily be reconciled with the nebular hypothesis. Referring to Isaac Roberts's photographs of the Andromeda nebula in 1887, which had galvanized William and Margaret Huggins into trying to photograph the corresponding spectrum, Chamberlin wrote, "The photographs of the nebula of Andromeda, that were hailed with such delight on their first appearance as exemplifying the Laplacian hypothesis, appear upon more critical study to support it only in vague and general terms, if indeed they lend it support at all."[6]

Mounting the most serious challenge to Laplace's hypothesis in more than 100 years, the geologist–astronomer duo suggested that nebulous material emanating from a young sun in long spiral arms would condense into lumpy knots and then into "planetesimals," cool rocky fragments which would slowly coalesce into planets and satellites. In 1902, Hale, while not supporting Chamberlin and Moulton's theory in all its details, noted in a semi-popular article that their criticisms of the nebular hypothesis appeared valid. He added that the spiral nebulae, which had been photographed in great numbers by astronomers at Lick Observatory in California, might represent the Chamberlin–Moulton type of solar nebula in its early stages of evolution.

American astronomers embraced the Chamberlin-Moulton theory with more enthusiasm than did their European counterparts. The theory enjoyed a certain popularity until the late 1920s, and one of its components, the planetesimal hypothesis for the formation of the Earth, is still well regarded. However, the theory suffered from some serious flaws, and early in 1907 Chamberlin himself admitted that most spirals probably were too large to be each the progenitors of a single sun and planetary system.

Hubble certainly was well acquainted with the Chamberlin–Moulton theory, for he explained and defended it to some astronomy students at Oxford some years after he had left Chicago. More importantly, his friendship with Moulton allowed

301

him later to re-establish contact with the astronomical community in Chicago after several years' absence at Oxford studying law.

In the fall of his final year at Chicago, Hubble shifted his emphasis to the humanities and pored over his textbooks. To win a Rhodes scholarship he would have to pass a qualifying examination in mathematics, Latin and Greek, with the latter subject traditionally proving the most difficult. That hurdle over, he solicited letters of recommendation, including one from his physics teacher, Millikan, and submitted to an interview. After the New Year he heard the news: he had been selected as the Illinois recipient of the Rhodes scholarship. The scholarship would pay for him to study the subject of his choice for three years, at whatever Oxford college would admit him. In Hubble's case, his father, who still insisted on his studying law, determined the choice of subject.

Hubble spent the summer of 1910 with his parents and siblings in Shelbyville, Kentucky, where they had moved in connection with John Hubble's insurance work. John was not well: he suffered from recurrences of malarial fevers and undiagnosed kidney disease. Hubble's brother Bill also came home from university for the summer, so the whole family was reunited for what was to be the last time.

In September, Hubble sailed for England and Queen's College, Oxford. He was 21, tall, good-looking, and outwardly confident. His letters home and observations by visitors and classmates, however, reveal that he lacked an inner compass. He affected a British accent, awkwardly and imperfectly, to the amusement of fellow Americans in his college, and began wearing a cape and carrying a cane. He wrote to his mother that he yearned to find some meaningful life's work: "I sometimes feel that there is within me, to do what the average man would not do, if only I find some principle, for whose sake I could leave everything else and devote my life."[7]

Adding to his struggle to find himself, Hubble had to keep some aspects of his life at Oxford secret. He attended church services only infrequently and often joined his classmates in drinking, despite an explicit promise to his father not to touch alcohol. And while he faithfully pursued studies of Roman and English law, he avoided mentioning to his parents that he had made friends with Herbert Hall Turner, Oxford's

Savilian Professor of Astronomy and director of the University Observatory.

By passing a preliminary exam early, in December of his first year at Oxford, Hubble was able to shorten the time needed to complete his jurisprudence degree from three years to two. By the fall of 1912, he had the freedom to attend lectures in whatever subject he wished. Curiously, he did not choose astronomy, but began a degree course in literature, then switched to Spanish.

A few months before the end of his scholarship years at Oxford, Hubble learned that his father had passed away. He made plans to return to his family, now living in Louisville, Kentucky, to support his mother. Perhaps he was thinking, too, that it was not too late to become an astronomer.

Graduate study in astronomy

John's illness had diminished the family income during the last months of his life. Henry, Hubble's "impractical" older brother, lived with his mother and brought home a modest income from his job as an insurance inspector. Bill still pursued his agriculture degree in Wisconsin. Thus, after John's death, the family faced straitened financial circumstances. The situation did not immediately improve with Hubble's return to the United States, for he proved to have less flair for managing money than he thought. He left his friends in England the impression that he had passed the Kentucky bar exam and was practicing law, while in fact he merely did some work translating legal documents.

At the start of the school year in the fall of 1913, Hubble took a job teaching Spanish, physics, and mathematics and coaching boys' basketball at a high school in Indiana, just across the Ohio River from his home in Louisville. He enjoyed unexpected success with the basketball team, taking its members to the state championships, where they took third place. He was also quite popular with the students, who found his English manner-isms intriguing—in contrast to his family members, who found them bizarre. But his heart was not in teaching, and former students recalled that he had difficulty making physics and mathematics accessible, and turned to reading his own astron-omy books at every opportunity.

In May 1914, Hubble contacted his former professor, Moulton, asking about financial assistance if he should enter graduate school in astronomy at the University of Chicago. His brother Bill was about to enlist in the army, and would thereafter support the family, forgoing his own dreams and independence. Hubble felt free to move away, and in fact his move back to Chicago marked the beginning of a gradual but eventually complete separation from his mother and siblings.

By the time Hubble made his inquiries about graduate study in Chicago, Hale was long gone from Yerkes Observatory and was monitoring Shapley's and van Maanen's work while carrying out his own solar research at Mount Wilson. Edwin Frost, to whom Hubble also wrote about opportunities for graduate study, had taken his place at Yerkes. On Moulton's recommendation, Frost promptly arranged for a tuition scholarship for Hubble.

Just before he began his graduate instruction at the University of Chicago—which consisted largely of research projects at Yerkes Observatory, rather than coursework in the city—Hubble had the good fortune to attend the August 1914 meeting of the American Astronomical Society. The meeting was held on the campus of Northwestern University in Evanston, where, as a high school student, Hubble had set a state record for high jump. There Hubble heard a landmark presentation by the astronomer Vesto Slipher on the amazing motions of some 40 spiral nebulae, all but a few of which he found to be receding from the Sun at speeds as high as 1100 kilometers per second. Slipher noted these radial motions of galaxies in the same way that Huggins had recorded the radial motions of stars, by looking for a shift to the left or right in the distinctive pattern of spectral lines in the astronomical object compared to the laboratory spectra (see chapter 6, figure 6.4). Slipher carried out his investigation at Lowell Observatory in Arizona, under the guidance of Percival Lowell of *Mars as the Abode of Life* fame. In part, Slipher's aim was to look for motions within nebulae, as opposed to motions of the nebulae themselves, that might support or refute elements of the Chamberlin–Moulton theory. Even after discovering the spirals' redshifts, Slipher did not give up on the idea that they might be single stars enveloped by nebulous matter. But other astronomers focused on a different aspect of his work: they welcomed his findings as strong evidence that the spirals, whatever they were, did not form part of the Milky Way system, and might in fact be island universes.

The unusual standing ovation Slipher received at the conclusion of his talk at the American Astronomical Society meeting surely helped direct Hubble's thoughts to the mysteries of the spiral nebulae as he considered possible topics for his doctoral dissertation. The spiral nebulae had never been resolved into stars, although some astronomers were aware of stellar spectra collected from the larger, nearer nebulae such as the Andromeda nebula. The velocities of the nebulae were unlike those of any star or cluster in the Milky Way system, implying they lay at a great distance: with such high speeds, any nebulae that were part of the Milky Way system would eventually find their way out. Beyond these facts, no one knew for certain what they were or how far away they might be. Any headway made in answering these questions would put the wise investigator in the company of such great astronomers as William Herschel and William Huggins. Here was fertile ground for Hubble to stake out.

As a lowly graduate student, Hubble could not hope to carry out a major observational program with the 40-inch telescope at Yerkes; he was granted access mostly in his capacity as a research assistant. For his own projects, he resourcefully commandeered a little-used but high quality 24-inch telescope, originally intended for solar work. He fitted it with a camera and began a photographic survey of faint nebulae.

Hubble could not have consulted with Frost about the results of his photographic work, for the unfortunate man was fast losing his eyesight to cataracts. But Hubble could and did turn to a senior astronomer on the staff for help, one of the greatest observational astronomers of his time. Edward Emerson Barnard, a fellow southerner from Tennessee, had grown up in poverty and received only two months of formal schooling. He had learned portrait photography as a trade and eventually found a way to combine this skill with his deep love of astronomy. By the time Hubble met him at Yerkes he counted the Gold Medal of the Royal Astronomical Society among his many awards and medals. He was widely respected for his recent photographs of the Milky Way, including some of "dark nebulae," pointing to the existence of obscuring clouds of material in interstellar space.

For his dissertation research Hubble canvassed the night skies with the 24-inch telescope, capturing the faint nebulae photographically with long exposures. Most of these nebulae were too faint to study spectroscopically, since obtaining a spectrum involves further weakening the light by dispersing or spreading it out through prisms. Information on physical conditions in the nebulae was not Hubble's primary goal, in any case. On a more basic level he wanted to try to characterize them by form and brightness, and to explore their apparent tendency to cluster together in space. In all, Hubble discovered more than 500 previously unknown nebulae of various kinds, the work of thousands of hours at the telescope and in the darkroom.

In the fall of 1916, as Hubble finished up required coursework on campus, he came to the attention of Walter Adams, Hale's right hand man at Mount Wilson and himself a former student of Frost. After a visit to the Chicago campus, Adams wrote to Hale that Hubble might make a fine addition to the staff at Mount Wilson, where the 100-inch telescope would soon be unveiled.

Hale followed up on Adams's suggestion, and in November 1916 offered Hubble the much-coveted position, pending the completion of his dissertation. But in April 1917, Hubble took a remarkable step, considering his life-long desire to become an astronomer and the allure of the world's largest telescope. He asked Hale to defer the start of his Mount Wilson tenure, a request that Hale granted. Hubble had decided to apply for a commission in the Officer's Reserve Corps. He would report for duty on 15 May.

Military service and a stay in Cambridge

From May to August 1917, Hubble followed a training course at Fort Sheridan, where he attained the rank of Captain in the infantry. He then served nearly a year of active duty at Camp Grant near Rockford, Illinois, training 25 officers who in turn trained 600 men, and rising to the rank of Major in January 1918. Not until September 1918 was Hubble sent to France on overseas duty.

Little objective information is available on his subsequent experience of war. It appears that Hubble may have exaggerated his combat experience, for the stories he later told do not mesh with the military record attached to his honorable discharge. Hubble claimed to have seen action and to have been knocked unconscious by a bursting shell. He awoke unattended in a military hospital, dressed himself, and left without communicating with anyone. But he also confessed to Frost, back at Yerkes, that he "barely got under fire."[8]

Perhaps the disappointment of missing combat after preparing for it for so long left Hubble reluctant to return immediately to the United States. He decided to serve a few months with the occupying forces on the continent. Then, during the summer of 1919, he lived in Cambridge, England, while supervising American army students in British universities.

During his stay in Cambridge Hubble might have crossed paths with the astronomer James Jeans, a thick-set man with a special talent for playing the organ. Jeans had just published a theory of the nebulae that would play a very big role in Hubble's future research. However, Hubble appears not to have digested Jeans's theory until several years later. He did take advantage of his residence in Cambridge to sit in on the lectures of Sir Arthur Eddington, who had elucidated the motions of stars in Kapteyn's "star streams," and to cultivate the friendship of Cambridge astronomer Hugh F Newall, a solar astronomer and friend of Hale. But it is difficult to understand how Hubble could have willingly delayed his return to the United States and the start of his position at Mount Wilson, when Shapley was rocking the astronomical community with his bold ideas about the size of the Milky Way galaxy and the off-center location of the Sun. Unless, perhaps, Hubble shied from the competition he would inevitably face once he joined the Mount Wilson staff as a junior member.

Early Mount Wilson years

On his way to Mount Wilson in late summer 1919, Hubble stopped for a visit at Lick Observatory atop Mount Hamilton, southeast of San Francisco. By the late 1910s, the small number of professional astronomers pursuing observations of the nebulae were concentrated at the Lick and Mount Wilson observatories in

California and at Lowell Observatory in Arizona, where Slipher carried on his radial velocity program. Curtis, the chief expert on spiral nebulae at Lick, remained there until shortly after the Great Debate with Shapley in 1920. If only it were Lick, and not Mount Wilson, with the superior telescopes, Hubble might have preferred to establish himself there. Both his scientific interest in the nebulae and his conservative approach would have fit nicely with the outlook of Lick astronomers—a cautious outlook Shapley referred to, somewhat derogatorily and unfairly, as the "Lick state of mind."

On 3 September, Hubble presented himself at the Mount Wilson offices, housed in a white neo-classical building on Santa Barbara Street in Pasadena. He joined a dedicated and hard-working staff. Shapley was not the only one rushing from data gathering to data analysis and back to more data gathering; astronomers who worked there at the same time recalled an intense level of activity and a memorable "spirit of research." Hale, the observatory's guiding spirit, still appeared at his Santa Barbara Street office, although he frequently traveled to Washington; it would be another four years before ill health forced him to stay away from the hub of activity he so much enjoyed.

Hubble lived in Pasadena boardinghouses and settled finally in a house he shared with a Caltech humanities professor and a seismologist employed by the Carnegie Institution, Mount Wilson's parent institution. It was the living on the mountaintop, however, that provided a pleasant sense of family and ritual. The "Monastery" building of Hubble's day housed up to a dozen astronomers in dormitory style rooms. At the other end of the building from the rooms stood a library with a fireplace and view over the San Gabriel valley, and a dining room, to which astronomers and their assistants were summoned by a bell.

Hubble's first opportunity to observe with the 100-inch telescope arrived in mid-October. The massive telescope, weighing more than 100 tons, was mounted under a sheet-metal dome 100 feet high and 95 feet in diameter. Electric motors controlled the opening and closing of the dome shutter or "slit," the rotation of the dome to allow the slit and telescope to view different parts of the sky, and the motion of the telescope itself as it compensated for the Earth's rotation.

Hale and the Mount Wilson astronomers had waited a long time for the installation of the 100-inch. The financial backer, John D Hooker, had made his original bequest in 1906. The Saint-Gobain glassworks in Paris had cast the glass disk to back the telescope's mirror in August 1907, and had shipped it across the Atlantic. After some uncertainty over the disk's usability, due to bubbles that had formed during annealing, work had begun on grinding and polishing it. Then the figured and silvered disk had arrived at the mountaintop for a first attempt at mounting in July 1917. The telescope finally saw "first light" more than ten years after Hooker's bequest, in November 1917.

In 1919, the year Hubble earned his first "run" with the 100-inch, janitor Milton Humason was promoted to night assistant thanks in part to Harlow Shapley's support. Humason, a down-to-earth but conservative character, quickly fell under Hubble's spell, addressing him from the first as "Major." Humason appreciated Hubble's care for the telescope and photographic equipment and his well thought out observing plan. Their working relationship got off to a strong start one night when clouds halted Hubble's observing and the astronomer, to Humason's surprise, asked if he could join Humason and members of the work crew in a game of cards.

In a letter to Barnard at Yerkes early in 1920, Hubble wrote that his plan was to learn all he could about the galactic nebulae — that is, the nebulous objects scattered among the stars and clusters of our own Milky Way system. To push his plan forward as quickly as possible, Hubble did not rely solely on his allocation of time at the 100-inch, but also used the 60-inch and a smaller telescope with a wide-angle lens, suitable for photographing large swaths of sky. But in the same year, Hubble also re-wrote his hastily-completed doctoral dissertation, at the insistence of Yerkes director Frost, and this activity prodded him to think as well about the non-galactic nebulae, those usually called spirals.

Hubble emphasized in the published version of his dissertation that the high velocities of the nebulae moving generally away from the Milky Way made it unlikely that these objects could in any sense pertain to the local system. He favored the hypothesis "that the spirals are stellar systems at distances to be measured often in millions of light years." However, he noted, "Extremely little is known of the nature of the nebulae, and no significant classification has yet been suggested; not even a precise definition has been formulated."[9] Thus both galactic and spiral nebulae filled his thoughts as he returned to civilian life and began formulating a research program.

Hubble was still enjoying the heady adventure of his first year with the world's largest telescope when, at the observatory site one evening, he met the woman who would become his wife. Grace, the sister-in-law of a visiting colleague from Lick Observatory, was married at that time, so both she and Hubble may have checked their enthusiasm at their first encounter. She later recalled feeling profoundly impressed by his good looks and serene detachment—a detachment which others viewed in a less positive light as aloofness.

Grace, petite and vivacious, was the same age as Hubble. She was born to a wealthy California businessman and his wife, the Burkes, and attended a private girls' school in Los Angeles. She graduated from Stanford University in 1912. Astronomers who knew her later described her seriously as an "intellectual giant."[10] She had concentrated her studies in English, and easily cultivated the friendship of prominent artists and writers.

From her diary entries and from remarks made by her friends, it appears that Hubble drew strength from her cool self-possession—a trait that he cultivated, but that she apparently expressed naturally. She, on the other hand, enjoyed his intellectual companionship and the opportunity to help manage his public image and career.

A year after Grace met Hubble, her husband, a geologist, died in an accident while attempting to retrieve a sample from a coal mine near Sacramento. The couple had had no children. Shortly thereafter, Hubble began a discreet courtship of Grace, away from the observatory, usually visiting her at her parents' home in Los Angeles.

Classifying the nebulae

In 1922, the International Astronomical Union (a member of the International Research Council, the association Kapteyn had worked to keep open to German scientists after World War I) appointed Hubble to its 14-member Commission on Nebulae.[11]

Members of the Union and its various Commissions were to meet in Rome, at the IAU's first general assembly. Hubble and fellow Americans Slipher, Curtis and Harvard's Solon Bailey looked forward to Slipher assuming the presidency of the Nebulae and Star Cluster commission following this meeting, replacing the French astronomer Guillaume Bigourdan.

Hubble had been preparing a classification system for the nebulae that he hoped the Commission would formally adopt, and he had explained it in a letter to Slipher in advance of the meeting. Slipher was favorably disposed to it, but in Rome four other Commission members, including president Bigourdan, produced systems of their own. None emerged a winner at the meeting. Undaunted, Hubble submitted for publication in the *Astrophysical Journal* a paper that included a description of his classification system. His assertiveness served him well, for the classification system turned out to be one of the four scientific achievements that he is best known for.

The paper, which was submitted in May 1922, the same month as the meeting, bears the title "A General Study of the Diffuse Galactic Nebulae."[12] The fact that it contains a general classification system as well as a treatise on galactic nebulae, as heralded by the title, suggests that Hubble inserted his broader classification system into a separately conceived paper as soon as he realized that the IAU deliberations were mired in academic debates.

Despite its odd mix of subject matter, the paper is a good example of Hubble's scientific and literary style. Drawing on his knowledge of the history of astronomy to provide some context for the non-specialist reader, he traced the evolution of nebular classifications, beginning with that of William Herschel. He then clearly articulated the advantages of his own system. In the second, larger section of the paper, he methodically laid out the evidence in an investigation of the physical nature of nebulae within the Milky Way galaxy. Important conclusions followed, seemingly from purely inductive reasoning. His straightforward, self-assured tone gives an authoritative ring to his analysis.

Hubble noted that William Herschel's "penetrating genius" had perceived a distinction between the intrinsically nebulous nebulae—the planetary and diffuse nebulae—and the stellar

311

nebulae, those that only appeared nebulous because even the finest telescopes could not resolve them into stars. In the years that followed, Hubble reminded his readers, Lord Rosse's great telescopes had apparently resolved "nebula after nebula," so that astronomers of the next generation abandoned the concept of a nebulous fluid. John Herschel, William's son, resorted to classifying the nebulae according to the difficulty of resolving them. The introduction of spectroscopy and photography in the mid-nineteenth century had restored William Herschel's idea of nebulous fluid in the form of gas or dust, at least as applied to some of the nebulae. Astronomers then no longer viewed all nebulae as either truly nebulous or as unresolved stellar clusters: there might be some of each kind. In particular, the nebulae lying in the zone of avoidance—that is, regions of the sky far from the dense groupings of stars in the Milky Way—could be suspected of being unresolved stellar clusters, or what Hubble called non-galactic nebulae.

The galactic nebulae, those belonging to the Milky Way system, could now be divided into two groups, Hubble suggested. The so-called planetary nebulae, such as the "nebulous star" in Taurus studied by Herschel and the Cat's Eye nebula studied spectroscopically by Huggins, clearly consist of dense nebulous material closely surrounding a star. Indeed, as we know now, the planetary nebulae are formed from a shell of stellar atmosphere ejected from the central star in its dying phase. The second group consists of diffuse, more or less widespread nebulosity, which might be either luminous or dark. The horsehead nebula in Orion (chapter 2, figure 2.9), which Hubble's friend Barnard had recognized and described for the first time in the 1910s, is perhaps the best-known example of a dark nebula. Barnard described one such dark nebula as "a drop of ink on the luminous sky."[13] The luminous diffuse nebulae include such obvious examples as the milky haze around the Pleiades asterism, visible through a telescope of at least 6 inches, and the Great Nebula in Orion (chapter 2, figures 2.7 and 2.2).

The most interesting of the galactic nebulae, from Hubble's point of view, were the luminous diffuse examples. Since the eighteenth century at least, astronomers had wondered what made them glow. Was the "active agency"—the source of light—outside the nebulae, or did they shine under their own power? From a survey of the spectra of some 60 diffuse nebulae, a distinction emerged, Hubble noted. Nebulae associated with very hot stars showed emission lines corresponding to elements in the nebular material, while nebulae associated with cooler stars, or the most distant parts of nebulae surrounding hot stars, showed absorption lines. The latter category of diffuse nebulae might be called reflection nebulae, since the light we see comes from the stars and reflects off dust or molecules. The emission nebulae were more difficult to understand in the early 1920s. Hubble demonstrated a sound intuition when he said of them, "It seems more reasonable to place the active agency in the relatively dense and exceedingly hot star than in the nebulosity, and this leads to the suggestion that the nebulosity is made luminous by radiation of some sort from stars in certain physical states. The necessary conditions are confined to certain ranges in stellar spectral type and hence are possibly phenomena of effective temperature."[14] Hubble's guess was on target. Emission lines in nebulae arise from ultra-violet illumination by a hot star nearby, and subsequent re-radiation by the gas at longer wavelengths.

Summarizing his arguments relating to the galactic nebulae, Hubble maintained that their source of luminosity is the radiation from their associated stars. "The nebulosity has no intrinsic luminosity," he wrote, "but either is excited to emission by light from a star of earlier [hotter] type or merely reflects light from a star of later [cooler] type."[15] At last, the mystery of the galactic nebulae, at least, had yielded to the powers of the telescope, camera and spectroscope. Further study of the galactic nebulae, Hubble remarked dismissively, could "well be left to special investigators."[16] Indeed, this paper of 1922 would be Hubble's last one to focus so deeply on the galactic nebulae. Having made some order among the planetary, luminous diffuse and dark diffuse nebulae, Hubble intended in future work to focus his investigative powers on the more elusive non-galactic spirals.

The physical nature of the fainter, non-galactic nebulae posed much more of a challenge than the relatively bright galactic nebulae. No doubt the unsatisfactory state of theories concerning the spiral nebulae, which had been underscored by the confusion of the Curtis–Shapley debate two years earlier, attracted Hubble to the problem. The classification system that he appended to the 1922 paper on galactic nebulae began a fresh attack on the problem.

Hubble recognized the non-galactic nebulae from their location in the sky and their lack of association with individual stars of our own system. These nebulae, in contrast to the planetary and diffuse galactic nebulae, avoided the concentration of stars in the Milky Way. Extremely cautious about drawing inferences, Hubble refused even to refer to the non-galactic nebulae as external systems—despite the fact that he had done so in his doctoral dissertation—or to be drawn into the island universe controversy. He wrote, "There appears to be a fundamental distinction between galactic and non-galactic nebulae. This does not mean that the latter class must be considered as "outside" our galaxy, but that its members tend to avoid the galactic plane and to concentrate in high galactic latitudes [i.e. far above or below the plane of our galaxy]."[17]

The only term for non-galactic nebulae in use up to that point in the twentieth century, as Hubble emphasized, was "spiral." The accepted wisdom seemed to be that the faint, fuzzy non-galactic nebulae of indeterminate shape were simply too distant for astronomers to detect spiral structure in them. The spiral structure was evident in a number of well-known cases, such as M51, which Lord Rosse had drawn by hand, or the Andromeda nebula. But Hubble, who had by then examined long-exposure photographs of thousands of nebulae, knew that while many faint and small non-galactic nebulae were unmistakably spirals, some of the larger and presumably nearer non-galactic nebulae showed no traces of spiral arms.

Hubble proposed a classification system in which the non-galactic nebulae ranged from the "globulars" to spirals in form. His "globulars" included globular clusters, which we now view as appendages to the Milky Way and other galaxies, as well as what we now know are spherically shaped galaxies. The gap between the two well-defined categories of globular and spiral was filled with elongated nebulae, which Hubble called "spindles" and "ovates." He surmised that the spindles, with their bulbous middles and long tapered ends, were spiral galaxies seen edge-on. The ovates, or egg-shaped, presumably took on the shape of globulars when seen from one side (see figure 9.2).

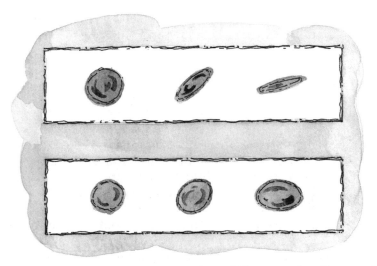

Figure 9.2 Galactic shapes as seen from different angles. Top panel: views of a disk galaxy seen face on (left), edge-on (right) and from an intermediate angle (middle view). Bottom panel: views of an ellipsoidal or egg-shaped galaxy from three different perspectives. (Credit: Layne Lundström.)

He would subsequently revise and enlarge on this system of 1922, but the hastily published outline marked the beginning of a career devoted to elucidating the nature of the enigmatic non-galactic nebulae and trying to account for these different forms.

Jeans' nebular theory

Sometime between May 1922 and April 1923, after his return from the IAU meeting in Rome and the publication of his classification system, Hubble must have studied the nebular theory developed earlier by Jeans at Cambridge, for the theory began to influence his research.

Jeans began with the premise at the heart of Laplace's nebular hypothesis, that the solar system originated out of a rotating nebulous mass of gas, but he sought to apply this premise to the origin of the spiral nebulae, which he believed were island universes like our own galaxy.

315

The nebulous mass in Jeans's theory—which, he admitted, came into existence "in an entirely unknown way"—contracted under the influence of gravity.[18] Initially it assumed a spherical form, but distortions from nearby masses or nebulae and further contraction conspired to flatten the nebula somewhat. Eventually it would take on a "lenticular" or lens-like form, bulging in the middle and tapering at the edges. Jeans compared this theoretical form to photographs of certain extra-galactic nebulae seen edge-on.

Further contraction of the nebula would lead to matter being thrown off its equatorial rim, "much in the same way in which water would gradually drip over the edge of a slowly shrinking cup," Jeans explained.[19] The ejected matter would form a chain of small nebular "satellites," which would cool and condense into clusters of stars. This chain of condensing material could be compared to the arms of spirals, Jeans suggested. The process of the central nebular mass ejecting matter would continue, with material streaming outward along the arms. Overall, the nebula would appear more distinct at the edges, where clusters of stars would form first, and more nebulous in the middle.

At least one other scientist besides Hubble, a geophysicist and mathematician at Cambridge, took note of Jeans' theory. In 1923, Harold Jeffreys suggested, though rather vaguely, that the spiral nebulae might represent the end point of Jeans' nebular evolution, while the "lenticular" nebulae, those Hubble called spindles, represented an earlier stage.

Whether Hubble learned the details of Jeans' theory from such published opinions or from Jeans himself, he wasted no time incorporating the theory into his scientific agenda. In April 1923, Hubble mentioned it explicitly in a letter to Slipher at Lowell Observatory. He wrote, "I have been trying to construct a classification of non-galactic nebulae analogous to Jeans' evolution sequence but from purely observational material. The basis is a distinction between amorphous nebulosity and the granular beaded arms of spirals." He noted that the amount of material in the various "amorphous" and "spiral" nebulae he had looked at seemed to be roughly equal, judging from their luminosities. Thus it was "quite possible to conceive of them as representing different stages of an evolutionary sequence."[20]

In accordance with this effort to relate the non-galactic nebulae to Jeans' theoretical forms, Hubble directed his attention to some of the more prominent non-galactic nebulae, trying to see individual stars, which had never been resolved in spiral systems, in their outskirts. One such nebula is the elliptical galaxy M87 in the constellation Virgo, a conspicuous member of a vast cluster of galaxies. Between 1920 and 1923, Hubble collected images of this nebula with the 100-inch telescope. These images, far better than any that could have been taken before, tantalized him. Extending in a line from the nucleus of the nebula Hubble saw "a remarkable chain of nebulous objects," a series of five small round nebulae. He described them as "almost stellar condensations." He also noted—again, in conformity with Jeans' theory, although he did not mention Jeans in his 1923 report on M87—that faint stars seemed to "cluster about the outskirts of the nebula," just outside the central regions of amorphous nebulosity.[21] Here might be evidence of "condensations" at the outer limits of the nebula, as predicted by Jeans. Then again, the bright points might be unresolved clusters of stars rather than single stars embedded in some remnant nebulosity, so Hubble expressed himself cautiously. He was right to be cautious—those bright points were, in fact, globular clusters attendant upon the M87 galaxy. The situation tugged at his most basic tendencies: his ambition to discover observational clues to the physical nature of the nebulae, and his desire to investigate the phenomena without theoretical bias and hence without the possibility of being led astray if the theory proved incorrect.

The M87 question was still unsettled that summer when he obtained a different sort of clue. Comparing photographic plates of the irregularly shaped nebula known as NGC 6822—a cloudy blob in the constellation Sagittarius, discovered by his friend Barnard in 1884—Hubble thought he saw brightness changes associated with variable stars. If real, the light variations would not only signal the presence of individual stars, but might also afford an estimate of the distance to the nebula, based on a comparison of similar variable stars at known distances.

In July, Hubble dashed off a letter to Shapley, with whom he had maintained a civil if not warm relationship after the latter became director of the Harvard College Observatory. Hubble wanted to know if Shapley could provide comparison plates from an earlier epoch, from Harvard's archives. Shapley could, and published his own analysis of the plates, suggesting that

NGC 6822 lay more than a million light-years away. Here, for any who paid attention, was a hint that the barrier of inter-galactic distances would soon be breached.

Breakthrough

Three months later, in October 1923, Hubble was examining photographs of the outer sections of the Andromeda nebula—a true spiral—searching for novae. Several novae were already known, but a larger sample might shed light on the question of which novae were comparable to those in the Milky Way: the 1885 example of S Andromedae, which had become almost as bright as the entire Andromeda nebula itself, or the more numerous, less spectacular cases that Curtis had pointed to in the "Great Debate." Once that uncertainty about the novae was resolved, they could be used as distance indicators using the "faintness means farness" principle.

Three star-like points of light on the photograph caught Hubble's eye, and he marked them with a pen and the letter N. But further examination of plates of the same area in Mount Wilson's own archives caused him to return to one of those markings with great excitement. His "nova" had dimmed and brightened again, proving it was no nova, but a variable star. The photograph bears witness to his state of mind: he crossed out the "N" and wrote, "VAR!" for variable.

Some painstaking work lay ahead, as Hubble estimated the brightness of the variable as it appeared on archived and new plates, and plotted the results as a function of the date on which the exposure was taken. As the variable's lightcurve unfolded, Hubble must have felt his excitement mount. The curve followed the standard pattern of Cepheid variables, with a relatively fast rise to maximum and a slower return to the dimmest point. Over the three-night period of 5, 6, and 7 February 1924, Hubble caught one of the Cepheids in the act of brightening by more than a magnitude, firmly establishing the most critical phase of the lightcurve and confirming beyond doubt that the variable was a Cepheid type.

On February 19, Hubble wrote to Shapley. "You will be interested to hear that I have found a Cepheid variable in the

Andromeda nebula (M31)," he announced in the much-quoted letter.[22] We know from Cecilia Payne-Gaposhkin's memoir that Shapley realized immediately that his conception of the universe had suffered a blow. The detection of individual stars of a familiar type in the Andromeda nebula refuted at once any idea of the spirals as truly nebulous, unresolvable entities. Furthermore, the period–luminosity relationship and Hubble's estimates of the Cepheid's apparent magnitudes put the nebula at a distance of some 825 000 light-years, well beyond the Milky Way's most remote globular clusters.

Shapley would now have to accept that the spirals were large systems of stars, perhaps even comparable to the Milky Way. In his response to Hubble, however, Shapley merely cautioned him about a number of technical difficulties that might have led him astray in his analysis of the data. But Shapley's cautionary note may have been just the reaction Hubble sought to elicit as he considered whether or not to publish his findings. Van Maanen's results on the rotation of M101 still commanded respect for the view of spirals as nearby objects. If Shapley had given up on van Maanen's rotations upon seeing Hubble's data, Hubble might, one can guess, have considered publishing immediately. But in 1923 he decided to wait, and gather more evidence.

As Shapley digested the news, Hubble and Grace were preparing for their wedding. They were married 26 February 1924 by the Burke family priest, then repaired to a cottage her parents owned on the northern California coast near San Francisco. Hubble's relatives were not part of the picture. He told his in-laws almost nothing of his background, and never introduced Grace to his mother or siblings. One of Hubble's sisters evidently reconciled herself to his abandonment of the family; "great men have to go their own way," she told an historian in a 1971 interview.[23] And while Grace's father once pumped an old friend of Hubble, whom he chanced to meet, for information on Hubble's youth, Grace was determined that only Hubble's own version of his life should be known outside the family.

The Hubbles's honeymoon in Europe—in part, an extended paid vacation for Edwin—was Grace's first visit to the continent. The couple visited the tombs of some of Hubble's heroes: Isaac Newton's and John Herschel's in Westminster Abbey, and

Galileo's in Italy. The trip held all the more enchantment because Hubble's as yet unpublished news of Cepheids in the Andromeda nebula brought him unexpected fame, at least among astronomers in the know. The Hubbles dined with prominent scientists at Oxford and Cambridge, and the Royal Astronomical Society hosted a dinner for Hubble, after which he discussed his recent work on the nebulae.

Back in Pasadena in May, the couple settled into an apartment near the Caltech campus while Hubble resumed his observing and data analysis. The original motivation behind his search for novae and distinct individual stars in the outer regions of nebulae was his classification scheme for the non-galactic nebulae. He made quick progress refining this classification scheme. He may have been spurred by conversations with Jeans, who visited Mount Wilson in the summer of 1924.

In late July, Hubble wrote again to Slipher. As president of the Commission on Nebulae (now renamed the Commission on Nebulae and Star Clusters), Slipher collected proposals for ideas to be discussed at the next IAU general assembly, to be held in July 1925 in Cambridge, England.

This time Hubble's plan called for nebulae to be divided into spirals and "ellipticals" rather than "amorphous" types, with the ellipticals further subdivided according to their degree of flattening. The spirals could be arranged in a sequence, too, Hubble suggested. At one end of the sequence the spirals had large, bright central cores and an envelope of amorphous nebulosity, while at the other end of the sequence the central region was smaller and fainter, and was surrounded not by nebulosity but by distinct spiral arms.

Hubble clearly implied that the nebulae evolved in time from one end of the sequence to the other. He wrote, "The gap between the two extreme forms is well filled [by examples of nebulae seen on photographic plates] and a series is readily constructed in which the granular spiral arms seemingly grow at the expense of the amorphous region, unwinding as they grow."[24] Hubble called the nebulae with large amorphous centers "early" types, and those with distinct arms reaching all the way to the center "late" types.

Slipher distributed copies of Hubble's letter to other members of the Commission. Shapley, who was privy to the exchanges of information with Slipher even though he was not added to the Commission until just before the meeting, needled Hubble about his classification nomenclature. "Non-galactic nebulae" seemed too neutral a term, considering that Hubble himself had shown these objects to be stellar systems comparable in nature, if not in size, with the Milky Way. Shapley suggested to both Hubble and Slipher that they be called "galaxies." Hubble agreed to drop the term "non-galactic" but insisted instead on calling them "extra-galactic" nebulae. It was not until Hubble's death that the term "galaxies" came to dominate usage.

Harvard's Solon Bailey had a more substantive and, in hindsight, entirely appropriate criticism of Hubble's scheme. Though Hubble always stressed that he had developed his classification scheme from purely observational considerations, Bailey noted that the terms "early," "middle" and "late" clearly pre-supposed an evolutionary order that was still speculative. This basic criticism was to resurface at the 1925 meeting, which Hubble did not attend.

In the meantime Hubble, guided by Jeans' theory, was forging ahead with his attempt to isolate individual stars in the outer edges of some of the more prominent nebulae. By August 1924, Hubble had amassed data on so many new variable stars in NGC 6822, the spiral galaxy M33 and the Andromeda nebula that he felt Shapley must face up to a new conception of the nebulae. "The straws are all pointing in one direction, and it will do no harm to begin considering the various possibilities involved," he wrote. Shapley could find nothing to quibble with this time. He wrote back, "I do not know whether I am sorry or glad to see this break in the nebular problem. Perhaps both. Sorry because of the significance for the measured angular rotation, and glad to have something definite and interesting come to hand."[25]

Hubble of course informed Jeans, also, of his discovery of Cepheids in extra-galactic nebulae, and Jeans passed the word to Henry Norris Russell, Shapley's former advisor at Princeton. Such hot news — word of a discovery that effectively settled the Great Debate argument over the nature of the spiral nebulae — could not be kept under wraps for long. The *New York Times* even carried an article on page 6 of its edition of 23 November 1924, reporting that "the results are striking in their confirmation of the view that these spiral nebulae are distant stellar systems."[26] But Hubble shrank from publishing his findings because of the "flat contradiction," as he put it, with van Maanen's rotations.[27]

Ironically, Jeans, who had formerly liked van Maanen's measurements of rotation in spirals because they seemed to vindicate his theory of nebulae beginning as rotating masses of gas, now lent the great weight of his support to Hubble, who had shown that the nebulae could not be as close as van Maanen implied. Hubble had not demonstrated that the spirals rotate, but his identification of individual stars at the outer edges of spirals accorded very well with another aspect of Jeans's theory. "Van Maanen's measurements have to go," Jeans wrote to Russell, voicing what Hubble could only hope would become a common sentiment.[28]

A few months later it was Russell who formally made Hubble's news public. The American Astronomical Society and the American Association for the Advancement of Science held a joint meeting beginning in late December 1924, in Washington, DC. Hubble did not attend, but he mailed his manuscript, "Cepheids in Spiral Nebulae," for Russell to present on 1 January 1925. The wider astronomical community learned then that Hubble had found a dozen Cepheids in the Andromeda nebula and almost twice as many in the nearby spiral M33. Assuming, as usual, that interstellar matter did not dim the stars, and assuming "uniformity of nature" so that the Cepheid period–luminosity relation held for the newly found Cepheids, the variables indicated that these two spirals lay at a distance of about one million light-years. Hubble's supporters, particularly Henry Norris Russell, were jubilant about the "unequivocal" nature of his findings for the island universe hypothesis.[29]

Even astronomers who had favored Shapley's "big galaxy" arguments in the Great Debate, and who therefore would have doubted that the spirals lay outside the Milky Way, were obliged to take Hubble's findings seriously, since he was applying Shapley's own Cepheid period–luminosity relation. Some astronomers remained confused by the contradiction implied by van Maanen's rotations, but by and large the community regarded Hubble's work as a breakthrough by "a young man of conspicuous and recognized ability."[30] Hubble had made his mark in the most dramatic way one could imagine, pushing back the limits of the known universe and putting the centuries-old notion of island universes on a scientific basis.

A return to classification

In other papers published in the 1920s, Hubble solidified the case for extra-galactic nebulae as stellar systems comparable to, or somewhat smaller than, the Milky Way, lying at distances measured in millions of light-years. He enlarged the number of known Cepheids and novae in these remote systems and, as the data accumulated, he repeatedly pointed to what he called the principle of the uniformity of nature. The period–luminosity relation for Cepheids "functioned normally" in the ever more distant nebulae that he studied, and the distances it gave were consistent with those he derived from other criteria, such as the luminosity of the brightest stars in any given system. The principle of the uniformity of nature could therefore be used as a guide in "extrapolations beyond the limits of known and observable data," Hubble suggested, and speculations based on such assumptions as the universal validity of the period–luminosity relation were legitimate unless they led to self-contradictory conclusions.[31] Using the principle of the uniformity of nature, Hubble planned to explore ever-deeper regions of space, using the better-studied, nearer nebulae as stepping-stones to more distant worlds.

The 1920s saw the Hubbles put down roots in San Marino, near Pasadena. Grace's well-to-do parents paid for the construction of a house near the magnificent estate of Henry Huntington, now known as the Huntington Library, Art Collections and Botanic Gardens. The Hubbles hired an architect to design a house reminiscent, in its proportions and plan, of apartments in the Palazzo Vecchio in Florence. Hubble's newly increased salary made it possible for the couple to choose furnishings for their house on their trips to Europe.

The Hubbles' lives were not as easy as might appear, however. As their closest friends and neighbors knew, they tried to start a family when Grace was in her late thirties. Sadly, she delivered a premature, stillborn baby. The couple remained childless thereafter.

Hubble was not successful at getting the IAU to endorse a revised classification system for extra-galactic nebulae when the Commission met in Cambridge, England, in 1925. In 1926, he went ahead and published it anyway, as a paper entitled "Extra-Galactic Nebulae." Along with the paper read by Russell at the 1924/25 meeting in Washington, announcing the distance to the Andromeda nebula, this is one of Hubble's most famous publications.

Although many of Hubble's peers viewed with suspicion his identification of galaxy types with stages in an evolution described by Jeans, his categorizing of galaxies—his recognition of salient features—has stood the test of time. As before, he divided the multitude of observed forms into ellipticals— formerly called the "amorphous" types—and spirals. A small number of nebulae, lacking any symmetry, fell into the category of irregulars (see figure 9.3).

The ellipticals, which he designated by the letter E, appeared completely nebulous. They range from perfectly round forms, which he labeled E0, to flattened ovals with a long axis three or four times the short axis, which he designated E7. The spirals or S forms consisted of nebulous nuclei (bulges, in today's nomenclature) surrounded by spiral arms in a flat disk. These too he subdivided according to distinct features. The "early" spirals, Sa types, have a large nucleus or bulge, and nebulous arms tightly coiled about it. "Middle" or Sb spirals have a smaller bulge and somewhat more loosely wound, clumpy arms. "Late" or Sc type spirals have a small bulge and uncoiled arms that appear grainy rather than nebulous.

Hubble also distinguished the spiral-barred or SB types, which resemble their normal spiral counterparts, except that the arms appear to emerge not from the bulge but from the ends of a bar across the central region. These are similarly labeled SBa, SBb, and SBc according to the size of the bulge and the appearance of the arms. The irregular galaxies often appear grainy (stellar rather than nebular) but they lack distinct arms or rotational symmetry.

Hubble emphasized that the two types of regular nebula, elliptical and spiral, formed a sequence, with ellipticals merging into spirals. Furthermore, he conceived of the sequence as representing evolution in time as well as structure. He could

324

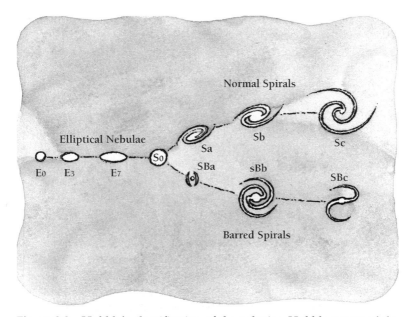

Figure 9.3 Hubble's classification of the galaxies. Hubble arranged the different types of nebulae he distinguished in a sequence, which he believed corresponded to an evolutionary sequence. At the left are the elliptical nebulae, more or less egg-shaped, without spiral structure. The more round galaxies he called E0, and the more "squashed" E7. At the right are the normal spirals (top branch) and the barred spirals, in which the spiral arms emerge not from the center of the galaxy but from a prominent bar running through the center. The spiral (S) and spiral-barred (SB) galaxies are designated also by the letters a, b or c, according to various morphological features such as the degree to which the arms appear "unwound." (Credit: Layne Lundström.)

find no examples of nebulae exactly at the junction of the two segments of the sequence: arms appeared fully formed in the spirals, or not at all in the ellipticals. This he took to imply that the progression from structureless ellipticals to spirals with arms occurred rapidly. He painted a vivid tableau of the majestic blossoming of galaxies: the flattened blobs of the E7 type of elliptical nebula gave way to systems with "a large nuclear region similar to E7, around which are closely coiled arms of unresolved nebulosity. Then follow objects in which the arms appear to build up at the expense of the nuclear regions and unwind as they grow; in the end, the arms are wide open and the nuclei inconspicuous."[32]

His description conveys a sense of Darwinian or biological evolution from the simple form of the ellipticals to the more complex form of the spirals. Even when he spoke in physical rather than biological terms, he underscored the unity of the extra-galactic nebulae. He noted that "the entire series can be represented by the various configurations of an original globular mass, expanding equatorially."[33] Jeans had also borrowed from the biological sciences. In *Astronomy and Cosmogony*, published in 1928, he wrote that the extra-galactic nebulae could be viewed as "astronomical plants belonging to the same species."[34]

Hubble again insisted that he had developed the classification system purely on the basis of observational criteria. However, he drew his readers' attention repeatedly to the close agreement between Jeans' theory and the characteristics he observed, such as the resolution of the outer regions of the spirals into stars, while the central region remained nebulous. This attempt to connect theory and observation put off some of Hubble's fellow astronomers.

Hubble's peers were right to be concerned, particularly about the direction of the proposed evolution from simple to complex. Indeed, in 1927 the Swedish astronomer Bertil Lindblad challenged that aspect of Jeans' theory on at least two counts. Lindblad did not agree that the bulges of spirals were truly nebulous, devoid of stars, and secondly he argued that flattened or disk forms would, over time, acquire a more globular shape — just the opposite of what Jeans proposed.

Furthermore, we understand now that the present appearance of galaxies is due to both "inherited" characteristics, such as the density of the cloud from which they formed, and environmental characteristics, such as the number of encounters they have had with other galaxies. Neither Hubble's nor Jeans' scheme takes galaxy interactions into account, but the idea was around as early as the 1920s. Another Swedish astronomer, Knut Lundmark, who had spent two years at Lick and Mount Wilson in the early 1920s, conducted a statistical study of galaxies. The study convinced him that galaxy encounters with their nearest neighbors in space are important factors in their evolution.

However, despite the problems with its theoretical under-pinnings, Hubble's classification scheme was, and still remains, a useful one. The morphological features that Hubble based it on do relate to the age and evolutionary history of galaxies. The scheme is appropriate for objects Hubble could observe with the 100-inch telescope: galaxies some 10–14 billion years after their formation.

The Milky Way and the Andromeda nebula: cosmic twins

A monumental paper Hubble finished in December 1928 con-solidated much of what the 40-year-old astronomer had learned since joining the Mount Wilson staff. At this point he had made two of the four major advances he is known for: he had clarified the relationship between the Milky Way galaxy and the spiral nebulae with his discovery of Cepheids in several extra-galactic nebulae, and he had laid the groundwork for theories of the nature and evolution of galaxies with his classification system. He drew on all his previous research in this landmark paper of 1928. A close examination of about 350 photographs taken with the 60- and 100-inch telescopes over a period of 18 years informed a study of the Andromeda nebula, the best known of the extra-galactic nebulae and a source of wonder for at least ten centuries.

The Andromeda nebula is some one million light-years dis-tant, Hubble found. (He was off by a factor of two—current measures put it at about two million light-years.) The outer regions of the spiral arms consist of "swarms of faint stars," while the central region of nebulosity had not yielded to the resolving power even of the 100-inch.[35] Hubble judged the Andromeda nebula to be somewhat smaller than our galaxy, while in fact it is larger; the error, like the error in the distance to the nebula, arose from his use of Cepheid variables as distance indicators, and the unsuspected difference in absolute magnitude between true Cepheid variables and similar W Virginis type stars (see chapter 8).

Hubble had discovered 63 new novae in the Andromeda nebula, adding to 22 already known. These all exhibited characteristics similar to galactic novae. Hubble concluded that

S Andromedae, the bright 1885 "nova" that had caused so much confusion in the great debate between Shapley and Curtis, was "a rare and peculiar type."[36] He was correct. We now call this type a supernova. In a supernova explosion, the last stage in the life of a massive star, it is not just a layer of the star's atmosphere that is blown off, as in a planetary nebula or in a nova explosion. The entire star explodes and a violent shock wave squeezes the light and energy out of the stellar material.

In a concluding section of the paper Hubble compared the Andromeda nebula to our galactic system. He drew on recent studies of the Milky Way system by Frederick Seares, Shapley's teacher at Missouri and a Mount Wilson astronomer. Hubble noted that the number of stars per unit volume of space is about the same in the Andromeda nebula's spiral arms as in the neighborhood of the Sun. The Milky Way clearly has an extensive disk surrounding the bulge or nuclear region; Seares's research suggested it is a "very late type spiral" in Hubble's classification, with loosely coiled spiral arms.

At last, Hubble's paper indicated, astronomers had found a way to hold a mirror to our own stellar system. We may never see our galactic home from the outside, but solid evidence, including that which Hubble presented in this paper, supports the view that the Milky Way is a majestic spiral not unlike the great nebula in Andromeda.

Hubble's expanding universe

The Dutch city of Leiden played host to the IAU's third general assembly in July 1928, a meeting at which Hubble presided over the Commission on Nebulae and Star Clusters. The city, nestled along the Old Rhine River not far from The Hague, flourished the sixteenth century as a textile center, and was home to the painter Rembrandt in the seventeenth. Its university was founded in 1575, and the university observatory in 1638. A more charming and historic setting could hardly be found for a discussion of current questions pertaining to the extra-galactic nebulae. For Hubble, it would also be the setting for personal contacts that would prompt an important shift in his research focus.

At the time of the IAU meeting, Willem de Sitter, one of Kapteyn's former students, directed the university observatory. Mild-mannered and dignified like his teacher, de Sitter was then in his mid-50s. Ejnar Hertzsprung, Kapteyn's erstwhile son-in-law and the promoter of Antonia Maury's "c" classification for giant stars, acted as adjunct director. Jan Hendrik Oort, another of Kapteyn's students, stood next in line to take over the reins at the observatory. He had come to Leiden in 1924, and had already made a name for himself as an astronomer of wide-ranging accomplishments.

De Sitter had played a key role in the dissemination and elaboration of Einstein's ideas on relativity. A mathematician before Kapteyn converted him to astronomy, he was one of a small number of astronomers who could follow Einstein's work and ponder the implications of relativity for observers. During World War I he served as a vital link between Einstein in Berlin and the scientific community in the allied countries. Einstein sent his papers to de Sitter—Holland was a neutral country—and de Sitter forwarded the papers to Eddington, secretary of the Royal Astronomical Society. Both Eddington and de Sitter helped "popularize" Einstein's work for the astronomical community.

It was not easy for astronomers to grasp the ramifications and potential applications of Einstein's theory. The general theory of relativity describes the properties of matter, gravitational forces, and radiation. We can think of the so-called field equations of general relativity as the "laws" that matter and radiation must obey, while the many possible solutions to the field equations represent the actual behavior of the universe. Similarly, the speed limit and other traffic laws specify how cars may move through a city, but many "solutions" are available to law-abiding drivers searching to go from point A to point B, including variations in route and speed. The challenge posed by Einstein's equations was to find solutions that represent the real universe.

Einstein's theory survived a big test in 1919. His theory predicted that the gravitational field of a massive object such as a star—that is, the effect of gravity in the space immediately surrounding the star—would attract light. A ray of light passing close to a star would be bent, deviated from its straight-line path. An eclipse of the Sun in 1919 had provided an opportunity to check Einstein's prediction. Eddington was one of several astronomers who led eclipse expeditions to test Einstein's theory, and his results were the best-known verifications of the light-bending effect.

Einstein discovered possible solutions to his own equations. His first solution described a static universe, neither expanding nor contracting, in existence for ever. Such a solution seemed attractive and reasonable at the time—who thought of the universe as expanding or contracting?—but it was a bit of a fudge. Einstein's field equations actually pointed to an expanding universe. To prevent such seemingly aberrant behavior in the solution, Einstein inserted a factor, the cosmological constant, into his field equations. Mathematically, the cosmological constant prevented him from obtaining the unwanted expanding-universe solution.

De Sitter also found a solution to Einstein's field equations. The universe corresponding to his solution was static and, bizarrely, devoid of matter or radiation, except for hypothetical test particles. The only way he could argue for the correspondence of his solution with reality was to point to the fact that space is mostly empty and that the formal mathematical solution was only approximately like the real universe. But a feature of de Sitter's solution that was to make it very interesting to astronomers, despite its unphysical lack of matter and radiation, is that it allowed for redshifting of light from moving test particles in the otherwise empty universe.

Einstein and de Sitter formulated their unsatisfactory solutions in the 1910s. During the 1920s, discussions of the general nature of the universe hit a kind of stalemate. The problems seemed nearly intractable, and other areas of physics drew theorists' attention. Furthermore, few astronomers were in a position to relate the theory of relativity to observations of the stars and galaxies.[37]

Against this backdrop Hubble joined discussions of the so-called de Sitter universe at the IAU meeting in Leiden. According to a diary Grace kept, de Sitter encouraged Hubble to continue Slipher's measurements of the velocities of extra-galactic nebulae. Certainly de Sitter would have been interested in the redshifts of galaxies, which might correspond to redshifts in his solution to Einstein's equations. At any rate, Hubble returned to the

United States full of enthusiasm for a new observing program focused on the redshifts and distances of the nebulae.

Slipher had, in fact, exhausted the capabilities of his instruments in about 1926, after amassing radial velocity measurements for dozens of galaxies. A few, such as the Andromeda nebula, showed small blueshifts, indicating that their complex motion within the local group of galaxies was bringing them closer. Most of the spectra, however, came from more distant systems and were redshifted, indicating the galaxies or nebulae were receding. The highest velocity of recession Slipher found was 1800 kilometers per second.[38]

Hubble and night assistant Milton Humason joined forces after Hubble's return to Mount Wilson. As Humason later remembered it, Hubble already anticipated, from his discussions in Leiden, that he would find a relationship between a galaxy's distance and its velocity of recession.[39] With the great light-gathering power of the 60- and 100-inch telescopes, Hubble and Humason would be able to look for a velocity–distance relationship at larger distances and, presumably, higher velocities than they could have using only Slipher's data.

Already by January 1929, six months after the Leiden meeting, Hubble had significant results to communicate. "A Relation Between Distance and Radial Velocity Among Extra-Galactic Nebulae," another landmark paper, appeared in the *Proceedings of the National Academy of Sciences*.[40] Hubble had collected together redshift or velocity data for 24 extra-galactic nebulae for which he deemed the distances to be reliable. Cepheids provided the distances to the six closest galaxies, while the brightest stars gave distance indications for another 13. Finally, to gauge the distances of four fainter galaxies in a cluster in the constellation Virgo, Hubble, like Shapley before him, had to rely on some less sophisticated techniques. He assumed that all galaxies have roughly the same absolute magnitudes, and he applied the "faintness means farness" principle.

Hubble plotted the velocities of the nebulae, derived from their redshifts, as a function of their estimated distances. To his eye, the data points appeared to define a straight line, with the most distant nebulae at 6.5 million light-years receding at a rate of about 1100 kilometers per second (see figure 9.4 for a similar plot, made later).

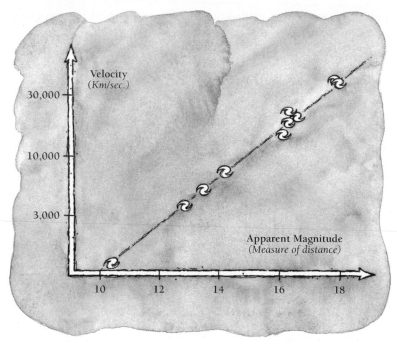

Figure 9.4 First hints of an expanding universe. Hubble plotted the velocity of galaxies (measured from the redshift of lines in the galaxies' spectra) against their apparent magnitude, an estimate of their distance. The result, similar to the straight-line plot shown here, showed that the galaxies are all in motion away from ours (except those which are members of our local group) and that the farther they are, the faster they are moving. This relation is just what one would expect if space is expanding in all directions. (Credit: Layne Lundström.)

The implications of his plot, simple as it is, were far-reaching. In simple terms, Hubble's plot could be interpreted to mean that the universe is expanding. As the space between all galaxies expands, the galaxies appear to fly away from us in all directions. Only those nearby, which form part of a group or cluster with our own galaxy, are exempted from the general pattern. An observer at any other location—in another galaxy outside our cluster— would see the same effect. The farther away the Galaxy, the more space there is to expand, and the more rapidly it appears to recede. The best analogy to illustrate the concept is that of dots drawn on the surface of a balloon. The dots represent galaxies, while the two-dimensional surface of the balloon, in this analogy, represents three-dimensional space. As the balloon is inflated, each dot moves farther away from every other dot. The most distantly separated dots are those which move away from each other the fastest.

From an historical point of view, the plot was important because he introduced numerical data into cosmological discussions that had been theoretical and abstract until then. From the cosmological perspective, he established the mathematical form of the velocity–distance connection: a simple linear relation. When, later, he was able to confirm this relation with more data (figure 9.4), his observations could be used to constrain or choose between competing cosmological theories.

The straight-line graph and the concept of the expanding universe brought to the fore the important question of the age of the universe. If the universe is currently in expansion, there must have been a time when all the galaxies were close together. Extrapolating backward in time, one could determine, from Hubble's straight-line plot, that the universe had been expanding for about 1.8 billion years. The problem with this result is that it conflicted with the age of the Earth and solar system. Even in 1929, geologists knew that the Earth must be more than 3 billion years old. Today, we know it is about 4.5 billion years old. The contradiction was largely resolved in the 1940s, when a revision of the Cepheid period–luminosity calibration led to improved accuracy in Hubble's values for galactic distances. Thereafter, the age of the universe as derived from the velocity–distance plot agreed more closely with the age of the Earth and solar system. But for many years the discrepancy provided an "out" for those who found it difficult to accept the notion of an expanding universe.

Lemaître and a renewal of interest in cosmology

In 1930, the cosmological stalemate over conflicting theoretical models of the universe came to an end in a most surprising way, with the revival of a "forgotten" paper by a Belgian cosmologist. The flurry of activity generated by this paper swept Hubble's research into the limelight.

Georges Lemaître, an unusual man who was both an ordained priest and a physicist, belonged to Hubble's generation. After serving as a soldier in the Belgian army during World War I, he returned to the Catholic University of Louvain, where he had earlier begun his studies. He obtained a PhD in mathematics and physics in 1920, and was ordained as an *abbé* in 1923. He then taught himself Einstein's relativity theory. In 1923, the Belgian government and the Belgian American Educational Foundation supported his post-doctoral stint at Cambridge University, where he deepened his understanding at Eddington's side and began to focus on the cosmological implications of relativity theory. The following year he began research with Harlow Shapley at Harvard and also at the Massachusetts Institute of Technology, working on a second PhD.

Lemaître attended the 1924–1925 meeting at which Russell read Hubble's epoch-making paper on observations of Cepheids in spiral nebulae. Lemaître realized then that Einstein and de Sitter's cosmologies, which had been conceived at a time when the Milky Way galaxy *was* the known universe, must be tested against observations of a universe in which galaxies, not individual stars, marked the contours of space. He set out to try to incorporate observational data, particularly Slipher's and Hubble's observations of receding nebulae, in a new cosmology.

Lemaître never saw any problem reconciling his religious faith with the rationalism of a scientist, and he demonstrated his willingness to contemplate the mysteries of the universe at their deepest level, from the perspective of a physicist. In 1927, he published in the *Annals of the Brussels Scientific Society* a remarkable paper on "A Homogeneous Universe of Constant Mass and Increasing Radius, Accounting for the Radial Velocities of Extra-Galactic Nebulae." He presented a concept that was radically new to theoretical cosmology: the physical expansion of the universe.

Not all the characteristics of Lemaître's proposed universe were radical. Like others before him, he envisioned a universe that was finite rather than infinite, but boundless, or without an edge. The surface of a balloon is often cited as an example of a finite but boundless space: the surface area of a balloon can be measured (even if it is expanding) but a line drawn on the surface can be extended indefinitely without encountering any edge.

Three-dimensional space may be finite but boundless like the two-dimensional surface of a balloon; it is just more difficult to visualize.

Lemaître showed that Einstein's static universe solution to the field equations of general relativity had a flaw: the solution was not physically realistic. Einstein described matter as uniformly distributed, but slight variations in this theoretical uniformity would cause the universe to lose its equilibrium and, rather than remain static, it would expand or contract. Thus, static solutions simply couldn't correspond to reality, Lemaître argued. His new perspective on the problem called for cosmologists to consider non-static solutions. In particular, he urged theorists to choose among the mathematically possible alternative solutions based on the kind of radial–velocity data provided by Slipher, Hubble and Humason—data that might indicate an expanding universe.

His own solution to Einstein's equations was an expanding universe that included both matter (unlike de Sitter's empty universe) and redshifts (unlike Einstein's static universe). He even predicted the linear velocity-distance relation that Hubble was soon to find among the extra-galactic nebulae, due to the expansion of space.

Cosmologists and astronomers did not, apparently, take any notice of Lemaître's paper when it first appeared; in their defense it could be noted that the *Annals of the Brussels Scientific Society* was a rather obscure journal. Lemaître himself did not promote his theory through personal contacts until he heard that Einstein and de Sitter had become dissatisfied with their own solutions to Einstein's field equations and had discussed the problem at a 1930 meeting of the Royal Astronomical Society in London. Then he gently reminded Eddington, secretary of the society, of his contribution to the field.

Eddington was embarrassed at having forgotten his former student's important analysis, which he found "brilliant."[41] He arranged for an English translation of the 1927 paper to appear in 1931 in the *Monthly Notices of the Royal Astronomical Society*, and attached a commentary of his own. Soon other "overlooked" papers began to be discussed, including some by the Russian Alexander Friedmann. Friedmann had also written about expanding space, although without relating his work to astronomical observations. After a period of stagnation in the 1920s, when theoretical cosmologists and observational astronomers knew little of each other's work, the 1930s began with an opening of communication and a surge of fresh ideas.

Even as Eddington prepared a translation of Lemaître's 1927 paper, Lemaître was tackling an even more challenging problem than solving Einstein's field equations. He was contemplating the origin and possible end of the universe itself. The expansion of the universe does not necessarily imply a beginning of the universe; for example, the universe might expand and contract repeatedly and indefinitely. Indeed, most astronomers and cosmologists recoiled at the thought of a singular moment in the history of the universe, when all came into being. But Lemaître ventured to contemplate such a moment, reflecting on the implications of an expanding universe for earlier and earlier times.

With what one historian of science called an "audacious notion," Lemaître speculated on the origin of the universe from what he called a primeval atom or super-atom, a nugget containing all that was or ever could be.[42] His primeval atom "decomposed" into a multitude of particles, like a radioactive atom decomposing into lighter atoms and elementary particles. The particles flew apart in an explosive moment of creation. For this reason Lemaître is sometimes known as the father of Big Bang cosmology, although the term "Big Bang" for the origin of the universe was coined later, after the theory had matured and had been elaborated on by others.

Eddington had earlier voiced the opinion of many scientists when he said, "Philosophically the notion of a beginning of Nature is repugnant to me."[43] But Lemaître's vision eventually persuaded cosmologists to consider seriously the physical state of the universe at early times. For example, if all the matter in the universe were more compact, the universe must have been much hotter in the past than it is now. In contrast to Jeans' suggestion that nebulae evolved from diffuse, amorphous states to condensed, organized bodies, Lemaître suggested that the evolution of physical bodies—whether of nebulae or of the universe itself—proceeded from an organized initial condition and led to a more fragmented, diffuse, low-energy state. "The origin of the world does not lie, apparently, in a primordial

336

nebula," wrote Lemaître, "but rather in a sort of primordial atom whose products of disintegration form the actual world."[44]

At the moment of creation, in Lemaître's cosmology, the simple primordial atom unleashed a fiery torrent of particles, which subsequently dispersed and cooled as the universe expanded. Most vivid was Lemaître's description of the cosmologist trying to piece together what happened. He wrote, "The evolution of the world can be compared to a display of fireworks that has just ended. A few red wisps, ashes and smoke. Astride one of the more cooled-off cinders, we watch as suns fade out, and we seek to reconstruct the vanished brilliance of the formation of the worlds."[45]

His paper ended with a speculation on the eventual end of the universe: "It is likely that the expansion has already passed a critical point and will not be followed by a contraction. In that case we cannot expect anything too exciting: the suns will cool, the nebulae will recede, ashes and smoke from the original fireworks will finish cooling off and dispersing."[46]

All of these new visions of the cosmos—theoretical formulations of an expanding universe, observations of redshifts of extra-galactic nebulae in all directions—caught the public's imagination in 1931, during the course of Einstein's visit to Pasadena and the frenzy of associated media reports.

Einstein and his wife Elsa stayed in January and February, allowing Einstein to discuss recent developments in relativity theory, astronomy and cosmology with Caltech faculty members and Mount Wilson staff members. Richard Tolman, a leading cosmologist on campus, was one with whom he had much to discuss. On the observational side, he particularly wished to meet Hubble.

On 4 February, Einstein spoke to a crowd assembled in the library of the Mount Wilson Observatory offices in Pasadena. There the one known as "the smartest man in the world" admitted that he had changed his mind about the state of the universe, and that he should not have constrained himself to a static solution to the field equations. Hubble's observations had played a key role in his review of the problem.

"New observations by Hubble and Humason [...] concerning the redshift of light in the distant nebulae make the presumption near that the general structure of the universe is not static," Einstein asserted. He added that the theories of Lemaître and Tolman "show a view that fits well into the general theory of relativity."[47]

Hubble enjoyed an unprecedented degree of media attention during Einstein's visit, worming his way to Einstein's side during photo opportunities and responding at length to reporters' questions about his own work. Curiously, amid the excitement over Einstein's revelations, he remained cautious about the interpretation of the redshifts as indications of the expansion of space. To the puzzlement of his friends and colleagues, he stayed clear of explicitly endorsing any cosmological theory.[48] In interviews with reports, as in his published papers, he did not wax philosophical about the expansion of the universe or the amazing speed of extra-galactic nebulae rushing away in all directions. His graduate student Allan Sandage, who started working with him in the 1950s, once remarked that Hubble was "of a poetic nature, he was an intellectual of a most profound type," but that he "didn't really open up" when it came to philosophical discussions.[49]

Some of the media attention may have gone to Hubble's head. In the 1930s, he was to have chaired the IAU's Commission on Nebulae and Star Clusters at a meeting in Cambridge, Massachusetts. But he failed to prepare a report or even to attend the meeting, leaving Shapley to assume the role. Other behavior that rankled with Mount Wilson Observatory director Adams included the fact that he treated his colleagues with something very like disdain, ignoring mail to be answered, and took extended vacations in Europe with full pay and with Grace in tow. Hubble's desire not to be bothered with his professional obligations and to shape his own public image seems even to have resulted in his breaking off contact with his family. When his mother, who had been cared for by his brother Bill, died in 1934, he failed to attend the funeral.

Adding to Hubble's difficulties maintaining good relations with his colleagues and family, the old conflict with van Maanen resurfaced in the mid-1930s. Hubble was determined to sweep away van Maanen's rotation measurements, a small but persistent thorn in his side. As early as 1925, Shapley had written to Seares that Hubble placed too much importance on being publicly vindicated. He wrote, "Hubble's attitude toward van Maanen disturbs me a little, because of my friendship for the latter: Hubble can so well afford to be generous as he has nothing to lose."[50]

Matters came to a head when Hubble examined his own photographic plates of the four major spirals van Maanen had studied and found no consistent signs of rotation at the level van Maanen had claimed to see. He wrote up his findings for publication, but observatory director Adams and editor Seares were aghast at the hostile way he expressed himself, and refused to publish the paper in the observatory's own series, *Contributions of the Mount Wilson Observatory*. In the end, Hubble published a report of his investigation in the *Astrophysical Journal*. Immediately following Hubble's paper, the journal published a note by van Maanen conceding that his earlier results might have been affected by systematic errors.[51] Hubble had succeeded in laying to rest the last serious argument against the spirals as extra-galactic star systems at great distances — but the victory added nothing to his reputation. As for van Maanen, his results were definitely in error, but to this day it is not clear how such a careful and well-intentioned observer could have been led so far astray.

The distribution of the nebulae

By the early 1930s, Hubble had made three major contributions to his field. He had put forward a valuable classification system for the nebulae that stimulated research on their origin and development. He had discovered Cepheids in extra-galactic nebulae and established their great distances, thereby putting our own Milky Way system in the right perspective. He had uncovered the linear relationship between the velocity and distance of extra-galactic nebulae, which most scientists interpret as evidence for the expansion of the universe.

The fourth of his major contributions, which relates to the ultimate fate of the universe, began to emerge in the mid-1930s. One of his more important papers of this period concerned the distribution of some 80 000 extra-galactic nebulae that he had identified on photographs taken with the 60- and 100-inch telescopes. Hubble imagined the volume of space surrounding our galaxy as a series of nested spheres, like the layers of an onion. To gauge the distribution of nebulae in radial distance from our galaxy, he counted the numbers of nebulae in each layer, assuming the nebulae in a given layer all had the same brightness. These counts showed that the nebulae are distributed isotropically: no region of extra-galactic space appears to have a higher density of nebulae than any other region. The nebulae do have a tendency to cluster, but no region or direction in space is distinguished by a concentration of nebulae or clusters. The isotropy of the universe is an important observational fact that cosmological theories must account for. At a basic level, it means that the universe has no center of expansion: all parts are expanding from all other parts.

Hubble used the counts to estimate the average density of matter in the observable universe, a quantity of much interest to theoretical cosmologists. The fate of the universe hinges on the density of matter in space. Above a certain density limit, gravity will ensure that all matter comes together again in what is popularly called the Big Crunch, the antithesis of the Big Bang. On the other hand, if the density of matter is in fact below that limit, gravity will not overcome the forces of expansion, and the universe may go on becoming larger and more dilute for ever.

Hubble described his major accomplishments in his famous monograph, *The Realm of the Nebulae*, which appeared in 1936. Some of the book is quite technical; he thought of it as "a cross between popular lectures and a textbook."[52] He wrote, as always, in a spare and straightforward style, minimizing his personal role in developments. Summing up a description of his classification system, he intoned dispassionately that "Order has emerged from apparent confusion, and the planning of further research is greatly simplified."[53] He discussed his discovery of Cepheids in extra-galactic nebulae and other findings as though pure logic had led from one step to the other. He glossed over the vexations that usually attend research, such as the dilemma of van Maanen's spurious rotations of spirals.

Second World War

Hubble's last paper before the United States entered the war reported on the direction of rotation of the spiral nebulae. He had every reason to expect the spiral nebulae would be in rotation, albeit very slowly. In 1925, the Swedish astronomer Bertil Lindblad, who had spent the years 1922–1924 visiting Lick and Mount Wilson Observatories, addressed this question. Lindlbad determined that the Sun and other stars are rotating around a distant galactic center, while a sub-system of stars also belonging to the Galaxy but residing outside the disk, the "galactic halo" stars, rotate much more slowly, seemingly suspended in the space above and below the disk. His announcement marked the resolution of the long-puzzling phenomenon of star-streaming, described by Kapteyn in 1904; the solar system's participation in the disk rotation gives us a perspective on stellar orbits that results in the apparent division of stars into two oppositely directed streams. Lindblad also confirmed the location of the galactic center at the point outlined by the globular clusters, as Shapley had shown. Then in 1927, the Dutch astronomer Oort, in trying to verify Lindblad's work, showed convincingly that the stars of the disk rotate differentially — that is, the stars closest to the center orbit fastest.

By the late 1920s, then, it was clear that the Milky Way is in rotation. Searching for an effect of similar magnitude in the distant nebulae, Hubble looked for small shifts in spectral lines from the spiral disks — a redshift from the side rotating away, and a blueshift from the opposite side of the nebula, which rotated toward the observer. He found the effect he was looking for, and the direction of rotation confirmed his guess: the arms trail the faster-moving inner parts of the nebulae.

In the late 1930s and early 1940s, Hubble had fairly well exhausted the capabilities of Mount Wilson's 60- and 100-inch telescopes in the pursuit of faint, distant nebulae. He waited now for Hale's last *tour de force*, a venture many had thought impossible: the completion of a telescope 200 inches in aperture.

The first contracts for building the 200-inch, which was destined for Mount Palomar, 50 miles north of San Diego, were signed in 1928. The Corning Glass Works company in western New York State began casting test disks of Pyrex shortly thereafter, exploring the properties of this new type of glass in ever-larger samples. The 200-inch blank was cast in December 1934, and after a 10-month cooling period, mounted in a railroad car and sent off to Caltech's optical shop for grinding and polishing. Hale died in 1938, before the job was completed; not until 1948 was the massive telescope mounted under its dome on the mountain and more or less ready to take data.

Hubble filled his free time with political activism. He became convinced that the United States must enter the expanding war in Europe, particularly after Paris fell to the Germans in 1940 and his beloved England faced a great threat. "We must face a long, tough, bitter war, but the sooner we start, the less terrible the price," Hubble argued in a speech to fellow citizens of Pasadena on Armistice Day, 1941, shortly before the attack on Pearl Harbor.[54] Hubble was in a position to influence opinion, at least locally where he was well known as a Mount Wilson astronomer of international repute. He had gained membership in the National Academy of Sciences, one of the youngest scientists to do so. He served on the elite board of trustees of the Huntington Library, Art Collections and Botanical Gardens, replacing Hale, who had played a key role in the founding of Huntington's institution. In 1940, he won the Gold Medal of England's Royal Astronomical Society. Besides these marks of distinction, Hubble enjoyed a high status among members of Grace's friends in the literary and artistic community. He parlayed this status into opportunities to give speeches and write articles in support of presidential hopeful Wendell Wilkie, an anti-isolationist, who eventually ran a doomed campaign against Franklin Roosevelt.

In the spring of 1942, Hubble received an urgent invitation to head the exterior ballistics program at the Aberdeen Proving Grounds in Maryland, just south of the Chesapeake Bay. Hubble professed not to know much about ballistics, and later claimed that when he first heard of the program, he "rushed to a dictionary" to look up "ballistics," and then turned to an encyclopedia.[55] But he accepted the job, knowing that it was his broad background in physics, surveying, optics and statistical analysis that would be pertinent. His department, which would include both men and women — Hubble called them "computer girls who had majored in mathematics"[56] — would study the effect of gravity and air resistance on the flight of bullets and rockets launched from all types of guns and rocket launchers used by the military. The term "exterior ballistics" emphasizes the distinction between the aerodynamical problems of flight within and outside of the barrel of the gun or rocket tube.

After his annual fishing trip to Colorado with Grace, Hubble departed by train for the East Coast and his position at Aberdeen. He left behind some of the darkest skies Mount Wilson astronomers

342

had ever seen, thanks to the blackouts in Los Angeles. The opportunity to study faint sources was a rare one; as a *Time* magazine article reported in a piece about Hubble, the night sky above Mount Wilson had turned from black to grey as the city had grown and the lights had spread. Hubble called the city at night the "Los Angeles Nebula" with a dense nucleus of light in the downtown district.

Hubble's colleague Water Baade, a naturalized citizen originally from Germany, took advantage of the lack of light pollution to study spectroscopically the Andromeda nebula and two of its companions, the elliptical galaxies M32 and M110. Contrary to Jeans' nebular theory, Baade found that the companion galaxies and the central bulge of the Andromeda galaxy itself could be resolved into stars. However, the bulge stars and disk stars exhibited different kinds of spectra. Ultimately, Baade's wartime research showed that galaxies are composed of at least two types of stars, which became known as "Population I" — the disk stars like the Sun — and "Population II" — the stars of the bulge and halo. The division into two types led astronomers to recognize the luminosity difference between true Cepheids, which are Population I stars, and the similar W Virginis stars, which are fainter and belong to Population II. Thus the wartime blackout enhanced the capabilities of the world's largest telescope and allowed astronomers at last to resolve puzzling discrepancies in the scale of the universe. Once it was understood that the W Virginis stars are intrinsically dimmer than classical Cepheids, the distances to the Andromeda galaxy and other nearby galaxies was doubled, and the size of our own galaxy fell into line with the size estimates of the spirals.

As Hubble immersed himself in the design and testing of ordnance *matériel*, Walter Adams and Vannevar Bush, then president of Mount Wilson's parent organization, the Carnegie Institution, were discussing the future leadership of the 200-inch telescope observatory at Palomar. The Carnegie Institution had made the grants for the telescope to Caltech, although Mount Wilson and Caltech had entered into a cooperative agreement over the construction and staffing of the new observatory.

Hubble fully expected to become observatory director on the basis of his high standing as an astronomer. But Adams stressed to Bush that Hubble, besides having no affinity for administrative work, had demonstrated what Adams called a "lukewarm attitude" toward the Carnegie Institution, arguing over his compensation and vacation, and not mentioning the Carnegie Institution as his employer in his publications.[57] Adams instead recommended the man who eventually got the job, the Caltech physicist Ira Bowen. Bowen had, like Hubble, studied physics with Robert Millikan at the University of Chicago, but there the similarities ended. Bowen, nine years younger than Hubble, was only 30 when, in 1928, he solved a problem that had interested William Huggins and had stumped astrophysicists since: the identification of the "chief nebular line," ascribed to a new element, "nebulium." The line is actually a closely spaced pair and the lines arise, under rare circumstances of temperature and density, from the common element oxygen. Bowen turned out to be a good choice to head the observatory. For one, he had enough diplomatic skills to manage Hubble.

Some of Hubble's friends thought he appeared worn out when he returned to Pasadena from Aberdeen in December 1945. Indeed, he must already have been suffering to some degree from heart trouble. In spite of any tiredness, however, Hubble continued to lecture to civic groups as he had just before the war. In 1946, his theme was that "a world authority armed with a police force" was the only solution to the problem of preventing war and its inevitable consequence, the ruin of civilization.[58] "War can be stopped when, and only when, we are ready to use physical sanction—when we are ready to use a police force bought at the price of a modicum of sovereignty," Hubble declared. In the world government that he envisioned, citizens would have the same power that Americans do over Federal authorities: "all of us, through the ballot, police the policemen," Hubble suggested.[59]

His speeches were not warmly received, just as his pro-war stance before the attack on Pearl Harbor had been out of sync with popular opinion. Perhaps it was to counter a diminishing of his status that Hubble engaged a publicity agent around this time. By 1948, he was no longer as vocal on political matters. He appeared on the cover of *Time* magazine in February 1948; the accompanying article, while mentioning his annual fishing trips, study of Chinese philosophy, friendship with Hollywood celebrities, and "civic activities," focused on his "flabbergasting discoveries" and did not mention his activism.[60]

344

New telescopes

Work on polishing the 200-inch telescope had stalled during the war, and resumed afterward with a new crew, which slowed progress. Finally, on 12 November 1947, the mirror was ready to be shipped from Pasadena to the top of Mount Palomar. Three diesel tractors combined forces to push the precious cargo up the winding road. Under the dome, engineers tested the support structure for the mirror with a concrete dummy, then removed it and began installing the mirror.

A dedication ceremony for the aptly named Hale Telescope was held in the summer of 1948, but the 200-inch behemoth still required adjusting and was not ready for scientific use. In the meantime, a much less impressive but very fine telescope, the so-called 48-inch Schmidt telescope, was completed and installed at the Mount Palomar site. This telescope had a mission: to photograph and map the entire northern sky down to a limiting magnitude of 22, more than two million times fainter than the 6th magnitude limit of the unaided eye. The resulting catalog became well known as the Palomar Observatory Sky Survey, and later served as a reference for pointing the Hubble Space Telescope.

The privilege of taking the first photographs with the 200-inch telescope, during a testing period in 1949, fell to Hubble. As expected, he found that nebulae four times fainter than could be seen with the 100-inch could be detected, which would put hundreds of millions more nebulae within telescopic grasp. Hubble's project for the 200-inch was, in fact, to count galaxies as a function of their apparent brightness to try to determine, more accurately than had been possible previously, the distribution of galaxies in space.

At his favorite fly-fishing spot in Colorado, on his summer vacation with Grace in July 1949, Hubble suddenly became ill. He had suffered a massive heart attack, and was lucky to recover, if slowly, in a hospital in Grand Junction, Colorado. Grace kept the news of his condition from the press. In the meantime, Humason and a graduate student, Allan Sandage, carried out his observing program at Palomar.

Hubble never resumed a full schedule of observing, and died at home on 28 September 1953, aged 63, of a cerebral thrombosis.

His student, Sandage, and his long-time observing partner, Humason, had carried out much of his observing for him in the final years. Sandage edited and published Hubble's posthumous work, the *Hubble Atlas of Galaxies*, a beautiful compendium of photographs of galaxy types.

Grace outlived her husband for 26 years. Although she kept diaries detailing her husband's life and career, and gave additional biographical material to the Huntington Library, these were not altogether helpful to historians of science. As Sandage noted, "known facts contradict part of the recollections" in those materials, and some of them "glorify him in ways larger than life."[61] Hubble's modern biographer, Gale Christianson, conducted extensive research to supplement the Hubble papers with more objective information.

Hubble's name, at least, is widely recognized today in association with the Hubble Space Telescope, a cooperative program of the European Space Agency and the National Aeronautics and Space Administration. The 2.4-meter (94-inch) telescope and its attendant cameras and spectrographs record observations from low earth orbit, above the distorting atmosphere. In addition to observing nearer objects such as star-forming regions in our own galaxy, users of the telescope, building on the foundation Hubble laid in observational cosmology, have plumbed the depths of space and time to uncover clues to galaxy formation and evolution.

Alan Sandage best encapsulated Hubble's contribution to the study of the Milky Way and other galaxies when he wrote, "What are galaxies? No one knew before 1900. Very few people knew in 1920. All astronomers knew after 1924," after Hubble's paper on Cepheids in the Andromeda nebula was read before the meeting of the American Astronomical Society. Galaxies, Hubble had shown, are the building blocks of the universe; as Sandage put it, "They are to astronomy what atoms are to physics. Each galaxy is a stellar system somewhat like our Milky Way, and isolated from its neighbors by nearly empty space. In popular terms, each galaxy is a separate island universe unto itself."[62]

10

THE MILKY WAY
REVEALED

"What we know is little, what we do not know is immense."

Pierre-Simon Laplace, 1827[1]

The Andromeda galaxy, which we perceive with our unaided eyes only as a faint oval nebula, is both neighbor and kin to our home galaxy, the Milky Way. Viewed with binoculars or a small telescope, the Andromeda galaxy distinguishes itself from the foreground stars of our own galaxy as a glowing ball of light encircled by a wide, thin disk. The light from the disk shines feebly compared to that from the central part of the Galaxy; the disk appears as insubstantial as a skirt of tulle, although the "skirt" is actually a retinue of billions of stars.

To appreciate the disk's whirlpool structure, we must use larger telescopes or capture the image with long photographic exposures. A photograph of the Andromeda galaxy taken with a short exposure shows mainly the glow from the center of the Galaxy. The same galaxy taken with a long exposure brings out the disk (see figure 10.1).

Such is the makeup of spiral galaxies: the visible light is concentrated in two components, a bright central bulge and a relatively faint disk composed of more or less prominent spiral arms. A third component, called the stellar halo, is mostly dark. The Milky Way follows the same basic plan.

If we could gain a vantage point outside the Milky Way — perhaps erecting a telescope on a planet orbiting one of the billions of stars in the disk of the Andromeda galaxy — we would take in at one glance the features of our galaxy that astronomers have had to uncover more painstakingly by mapping the skies in all directions and using light from the entire electromagnetic spectrum. We

Figure 10.1   Disk and bulge. The first view is a short-exposure image of the Andromeda galaxy, showing the bright central region. The dusty spiral arms show up in the second view, taken with a longer exposure. (Credit: Jason Paul Lisle, Sommers-Bausch Observatory, University of Colorado.)

would probably be dazzled by the glare from the Milky Way's central bulge; our view would not be as obscured by dust and gas as it is from our actual position within the Milky Way's disk. We would be able to trace the Milky Way's spiral arms, arcs of stars and bright star clusters delineated by dark inter-arm dust lanes. We might marvel at the proximity of the Large and Small Magellanic clouds, the brightest of several satellite galaxies hovering relatively near ours in our corner of the cosmos.

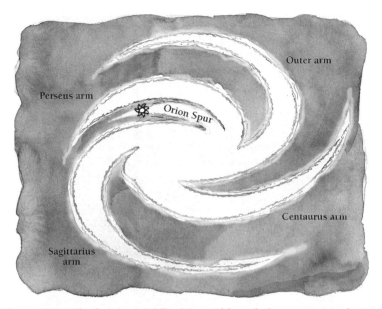

Figure 10.2 The four-arm Milky Way. Although they are not as distinct as illustrated here, four major spiral arms can be discerned in our Milky Way galaxy. Smaller arms or spurs also appear. The naming of the arms varies by source, but the spiral arm in which the sun resides is known as the Perseus arm, or, according to some authorities, the Orion spur of the Perseus arm. (Credit: Layne Lundström.)

The Milky Way galaxy is often depicted with four principal spiral arms (figure 10.2). Astronomers generally agree on the naming of the nearest two, the Perseus arm and the Sagittarius or Sagittarius-Carina arm. The Sun is located in the minor Orion arm, sometimes called the Orion "spur," which lies between the Perseus and Sagittarius arms. The third major arm is often called the Centaurus arm. A fourth arm goes by the name Outer arm, or sometimes Cygnus arm.[2]

The dimensions of the Milky Way galaxy

In 1930, when Lick Observatory astronomer Robert Trumpler proved that obscuring dust particles are widespread in the interstellar medium, our gauges of galactic dimensions began to attain modern standards of accuracy. Previously (as we saw in chapters 7, 8, and 9), estimates of stellar distances based on apparent brightness tended to run too high, because the unsuspected dust dimmed the visible light reaching us from the stars. Accounting for dust allowed astronomers to correct the scales of Kapteyn's and Shapley's stellar systems. Beginning in the 1950s, observations in the radio wavelengths that are not absorbed by dust allowed us to probe the outer limits of the Galaxy and to form a more complete picture of the Galaxy's size.[3]

The Milky Way has three major components (figure 10.3). The Milky Way's central bulge, whose mass is about 10 billion suns, is roughly 3000 light-years in diameter in the plane of the disk. It extends to about 1800 light-years above and below the

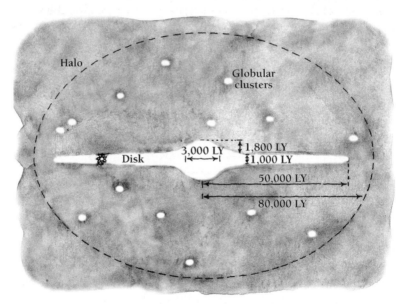

Figure 10.3 Components of the Milky Way. At the center of the disk of the Milky Way is the bulge, a bright region some 3000 light-years across where the stars are very densely packed. The visible disk itself, only some 1000 light-years thick, has a radius of about 80 000 light-years. Globular clusters fill the space around the disk, known as the halo. (Credit: Layne Lundström.)

disk; it is rather cigar shaped. The visible disk, on the other hand, with a similar mass, is flattened like a pancake. It is only 1000 light-years thick, and extends to 50 000 light-years from the center. A thin layer of invisible hydrogen gas continues beyond the limit of the stars in the spiral arms and forms a disk of about 80 000 light-years in radius from the center.

Globular clusters, spherical agglomerations of stars, populate a third component of the Galaxy's structure, a volume called the stellar halo. This space encloses the Galaxy like a bubble that is big enough to contain the disk. Any star or cluster found outside this volume, or roughly 50 000 light-years from the center of the Milky Way, is about as likely to "belong" to a nearby galaxy as to the Milky Way. A case in point is the faint Sagittarius dwarf galaxy, which lies only 50 000 light-years from the center of the Milky Way. At this distance, the gravitational pull of the Milky Way competes against the satellite galaxy's own gravity; the satellite system is being stretched apart, or "tidally disrupted," by its proximity to our more massive system.[4]

The halo

The halo is perhaps the least understood part of the Galaxy, even though, at first glance, it appears to be mostly empty.

Hundreds of globular clusters float in the Milky Way's halo, from the galactic center out to the edge of the stellar halo. (See chapter 2, figure 2.6, for an illustration of a globular cluster.) These systems contain hundreds of thousands or millions of stars — as many as in a small galaxy — in a compact volume only a few hundred light-years across, which makes them very luminous. Their beauty derives from their spherical symmetry: they are as round as dandelion puffballs, often with bright centers and more diffuse edges. Their orbits around the galactic center cause many of them to plunge through the plane of the disk, then to swing slowly out into comparatively empty space. Isolated or "field" stars in the halo may have come from the gradual dispersal of some globular clusters, early in the formation of the Galaxy.

The stars that populate the halo, either as field stars or as members of globular clusters, are generally much older than the stars in the disk. While some disk stars may be 10 billion years old, those in the halo are at least 13 billion years old, and that difference is significant. The halo stars formed almost exclusively from the primordial elements of hydrogen and helium, present when the Galaxy itself was young. The disk stars that we currently see, on the other hand, incorporated material from earlier generations of stars as they formed. Stars burn hydrogen and helium in their nuclear furnaces, creating more complex elements, and spew that material into interstellar space when they die. In this way, the stars in the disk, having formed more recently than the halo stars and in a more densely populated environment, have had a chance to enrich their chemical composition.

The advanced age and primitive composition of halo stars is not the only interesting aspect of the halo. While a substantial fraction of the matter in the disk is luminous—i.e., in the form of stars—most of the halo matter is dark. The halo stars, it turns out, contribute a meager fraction of the halo mass.

We can infer the presence of dark matter in the Galaxy through its effect on luminous matter in the disk and on the stellar component of the halo. Dark matter, whatever it may be, shares one fundamental characteristic with ordinary matter in dust, gas and stars: it exerts a gravitational force. Furthermore, the distribution of dark matter in space—how concentrated it is around the center of the Galaxy—affects the speed of stars and gas clouds in the disk as they orbit the center of the Galaxy. Charts of disk-component speeds, called rotation curves, allow us to determine the amount of unseen matter as a function of distance from the center of the Galaxy. The Milky Way's rotation curve shows that at least 90% of the matter composing it is dark, and that most of the dark matter must be in an extended halo that reaches beyond even the limit of the globular clusters, perhaps as far as a million light-years. Figuring out what the dark or missing matter is, is one of the great challenges astronomers currently face.[5]

The disk

Of the three main components of the Galaxy—central bulge, disk and halo—the disk, with its spiral arms, is the most visually complex. The spiral arms have been described in many beautiful ways: like pinwheels, or like cream poured into stirred coffee.

We know that the arms "trail" — that is, they follow the rotation of the Galaxy like the folds of a skirt winding around a spinning dancer. But exactly what the Milky Way's spiral arms would look like, if we could see them from a distance of millions of miles away, is an intriguing question.

In the visible wavelengths of light, the spiral arms look like concentrations of stars, leaving apparently little in the dark inter-arm lanes. Indeed, the mass of gas and stars per unit volume is enhanced in the arms, although stars and large amounts of dust and gas do collect in the inter-arm regions as well. The density of stars in the arms is only about 10% greater than in the rest of the disk, but the arms stand out because the stars there tend to be bright.

Besides the density enhancement in spiral arms, a factor that determines how prominent they look is the regularity of the spiral structure. At one extreme in this approach to classifying galaxies are the so-called "grand design" spirals, such as M51 and M81. These have a regular spiral structure, generally composed of two arms that one may easily trace from the center of the Galaxy to the outer limits of the visible arms. At the other extreme are "flocculent" spirals such as M63. These show fluffy, small-scale spiral structures throughout the disk (figures 10.4a and b). The Milky Way is probably of an intermediate type, with multiple arms and some "feathering" between adjacent arms.

The regularity of spiral arms appears to depend on various physical parameters. The proportion of mass in a galaxy's disk compared to the mass in the halo varies from one galaxy to another, so that one may speak of "light" or "heavy" disks. The proportion of mass in stars compared to that in gas may vary. The stars in the disk may appear settled, or "cool," or they may zoom about with a high degree of random motion, in which case astronomers call the disk "hot." According to recent theory, grand design spirals form in galaxies that have cool, light stellar disks and are relatively gas-poor. The Milky Way galaxy is evidently too rich in gas to form a stunning grand design spiral.

The origin of spiral arms is not understood in great detail. So much is clear: the spiral pattern is very likely the result of a ripple or wave that propagates through the disk. The wave, called a

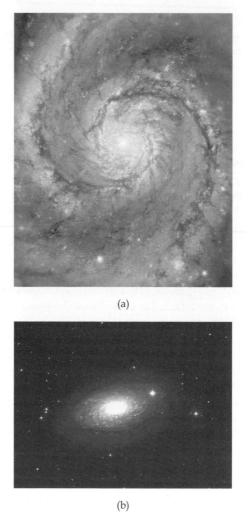

(a)

(b)

Figure 10.4 (a) "Grand Design" galaxy. The spiral arms are prominent and extend in an unbroken line from the center to the extremities in this example of a so-called "grand design" galaxy, M51, the "Whirlpool" galaxy. (The bright round patch shown in Lord Rosse's drawing at the end of one of the spiral arms (figure 6.2) is just off the top edge of this image.) (See color section.) (b) "Flocculent" galaxy. The spiral pattern of this galaxy is more patchy, or flocculent, than that of a "grand design" galaxy.

354

density wave, arises naturally from the interplay of forces in the disk. Both it and the stars and gas clouds that make up the disk rotate around the center of the Galaxy, but not at the same speed. The stars and gas clouds rotate at varying speeds depending on their distance from the center, with those closest in orbiting fastest. The density wave rotates at a slower rate than even the outermost stars do. The stars and gas clouds overtake the density wave, and this is what gives rise to the spiral arm pattern.

The situation is often compared to that of cars entering and passing through a knot of traffic on the highway, with the stars as vehicles and a peak in the density wave as the region of traffic congestion. As the stars pass through the peak in the density wave, they slow down and become more bunched together. The combined effect of stars at varying distances from the galactic center slowing down and entering a "traffic jam" creates the spiral arms.[6]

Our Sun passes through a spiral arm every 200 million years or so, and is currently headed for the Perseus arm—in another 140 million years. Some astronomers have put forward the interesting hypothesis that mass extinctions on Earth, like the demise of the dinosaurs, correlate with the solar system's passage through spiral arms.[7] In this scenario, the solar system takes a heavier pelting from comets and asteroids during a passage through a spiral arm, and the impacts on Earth lead to extreme climatic changes.

The spiral arms are delineated, on their trailing edges, by bright star-forming regions. That's because the gas in the galaxy's disk is compressed as it enters a spiral arm, and the compression leads to favorable conditions for star formation. The "Great Nebula" in Orion is at the star-forming edge of one such spiral arm in the Milky Way, the Orion arm. We have a great view of it from our Sun's position between the Sagittarius arm and the Orion and Perseus arms (see figure 10.2).

In the 1950s, astronomers used the Orion nebula and others of the same type, called HII ("H-two") regions or regions of ionized hydrogen gas, to trace the contours of the nearer spiral arms of the Milky Way. Since then, we have mapped more distant parts of the galaxy using the techniques of radio astronomy and using other features, such as molecular gas clouds, as tracers.

The Orion nebula (chapter 2, figure 2.7), is in many ways typical of the glowing pink nebulae lining the inner edges of spiral arms. The diameter of the gas cloud associated with it is more than 20 000 times that of the solar system. The mass of gas composing the cloud—mostly hydrogen—is enough to form tens of thousands of stars. Hundreds of young stars, formed within the last million years, already burn within this stellar nursery. Four of the young stars form a bright cluster, the Trapezium, that one can see with binoculars or a small telescope. These O and B type stars radiate intensely in the ultraviolet part of the spectrum, which allows electrons to separate from hydrogen atoms in the gas. When the electrons have an opportunity to recombine with hydrogen atoms, they emit red light. This fluorescence process gives the Orion nebula, and other HII regions, a characteristic pink cast on photographic images.

Associated with the Orion nebula is a dense cloud of molecular hydrogen called the BN-KL infrared nebula. Star formation there is at an even more primitive stage. The turbulent roiling of the cloud and the effect of the cloud's own gravity is causing hundreds of pockets of molecular material to condense and take shape as stars and embryonic solar systems. In 1993, the Wide Field and Planetary Camera 2 on the Hubble Space Telescope first captured images of these condensing stars and their associated proto-planetary disks in the Orion region. In 1995, the same camera found even more spectacular views of star formation in progress, in the Eagle nebula (also known as M16) (figure 10.5).[8]

While stars are forming in the spiral arms, and particularly along the trailing edges of the arms, older stars created in earlier epochs of star formation are dying. The Perseus arm of the Milky Way has provided notable examples of dying stars over the past few thousand years of recorded history. The event that led to the Crab nebula is a well-known instance.

In 1054, Chinese and Japanese astronomers witnessed the death of a star by supernova near the horns of the bull in the constellation Taurus. They described the appearance of a new or "guest" star about as brilliant as the full moon, visible in the daytime for about a month and visible at night for more than a year. Curiously, European astronomers seem not to have noted it. Native Americans in Arizona may have observed the

Figure 10.5 Star-forming region. This 1995 image from the Hubble Space Telescope shows that the Eagle nebula, associated with the open cluster M16, is home to great pillars of molecular hydrogen gas and dust, many light-years in length. Within these pillars, gas is condensing under the influence of gravity, forming new stars. At the ends of the pillars, radiation from stars that have recently turned on their nuclear burning is clearing away some of the pillar material and exposing dense globules where stars are at an earlier stage of formation. (Credit: NASA.)

supernova, according to one interpretation of pictographs found at White Mesa and Navajo Canyon.[9]

The visible output of the supernova declined below the threshold of naked-eye observations. Then, in the 1700s, European astronomers scanning the skies with telescopes found the supernova's gaseous remnant, without knowing what it was. Lord Rosse (see chapters 5 and 6) named the remnant the Crab nebula because of its resemblance to a crab's claw. In 1942, astronomers translated the Chinese accounts of a "guest" star into English and made the connection between the supernova and the glowing remnant. In the 1960s, the left-over star at the heart of the Crab was one of the first to be identified as a pulsar, a very dense, rapidly spinning star that emits a beacon of radio waves and x-rays.

Stars are born and die in the disks of spiral galaxies. But the disks play host to a rich variety of objects besides star-forming nebulae and supernovae. In the Milky Way's disk we observe stars in various stages of evolution; stars in pairs or triplets or in irregularly shaped galactic clusters; and planetary nebulae, the remnants of expired low-mass stars. We see large and small clouds of dust and gas, some in the form of filaments, knots or sheets. Our position in the plane of the disk, about two-thirds of the way out from the center of the Galaxy, grants us a good view of these diverse objects.

The bulge

The boundary between the galaxy's disk, comprising the arms and inter-arm dust lanes, and the amorphous central bulge is a blurry one. Lacking a clear-cut transition or precise definition, astronomers commonly refer to the bulge as the luminous mound in the middle of a galaxy that would be left over if one subtracted the disk. In the case of the Milky Way, most of the light from the bulge is contained within a radius of about 1500 light-years from the galactic center. The bulge becomes difficult to separate from the disk at about 10 000 light-years from the galactic center.[10]

The mystery of bulges is how they relate to galaxy formation. In spiral galaxies, did the bulges form first, before the disks? Or did they build up over time, under particular conditions? Is it a coincidence that bulges look like elliptical galaxies? Despite the difficulties associated with studying the dense heart of galaxies, astronomers are interested in what clues bulges might provide.

Bulges are a defining characteristic in the phenomenological system Edwin Hubble devised to classify galaxies — the "Hubble system" (see chapter 9). In the *Sa* category of spiral galaxies, the bulge is relatively large compared to the disk. As one progresses to the *Sc* category and beyond, the size of the central bulge decreases relative to the size of the disk. Other characteristics vary in tandem with the bulge-to-disk ratio—for example, a galaxy's current star formation activity appears to increase as the relative bulge size decreases, and the arms become less tightly wound.

This "sequence" does not necessarily imply an evolution in time (say, from large bulges to small ones), as was once thought. It does suggest that the Hubble type of a particular galaxy, and the relative size of the bulge, derive somehow from the physical conditions under which the Galaxy formed.

The Milky Way bulge is a good place to start studying bulges in general, because of its proximity. However, dust and gas in the plane of the disk dim the visible light from the central regions of the Galaxy by a factor of about a trillion. When we look to the bulge, toward the constellation Sagittarius, we see only that the band of light that we also call the "Milky Way" in that direction is thick with stars, and that the star clouds appear somewhat yellower than in other dense regions, reflecting a difference in average color of bulge stars.

In the 1940s, Mount Wilson astronomer Walter Baade discovered the first known of a few small tunnels into the bulge—lines of sight that happen to skirt the irregular concentrations of obscuring dust and gas. His line of sight from the Earth passes within 1800 light-years of the galactic center, and gives us a rare glimpse into the bulge in the visible wavelengths (figure 10.6). The view through "Baade's window," as it is called, is the subject of intense study. Baade himself used it to obtain a first measure of the distance from the Earth/Sun system to the galactic center. The current value is about 26 000 light-years.[11]

Infrared wavelengths travel through dust with less dimming, and so provide another perspective on the bulge. The wavelength of visible light is comparable to the scale of molecules and dust particles. This similarity of scale means the interstellar matter can dim visible light; infrared and radio wavelengths are longer, and they pass through space unimpeded. (The disadvantage of the longer wavelengths is that they generally provide a coarser view, or they may reveal a different aspect of the source.)

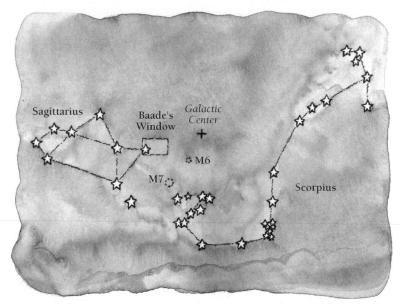

Figure 10.6 Baade's Window. Named for the astronomer who discovered it, Walter Baade, the "window" is a relatively unobstructed view deep into the heart of our galaxy. The sky in the direction of the constellations Sagittarius and Scorpius is full of dust and gas, but a line of sight through Baade's window penetrates to the rich star fields close to the galactic center. The "window" can be found near the spout of the "teapot," which many recognize in the constellation Sagittarius. (Credit: Layne Lundström.)

In 1990, NASA's Cosmic Background Explorer Satellite (COBE) took a photograph of the disk of the Milky Way in the near-infrared wavelengths (figure 10.7). The image was the first to show the bulge directly, as a bulbous thickening in the middle of the disk. The shape of our galaxy in this view is reminiscent of visible-light images of some distant spiral galaxies seen edge-on.

The COBE image shows the bulge in two dimensions. Studies of the kinematics of stars—how they move in complex orbits around the center of the Galaxy—provide some information on their distribution in three dimensions. An intriguing result that has emerged in the last decade is that the bulge is in the shape of a bar, albeit a "mild" bar, about twice as long as it is wide.

360

Figure 10.7 The Milky Way galaxy from inside. This view of our galaxy, taken in the infra-red portion of the spectrum using instruments on the Cosmic Background Explorer (COBE) satellite, shows our galaxy's thin disk of stars, and dust (which appears red or orange in this image) within the disk. Infra-red light penetrates dust and gas much better than visible light, so the image reveals much more of the central swath of the Galaxy than we could see with traditional telescopes. (See color section.) (Copyright Edward L. Wright. Used with permission.)

The long axis points nearly toward our Sun's position in the disk. If the Milky Way does have a bar, it is in good company. (No, the bar at the center of the Milky Way is not a good place to have a drink!) Recent research suggests at least one third of all spiral galaxies are barred, although the bar may be more pronounced when the galaxy is imaged using a colored filter.[12]

The galactic center

The very heart of the Milky Way, at the center of both the disk and bulge, is almost impossible to study in the visible wavelengths. Interest in this region has been high, however, since the 1950s. That's when astronomers found a strong, compact radio source in the direction of the constellation Sagittarius—a radio signal not associated with any known star or stellar remnant. The source is known as Sagittarius A*, pronounced "Ay-star." The unusual characteristics of Sagittarius A* have long fueled speculation that the center of the Galaxy harbors an exotic object called a black hole.

Early investigators in Australia came across Sagittarius A*
during the course of sky surveys with large radio antennae oper-
ating at wavelengths of 25 centimeters to 3 meters. In their reports
in scientific journals they described the source of the radio signals
with phrases such as "remarkably powerful" and "unusually
intense."[13]

The source lay in the direction of the center of the Galaxy
but, without knowing how far away it was, the investigators
and other astronomers were initially cautious about identifying
it with the nucleus of the Milky Way. They were tempted to do
so, however. At about the same time that Sagittarius A* was discovered, astronomers had found that some strong radio sources
in the sky corresponded to the cores of nearby galaxies.

By 1960, the consensus was that the source of the radio signal
coincided with the center of the Galaxy. The Dutch astronomer
Jan Oort, writing in a popular text of the late 1960s, said of this
"striking" source, "As far as we can judge it lies precisely in
the direction where the center of mass of our Galaxy is thought
to be, and is probably connected with this."[14] In 1971, Donald
Lynden-Bell, then at England's Royal Greenwich Observatory,
and Martin Rees at Cambridge University laid out persuasive
arguments for the existence of a black hole, surrounded by a
ring of dust and gas, at the center of the Milky Way and at the
origin of the Sagittarius A* source.[15] However, as late as the
mid-1980s, many astronomers doubted the existence of a black
hole at the center of our galaxy, and sought alternative explanations for the unusual characteristics of Sagittarius A*.

"Black hole" is the term usually applied to a massive star that
has collapsed under its own weight so that its gravity, which
arises from its mass, is unusually concentrated. Dust and gas
molecules and matter particles making up the star cannot
escape, and matter outside the black hole that comes within a
certain radius will be forever trapped by this extraordinary gravitational field. Even light waves coming from the collapsed object
are bent back into the gravitational well — hence the name "black
hole." Supermassive black holes, which is a more accurate term
for black holes at the centers of galaxies, are black holes that
formed not from a single star, but from the collapse of a giant
gas cloud or star cluster, or perhaps from the merger of two
galaxies with small black holes at their centers. Both kinds of

black hole may grow by accreting gas and stars, or by mergers with other black holes.

In the 1970s, astronomers understood the basic properties of black holes from a theoretical standpoint, and were still searching for evidence of their existence. Their reasons for suspecting that a black hole might lurk at the center of our galaxy and at the centers of other galaxies included strong theoretical arguments and some observations indicating that the central object was very compact.

The development of infrared and sub-millimeter imaging technology in the 1980s and 1990s has allowed us to part the curtains on the center of the Galaxy and the area around the supposed black hole, just as it provided the first view of the bulge itself. Since the mid 1990s, astronomers have exploited that view to build a substantial body of knowledge about the galactic center and to amass very strong evidence pointing to a supermassive black hole. They're still working on it, but the picture they are piecing together shows a cornucopia of odd and uncommon objects surrounding the black hole candidate at the center (figure 10.8).

If we were to approach the center of the Galaxy from somewhere deep in the bulge, say about 500 light-years from the galactic center, the first thing we would notice would be the very high density of stars—mostly rather cool, yellowish stars—and their increasing concentration as we traveled farther in. But stars are not the only denizens of the inner bulge. If we had the ability to see various types of molecules, we would also observe giant gas clouds as we sailed through them, warmer and denser than those elsewhere in the disk, and very turbulent. We would continually be buffeted by long filaments and arches of gas coming at us from all directions, ghostly shells of material cast off from supernova explosions. Looking up, out of the plane of the galaxy, we might observe giant bubbles shaped by magnetic fields, expanding out and away from the galactic center. If we were equipped with x-ray vision, we would notice around us an intense flickering light from several isolated sources. The x-rays probably arise from heated gas feeding stellar black holes or neutron stars.

If we chose our angle of approach just right, we could catch a ride on a fast-moving river of gas about 30 light-years long, emanating from one of the giant molecular clouds and pointed

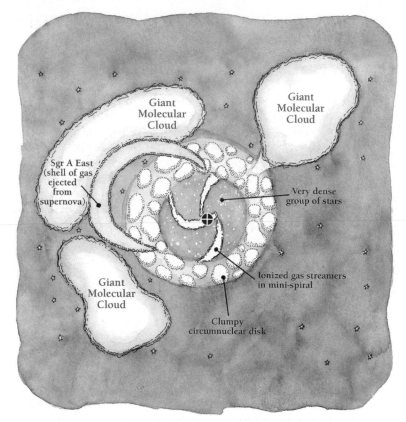

Figure 10.8 The galactic center. At the very center of our galaxy lies a black hole with a mass some 3 million times that of the Sun. Other unusual structures can be found in this innermost region of the Milky Way. A clumpy disk or ring of gas surrounds the black hole, and streamers of hot gas are apparently spiraling into the black hole area from the disk. A shell of gas, like that which might be ejected from a powerful supernova, looks like it is expanding into one of several giant molecular clouds—a hint of turbulent times in the relatively recent past. (Credit: Layne Lundström, adapted from Genzel *et al* (1995).)

in the general direction of the galactic center. Our ride might deposit us at a point about 25 light-years from the galactic center. There we would find ourselves in a vast rotating ring of gas and dust clouds. This is the circumnuclear disk.

The clouds within the circumnuclear disk are about a light-year across, on average, and they bob up and down as well as rotate around the galactic center at speeds of about 100 kilometers per second. The disk probably draws material in from some of the outlying molecular clouds, through the river or streamer of gas.

The circumnuclear disk has a distinct inner edge at about 5 light-years from the galactic center. Crossing this boundary, we would find that space appears to have been swept clear of much of the gas and dust that we noted in the disk and beyond. This region is sometimes called the central cavity. The cavity is not completely empty, however; it is only low in interstellar material. Hundreds of thousands of stars swarm in this space. Many are red giants and some, very close to the center and the Sagittarius A* source, are young massive blue stars.

Recent observations have also delineated three arc-shaped streamers of ionized gas in the central cavity. The streamers have the shape of a mini-spiral, and they appear to converge at the location of the proposed black hole like streams of bath water headed for the drain. These streamers are apparently funneling gas from the circumnuclear disk to the black hole, just as the 30 light-year-long streamer seen outside the disk may connect the disk to a source of material in one of the giant molecular clouds.

In the immediate vicinity of the suspected black hole, the density of stars is extremely high, possibly more than a million times that in the neighborhood of the Sun. The view would be spectacular, if blinding. It would also be quite instructive for a student of stellar structure and evolution. Space is so thickly studded with stars, at that level, that collisions between stars become inevitable. Stellar collisions and the mixing of stellar atmospheres are extremely rare outside this special zone of the Galaxy, and the kinds of stars that result are unknown to us in the far reaches of the Milky Way's disk.[16]

What of the black hole that we suppose is at the very center of the Galaxy, our journey's true heart of darkness? The black hole itself remains difficult to visualize and impossible to image with telescopes and detectors.

Within a certain perimeter of a black hole, called the event horizon, everything belongs to it. Outside the perimeter, stars or gas clouds can orbit a black hole as they can any other massive body. Contrary to the popular imagination, a black hole is not like a malevolent vacuum cleaner, capable of sucking up everything around it. However, any star whose orbit takes it within the event horizon will add to the black hole's mass — and that will enlarge the perimeter, albeit by a very small amount. So, as long as there are stars (or gas clouds) nearby to feed on, a black hole may grow by enlarging both its mass and its perimeter.

If we could hang back and watch a star approach the perimeter of a black hole, we would see the effects that an intense gravitational field has on space and time — effects described in Einstein's theories of relativity. The light from the star would become redshifted — that is, even if it emitted mainly in the visible portion of the spectrum, as the Sun does, it would appear to emit mainly in the infrared, then in the radio regime, and in practice the star would eventually become undetectable, as the wavelength of its emission became very long. The star itself, anyway, would be stretched into a long spike and ripped apart. As it neared the perimeter of the black hole it would encounter a rapidly increasing force of gravity that would pull it to pieces. And if the star pulsed regularly so that we could gauge the passage of time, we would have the impression that those pulses were delayed by the star's sinking into the gravitational well of the black hole — as though time itself slowed down for the star. These effects follow directly from the intense gravity of a black hole or supermassive black hole, and the intense gravity can arise from the collapse of an ordinary massive star, yet the net result is an object that seems to belong to another, weirder, universe.[17]

A few still doubt the existence of a supermassive black hole at the center of the Milky Way. However, recent observations have provided the best evidence yet for its existence.

In the mid-1990s, two teams of astronomers, one at the University of California, Los Angeles, and another at the Max Planck Institute for extra-terrestrial physics (MPE) in Garching, Germany, recorded the motions of stars careening around the very edges of the supposed black hole at the center of the Galaxy. Some of these stars are whipping around at speeds more than 1000 km per second. The motions of the stars are governed by the mass holding them in orbit, so the measurements allowed the researchers to calculate the mass of the dark matter or black hole. The most recent results imply a mass of about 3 million times the Sun's mass, inside a radius of that is at most like the scale of the solar system.[18]

As observations fill in more details, it may become possible to rule out alternative explanations for the proposed black hole — alternatives such as a concentration of neutron stars or other massive, low-luminosity objects. We may also gain a better understanding of the energy balance at the core of the Galaxy. At present, it seems that the black hole, while radiating powerfully in the long wavelength or radio part of the spectrum, is less bright than expected in the shorter wavelengths. It may be radiating energy less efficiently than predicted, or it may be "starving" — unable to satisfy its appetite for gas because of strong stellar winds keeping material out of its reach.

We have much yet to learn about the dark heart of the Galaxy. However, new observations both from the Hubble Space Telescope and from ground-based observatories have greatly increased our understanding and the case for a supermassive black hole. As one astronomer put it recently, "We have now moved from debating the existence of supermassive black holes to asking what regulates their formation and evolution, and how their presence influences, and is influenced by, their host galaxies."[19]

The Local Group — the Milky Way's neighborhood

The Milky Way is one of 10 billion galaxies in the universe — or maybe 100 billion... no one knows for sure. Viewed from a distance, the galaxies appear to link together to span space like soap suds fill a sink. They associate with other galaxies in groups and clusters that form extended sheets and surfaces of bubbles, outlining irregular dark voids.

The Milky Way belongs to the "Local Group" of galaxies, which in turn connects to other groups and clusters to form the "Local Supercluster." The Local Supercluster is dominated by the Virgo Cluster of about 2500 galaxies, some of which one can see with binoculars or a small telescope. The Local Supercluster has a diameter of at least 100 million light-years.

The Local Group is spread over a more modest 3 or 4 million light-years. About 35 galaxies that we know of can claim membership; the exact number depends on how one defines the boundaries of the Local Group. Most of the galaxies are relatively faint, like smeared-out globular clusters, and escaped notice for a long time. Astronomers discovered new "dwarf" or faint Local Group members as recently as 1990, 1994, 1997, and 1999, and the search continues.[20]

The Milky Way and the Andromeda galaxies, both large spirals, dominate the Local Group in terms of mass and luminosity. Most of the rest of the Local Group galaxies are companions to these, entrained by the larger systems' gravity. Even M33, the only other spiral in the Local Group, belongs to the Andromeda galaxy's domain.

Astronomers have a new outlook on dwarf galaxies since the early 1980s. Before then, it was easy to overlook them as insignificant members of the Local Group. But dwarf galaxies turned out to carry a surprising amount of dark matter. In fact, dark matter dominates their make-up. And while spiral galaxies like the Milky Way have a bright bulge in the middle and extended dark halos, the dwarf galaxies are dark even in their middles.

Some astronomers have now proposed a theory that gives dwarf galaxies center stage in the drama of galaxy formation. The early universe may have spawned very dark dwarf-type galaxies, rich in gas. These would have agglomerated to form larger systems, which became the spiral galaxies. The dark matter at the centers of the larger systems would have gravitationally attracted the gas, which eventually formed the luminous stars, leaving the halo regions dark. The Local Group dwarf galaxies that we currently see would generally be outlying remnants that took longer to participate in spiral-building.[21]

However the theory stands up to further investigation, it is clear at least that the Milky Way is currently merging with lesser galaxies such as the Sagittarius dwarf, and has stripped gas from other galaxies in the past. The Magellanic Stream provides evidence of an encounter about 100 billion years ago, for example. Astronomers have tracked a filament of hydrogen gas that is anchored in a cloud encompassing both the Small and Large Magellanic Clouds. As the Magellanic Clouds drifted near the Milky Way, the gas dragged behind, and now leaves a trail that arcs about one quarter of the way around our galaxy. Other evidence of past merger activity comes from observations by the Sloan Digital Sky Survey of galaxy detritus, left behind after the Milky Way had a "snack" on the Sagittarius dwarf galaxy a few billion years ago.[22]

Many astronomers now believe that mergers and interactions among galaxies help define their properties, so that a galaxy's "environment" is as important to its evolution as is the intrinsic factor of mass. The Local Group is too small to readily show the effect of environment, but studies of a number of larger clusters have revealed distinct patterns. Elliptical galaxies, stripped of most of their gas and populated by older stars, tend to congregate at the centers of rich clusters, leaving spirals at the edges. This segregation of galaxies makes sense if, as is supposed, mergers involving spiral galaxies lead to the formation of ellipticals. Spirals near the centers of rich clusters would have many opportunities to merge and transform, and so would be rarer than spirals in the quieter outer regions of the clusters.

Since the 1920s, when we first conceived of our stellar system as one of many island universes, our view of the Milky Way has undergone many transformations. We no longer think of it as lying at the center of the universe, nor even as a "continent" among islands. Our galaxy is large, but not outstandingly so; even within our small Local Group, the Andromeda galaxy is slightly more massive. Our galaxy has a supermassive black hole at the center, but not a particularly impressive one, and it is not stirring up enough gas around it to produce bright x-rays and other indications of activity. But this rather modest, temperate galaxy that we call home has enough wonders and curiosities, particularly in the bulge and halo, to keep astronomers busy for a long time. And our galaxy shares with others a mysterious origin that is somehow connected to dark matter, whose existence we were not even fully aware of until about 30 years ago. Considering that more than 90% of the universe consists of unknown dark matter, we may have to admit that the stellar system we have studied so long, and with so much success since the middle of the twentieth century, may add up to only a tiny part of the galaxy—as though we had been studying the foam on a breaking wave, and thought we understood the ocean.

NOTES

Notes for chapter 1

1. Laplace (1835) pp 550–551, author's translation
2. Caroline Herschel's phrase about minding the heavens appears in Herschel (1876) p 77
3. Heilbroner (1953)
4. Information on the Voyager spacecraft is available from a web site maintained by NASA's Jet Propulsion Laboratory: <http://www.jpl.nasa.gov/calendar/voyager1.html>
5. Versions of this story are told in, e.g. Condos (1997) pp 109–110, and Henbest and Couper (1994) p 4
6. Drake (1957) p 49
7. Wright (1750) p 177
8. Wright (1750) p 177

Note for chapter 2

1. Raymo (1982) p x

Notes for chapter 3

1. Quoted in Jaki (1990) p 88
2. Hughes (1951) p 4
3. Hughes (1951) p 4
4. Hughes (1951) p 8
5. Wright's journal, with the alterations attributed by historian Edward Hughes to Wright's friend George Allan, is reproduced in Hughes (1951). In the year 2000 I had the opportunity to see the original manuscript in the British Library. The third person pronouns appear in darker ink than the rest of the text, and the paper at those locations has been roughened by the erasures. The same hand apparently

clarified some of the individual letters in Wright's words, e.g., the "t" and the "s" in the word "indentors" in the entry for November 1729

6. Both quotes, Hughes (1951) p 3
7. Hughes (1951) p 3
8. Hughes (1951) p 3
9. For a description of a watchmaker's apprenticeship in this era see Weiss (1982)
10. Hughes (1951) p 4
11. Hughes (1951) p 5
12. "Lest the system..." Newton (1729) p 389. On the Creator's "reformation" of the solar system, see Newton (1730) p 402
13. Wright might have come to Derham's work first from his 1714 treatise, *Artificial clock-maker. A treatise of watch and clock-work, shewing to the meanest capacities, the art of calculating numbers to all sorts of movements ... With the antient and modern history of clock-work*
14. Title of document in Durham University Library's collection of Thomas Wright manuscripts, reference number GB-0033-WRM 6/7
15. I am indebted to Michael Hoskin (Hoskin, 1971, introduction) for illuminating the context for this remark by Wright
16. "Great Notice...Pavement Figure" Hughes (1951) p 5
17. Description in <http://www.bmz.amsterdam.nl/adam/uk/groot/paleis1.html>
18. Both quotes, Hughes (1951) p 5
19. All quotes, Hughes (1951) p 5
20. Hughes (1951) p 16
21. All quotes, Hughes (1951) pp 8–9
22. All quotes, Hughes (1951) pp 9–10
23. Quoted in W*right (1750) p xiii
24. "Sacred Throne of Omnipotence, Wright (1750) p 3; myriads ("meriads") of systems, p 9
25. "General Representation of Euclid's Elements..." Hughes (1951) p 7; "Universal Vicissitude..." Wright (1750) p xiii
26. Wright (1750) p 5
27. Hoskin (1971) p 3
28. Wright describes the divine presence as "full in the middle of ye principal scheme," but later discusses his "grand section or rather sector of ye creation," which clearly has one "end of the scheme comprehending the center, the other [end]...supposed to be the most remote from the focus of power." Wright (1750) pp 5–7
29. "near ye center...": Hoskin (1971) p 8; "proportionable removed... human eye": Wright (1750) p 9
30. Halley (1718). One of the reasons Halley ruled out precession of the equinoxes and refraction as sources of error is that the Moon had

occulted Aldebaran, as seen from Athens in 509 AD; this did not seem possible unless Aldebaran had shifted its position more than could be explained by precession
31. Quoted in Weiss (1982) p 20
32. Wright (1750) p 109
33. Hughes (1951) pp 12–13
34. A reproduction of this rare plate is in Dingle and Martin (1964) p 115
35. Wright (1742) p 75
36. Wright (1742) p 16
37. Wright (1750) pp 91–92
38. Wright (1750) p 20
39. Wright (1750) p 91
40. Wright (1750) p 31
41. Wright (1750) p 77
42. Both quotes, Wright (1750) p 43
43. Wright (1750) p 52
44. Wright (1750) p 92
45. Wright (1750) p 97
46. Wright (1750) p 100
47. Wright (1750) p 162
48. Wright (1750) p 117
49. Both quotes, Wright (1750) p 134
50. Wright (1750) p 134
51. Wright (1750) p 138
52. Wright (1750) p 137
53. Wright (1750) p 153
54. Wright (1750) p 162
55. Wright (1750) p 155
56. Wright (1750) p 177
57. Both quotes, Wright (1750) p 171
58. Hastie (1968) p 18
59. Hastie (1968) p 50
60. Hastie (1968) p 21
61. Hastie (1968) pp 49–50
62. Hastie (1968) p 42
63. Allen (1793a) p 127
64. Allen (1793b) p 215
65. Details of his retirement years emerge from his description of his house—see Allen (1793b)
66. Wright (n.d.) p 79
67. Wright (n.d.) p 26
68. Wright (n.d.) pp 10–11
69. Wright (n.d.) p 63

70. Hoskin, 2000. Personal communication
71. "wholly-unsuspected:" Wright (n.d.) p 8

Notes for chapter 4

1. Quoted in Panek (1998) p 149
2. Herschel's description of his meeting with Watson is quoted in Clerke (1895) p 22 and MacPherson (1919) pp 19–20
3. Quoted in MacPherson (1919) p 9
4. Quoted in Gingerich (1984) p 77
5. Quoted in MacPherson (1919) p 10
6. Herschel (1876) pp 10–11
7. Gingerich (1984) p 79
8. Readers may, to some extent, discover Herschel's music for themselves. Two CDs are available: one a collection of his works for organ, the other a compendium that includes two oboe concerti and a chamber symphony for strings and continuo. The organ music, "Pièces d'Orgue de William Herschel," played by Dominique Proust, an astrophysicist, is available from Disques DOM, 4–6 rue du Donjon, 94300 Vincennes, France. The orchestral works are available from Newport Classic, 106 Putnam Street, Providence, RI 02909, USA
9. Herschel (1876) p 8
10. Herschel (1876) p 20
11. Herschel (1876) pp 19, 24–5
12. Herschel (1876) p 26
13. Herschel (1876) p 33
14. Herschel (1876) p 34–5
15. Gingerich (1984) p 79
16. Herschel (1876) p 35
17. Gingerich (1984) p 79
18. Herschel (1876) p 38
19. Herschel (1876) pp 37–38
20. "Slender arms" in MacPherson p 17; "infinite satisfaction" in Gingerich (1984) p 79
21. Gingerich (1984) pp 79–80
22. Quoted in MacPherson (1919) p 42
23. Herschel (1876) p 50
24. Herschel (1876) p 43
25. Herschel (1876) p 44
26. Herschel (1784) p 441
27. Herschel (1784) p 441

28. Herschel (1784) p 441
29. Herschel (1876) pp 41–42
30. Quoted in MacPherson (1919) p 27
31. Herschel (1876) p 52
32. Herschel (1876) p 325
33. Herschel (1876) p 55
34. See, e.g., Herschel (1811) p 269
35. For a discussion of Herschel's copy of Wright's Original Theory, see Hoskin (1963) pp 115–116
36. Herschel (1784) p 449
37. "Nebulous ground" quoted in Herschel (1784) p 449
38. Herschel (1784) p 442
39. Herschel (1785) p 244
40. Herschel (1785) p 245
41. Herschel (1785) p 244. Note that although the term "island universe" was not yet in use, Herschel did compare our system to an island in the 1785 paper, p 247
42. Herschel (1785) p 219
43. Herschel (1876) p 57
44. Herschel (1876) p 147
45. In 1798, Herschel claimed to have discovered four other satellites of Uranus, but these four proved to be spurious
46. Herschel (1876) p 73
47. Herschel (1876) p 77
48. Herschel (1876) pp 78–79
49. Herschel (1876) p 178
50. All passages in Herschel (1789); concluding paragraph is p 226
51. Herschel quotes his own notebook entries and speculates on the nature of the self-luminous matter in Herschel (1791) pp 82–84
52. Herschel (1791) p 85
53. Herschel (1800) p 272
54. Herschel (1876) p 116
55. See Clerke (1895) p 146
56. Herschel (1817) p 302
57. This phrase appears as a section heading in the 1817 paper
58. Herschel (1817) p 309, p 304
59. "Faintness means farness" is Owen Gingerich's apt description of the common assumption
60. Herschel (1817) pp 309–311
61. See Hoskin (1963) p 175
62. Herschel (1817) p 326
63. Herschel (1817) p 327
64. Quoted in Armitage (1963) p 37

Notes for chapter 5

1. Herschel (1841) p 97
2. Struve (1895) p 9
3. Quoted in a history of the Tartu university given on the university's web site and in travel guides to Tartu, e.g. see <http://travel.lycos.com/> and search on Estonia and Tartu
4. Struve (1895) p 21
5. Felber (1994) p 14, author's translation
6. Quoted in Batten (1988) pp 46–47
7. Batten (1988) p 75
8. Felber (1994) p 199
9. Quoted in Kellner (1963) p 34
10. Felber (1994) p 72
11. Struve (1895) p 48, author's translation
12. Struve (1895) pp 48–49
13. Struve (1895) p 55, author's translation
14. Batten (1988) pp 125–127
15. Quoted in Burnham (1900) p xiv Pub. Yerkes Obs. vol 1, p xiv
16. Dick and Ruben (1988) p 119
17. Dick and Ruben (1988) p 119
18. Felber 1994 pp 107–108
19. Struve (1957) p 71
20. Bessel (1838) p 152
21. Dick and Ruben p 120
22. Herschel (1841) p 9
23. Quoted in Batten (1988) p 127
24. Struve (1847) p 67, author's translation
25. Struve (1847) p 49
26. Struve (1847) p 63
27. Struve (1847) p 67
28. Quoted in Batten, p 151
29. Quoted in Batten, pp 158–159

Notes for chapter 6

1. De La Rue (1861) p 130
2. The biography is referenced as Mills and Brooke, 1936. Lady Huggins' manuscript passed to a friend's brother, and after both he and his sister died, to the executor of her friend's estate, Charles E Mills. Mills and his collaborator C F Brooke, who confessed in the preface to having "no knowledge of the mysteries

of science," edited the partially completed manuscript and brought it out as a small book

3. Mills and Brooke (1963) p 7
4. Most of the objective information about Huggins' life and career comes from the PhD thesis of Barbara Becker—see Becker (1993) and references therein
5. See Gernsheim and Gernsheim (1958) p 720 for details of Daguerre's kit
6. See Mills and Brooke (1963) p 9
7. Quotations are all from Mills and Brooke (1936) p 17
8. Mills and Brooke (1936) pp 18–19
9. Huggins produced his own lunar photographs—see Becker (1993) pp 36–37
10. The first of these is Huggins (1856) pp 175–177
11. For an English translation and partial reprint of Kirchhoff and Bunsen's 1860 paper in *Annalen der Physik und der Chemie* **110** 161–189, see Farber (1966) pp 19–25; this quote, p 24
12. Farber (1966) p 24
13. Farber (1966) p 25
14. Moulton (1924) p 392
15. In 1834, for example, the positivist philosopher Auguste Comte mentioned the chemical composition of the stars as an example of a question that could never be answered by science. See, e.g., Serres *et al* (1975) p 301
16. See note 1 in this chapter
17. Huggins and Huggins (1909) p 6
18. Details of the experimental set-up are in a paper on the spectra of the chemical elements, Huggins and Miller (1864a), and in the paper on stellar spectroscopy, Huggins and Miller (1864b)
19. Huggins and Miller (1864b) p 413
20. Huggins and Miller (1864b) p 418
21. Huggins and Miller (1864b) p 433
22. Huggins and Miller (1864b) pp 433–434
23. Huggins and Miller (1864b) p 434
24. Huggins (1864) p 438
25. Huggins (1865) p 42
26. Huggins (1865) p 42
27. The chief nebular line was not identified until 1928. It is actually a closely spaced double line and is due to oxygen atoms in a physical state that is difficult to replicate in a laboratory
28. Huggins (1865) p 42
29. Discourse at Nottingham reprinted in Huggins and Huggins (1909); see p 503

30. Fournier d'Albe (1923) pp 154–155 and 172
31. Huggins (1868) pp 548, 549
32. Quoted in Becker (1993) p 213
33. Quoted in Becker (1993) p 213
34. Quoted in Becker (1993) p 215
35. Lockyer (1874) p 255
36. For biographical information about Margaret, see Bruck (1991) and Bruck and Elliott (1992)
37. Quoted in Becker (1993) p 269
38. Quoted in Becker (1993) p 273
39. Quoted in Plotkin (1982) p 323
40. Quoted in Plotkin (1982) p 326
41. For a reference to the women as Pickering's harem, see, e.g. Welther (1982) p 94
42. Davis (1898) pp 223–224
43. Quoted in Becker (1993) pp 404–405
44. Quoted in Becker (1993) p 272 n 106
45. Quoted in Becker (1993) p 272
46. Huggins (1889) p 40
47. His style here is so different from that of his earlier scientific papers, one wonders if Margaret influenced him to drop his normally formal and long-winded mode of expression. On the historical accuracy of the article, see Becker (1993) especially pp 5–7, 83–84, 324 and 416–419
48. Reprinted in parts in Huggins and Huggins (1909); see pp 5–6
49. Reprinted in Huggins and Huggins (1909) pp 523–539
50. Quoted in Becker (1993) pp 288–289
51. Quoted in Becker (1993) pp 291–292

Notes for chapter 7

1. Kapteyn (1909) p 46
2. The story is recounted in Krul (2000) pp 67–68
3. Paul (1993a) p 7
4. Both quotes: Paul (1993a) pp 14, 15
5. Paul (1993a) p 21
6. Paul (1993a) p 22, with corrections by Van Berkel and Van der Kruit (2000) p 372
7. Paul (1993a) p 46
8. Gill (1896) p xii
9. Quoted in Paul (1993a) pp 26
10. Henrietta seemed to think Kapteyn learned of Gill's work only through Gill's article, but Kapteyn met and corresponded with Gill

before Gill's article appeared, and Gill may have appealed rather directly to Kapteyn for help See Krul (2000) p 55

11. Paul (1993a) p 27
12. In February 1923, the writer of an obituary of Kapteyn, who signed himself only "J. J.," wrote in the *Monthly Notices of the Royal Astronomical Society* that Kapteyn had help from prison convicts. This story cannot be substantiated. See Krul (2000) p 73
13. In fact, it progressed unevenly until another generation of astronomers wrapped it up in 1970, and the atlas portion was in some ways superseded by later sky surveys that took advantage of advances in photographic technology
14. Paul (1993a) p 37
15. Krul (2000) p 62
16. Quoted in Jones and Boyd (1971) pp 35–36
17. Eddington (1938) p 169
18. Paul (1993a) pp 61–62
19. Seares (1922) p 233
20. Wright (1966) p 78
21. Quoted in Jones and Boyd (1971) p 240
22. Quoted in Krul (2000) p 77
23. Paul (1993a) p 75
24. Kapteyn (1914) p 153
25. Quoted in Paul (1993a) p 79
26. Paul (1993a) p 87
27. Quoted in Gingerich (2000) p 201
28. Quoted in Paul (1993b) p 210

Notes for chapter 8

1. H N Russell to H Shapley, 17 September 1920. Quoted in Smith (1982) p 89
2. Letter from H Shapley to J C Kapteyn, 6 February 1917. Quoted in Gingerich (1970) p 346
3. Shapley (1969) p 12
5. Quoted in DeVorkin (2000) p 104
4. Shapley (1969) p 11
6. Shapley (1969) p 22
7. Shapley (1909)
8. Landi, 2001. Reference Archivist, University of Missouri–Columbia. Personal communication
9. Shapley (1969) p 24
10. Kopal (1981) p 261

11. Shapley (1969) p 31
12. Shapley (1969) p 32
13. See DeVorkin (2000) p 257, and references therein
14. Lorett Treese, Bryn Mawr College Archivist (2001). Personal communication
15. Matthews (2001). Personal communication
16. Kopal (1986) p 158
17. See note 24 for a list of her papers
18. Shapley (1969) p 42
19. Shapley (1969) p 40
20. Shapley (1969) p 41
21. Mildred Shapley Matthews (2001). Personal communication. George Hamilton Combs was the author of several monographs, including *The New Socialism* and *The Call of the Mountains*
22. Shapley (1969) p 49
23. Shapley (1969) p 55
24. The five children were born in February 1915, March 1917, March 1919, June 1923, October 1927 (Mildred Shapley Matthews (2001), personal communication). Martha's papers appeared in *Ap. J.* **42** 148–162, 1915 (submitted 2 April 1915); *Ap. J.* **44**, 51–58, 1916 (submitted 3 April 1916); *Ap. J.* **45**, 182–188, 1917 (submitted January 1917); *Ap. J.* **46**, 56–63, 1917 (submitted April 1917); *Ap. J.* **50**, 42–49 (submitted September 1918) and 107–140, 1919 (submitted November 1918); *Publications of the Astronomical Society of the Pacific* **30**, 343–346, 1919; *Harvard Observatory Circular* No. 238, 1922; *Harvard Bulletin* No. 797 (1924); *Harvard Bulletin* Nos. 843, 844, 845 (1927)
25. That Shapley was thinking about the origins of the Milky Way, not just arrangement of stars, can be seen from his correspondence and minor papers. In 1918, writing to Hale, he suggested that stars in globular clusters were younger or had evolved more leisurely. In 1919, Shapley thought the Milky Way may have originated in the combination of two clusters and that it continued to accrete clusters. See Smith (1982) pp 63, 91
26. Shapley (1915)
27. Shapley (1969) p 53
28. See H Shapley to A Eddington, quoted in Smith (1982) p 60. Scutum would have been observable in summer and fall
29. Shapley (1917) p 216
30. Quoted in Smith (1982) p 32, and see also p 156
31. Shapley to Russell, 3 September 1917, see Berendzen and Hart (1973) p 53
32. Shapley to Russell, 31 October 1917, see Smith (1982) p 60
33. Smith (1982) p 36

34. Haramundanis (1996) p 177
35. Quoted in Gingerich (1970) p 347
36. All quotes are from Shapley to Hale, 19 January 1918, quoted in Smith (1982) p 62
37. Hale to Shapley, 14 March 1918, quoted in Smith (1982) p 68
38. 10 December 1917, Adams to Hale. The exchange is quoted in Smith (1982) pp 76–77
39. Shapley (1969) p 65
40. Shapley (1969) p 57
41. Hoskin (1976) p 169
42. H Shapley to H Russell, 31 March 1920, quoted in Smith (1982) p 80
43. Shapley (1969) p 78
44. For details about the debate, as opposed to the printed versions of the talks, see Hoskin (1976)
45. All quotes, see Hoskin (1976)
46. Curtis (1921) p 217
47. Both quotes, Hoskin (1976) p 174
48. Bok (1974) p 56
49. Baade (1963) p 9
50. Hoagland (1965) p 429
51. Haramundanis (1996) p 124
52. Haramundanis (1996) p 141
53. Haramundanis (1996) p 154
54. Haramundanis (1996) p 154
55. Hers was also, necessarily, the first astronomy PhD awarded to a woman at Harvard, but it was not the first astronomy PhD awarded to a woman in the United States—see Haramundanis (1996) p x
56. Hogg (1965) p 336
57. H Shapley to A van Maanen, 8 June 1921, quoted in Berendzen and Hart (1973)
58. Haramundanis p 209
59. Bok (1965) p 416
60. Shapley (1956) p 73
61. See http://www.nypl.org/research/chss/spe/rbk/faids/Emergency/index.html
62. Shapley (1969) p 127
63. Bok (1976) p 254
64. The history of the NSF, and quotes relating to it, are from Blanpied (1998)
65. Bok (1976)
66. Shapley (1969) pp 145–149
67. Shapley's passing of the torch to Needham and Huxley is revealed in a State Department memo of 29 August 1945 (501.PA/8–2945). I am

indebted to Dr. Gail Archibald of Unesco's Section on Engineering Sciences and Technology and Mr. Mahmoud Ghander at Unesco's Archive services in Paris, for their help Dr. Archibald kindly shared with me her presentation copy of a paper she presented at the XXIst International Congress of History of Science, Mexico City, 8–14 July 2001, on "Putting the 'S' in Unesco (1943–1945)."

68. Bok (1976) p 256
69. Shapley (1969) p 156
70. Shapley (1969) p 154
71. Shapley (1969) p 167
72. See Matthews, M S and Wilkening, L L (1982) *Comets* (Tucson: University of Arizona Press); Binzel, R P, Gehrels, T, and Matthews, M S (1989) *Asteroids II* (Tucson: University of Arizona Press); Kieffer, H H, Jakosky, B M, and Matthews, M S (eds) (1997) *Mars* (Tucson: University of Arizona Press); Lewis, John S, Matthews, Mildred S, and Guerrieri, Mary L (eds) (1997) *Resources of Near-Earth Space* (Tucson: University of Arizona Press)

Notes for chapter 9

1. Crommelin (1917) p 376
2. Biographical details such as these are from Christianson (1995) unless otherwise noted
3. The visibility of planets can be checked at <www.skyviewcafe.com>
4. Quoted in Christianson (1995) p 31
5. Edwin's feelings about his father are quoted in Christianson (1995) p 15
6. Quoted in Brush (1978) p 9
7. Christianson (1995) p 67
8. Christianson (1995) p 109
9. Mayall (1970) pp 188–189
10. Burbidge (1994) p 23
11. Christianson (1995) p 153. Note that Christianson gives the date of the 1922 meeting as July, not May. The Commission on Nebulae is now known as the Commission on Galaxies. The International Research Council became the International Council of Scientific Unions in 1931
12. Hubble (1922) pp 162–199
13. Barnard (1913) p 500
14. Hubble (1922) p 189
15. Hubble (1922) p 162
16. Hubble (1922) p 167

17. Hubble (1922) p 166
18. Jeans (1919) p 206
19. Jeans (1919) p 209
20. Hart and Berendzen (1971) p 112
21. Hubble (1923) p 262
22. E Hubble to H Shapley, 19 February 1924, quoted in Smith (1982) p 141
23. Quoted in Christianson (1995) p 22
24. Hart and Berendzen (1971) p 114
25. Exchange is quoted in Smith (1982) pp 118–119
26. Quoted in Smith (1982) p 141
27. Quoted in Christianson (1995) p 161
29. Quoted in Smith (1982) p 120
29. Quoted in Berendzen and Hoskin (1971) p 8
30. Christianson (1995) p 161
31. Hubble (1925) p 432
32. Hubble (1926) p 325
33. Hubble (1926) p 351
34. Smith (1982) p 151
35. Hubble (1928) p 103
36. Hubble (1928) p 145
37. North (1965) p 109
38. Trimble (1996) p 1076
39. Christianson (1995) p 188
40. Hubble (1929)
41. Kragh (1987) p 127
42. Kragh (1987) p 128
43. Kragh (1987) p 130
44. Lemaître (1931) p 406
45. Lemaître (1931) p 408
46. Lemaître (1931) p 410
47. Smith (1982) p 187
48. Christianson (1995) p 347
49. Christianson (1995) p 350
50. Smith (1982) p 131
51. Christianson (1995) p 233
52. Christianson (1995) p 246
53. Hubble (1936) p 56
54. Hubble (1954) p 59
55. Hubble (1954) p 64
56. Hubble (1954) p 69
57. Christianson (1995) pp 254, 306
58. Hubble (1954) pp 77, 75

59. Hubble (1954) p 72
60. *Time Magazine,* 9 February 1948, p 56
61. Sandage (1989) p 352
62. Sandage (1961) p 1

Notes for chapter 10

1. Quoted (in French) in Berry (1961) p 307; author's translation
2. Four-armed spiral models of the galaxy derive from that of Georgelin and Georgelin (1976). Bertin and Lin discuss the number of arms in chapter 6 of their book (Bertin and Lin, 1996). Various well-consulted sources disagree on the continuity and hence the naming of spiral arms. The authors of the textbook *Voyages* (Fraknoi, Morrison, and Wolff, 1997) show four major arms, three of which they name. These are the Carina, Perseus, and Cygnus arms. The authors of the excellent *Guide to the Galaxy* (Henbest and Couper, 1994) show an "outer" arm, the Perseus arm, the "Local" arm, the Sagittarius arm, and, further in, the Scutum-Crux arm, the Norma arm, and the 3-kiloparsec arm. The National Geographic Society publishes a map of the Milky Way showing the Outer, Perseus, Orion, Sagittarius, Carina, Crux, Scutum, Norma and 3-kiloparsec arms
3. A map of the distribution of atomic hydrogen gas in the Milky Way, showing a more or less global view of the spiral structure of the galaxy, was made in 1958. See Oort, Kerr and Westerhout (1958)
4. The discovery of the Sagittarius dwarf galaxy is reported in Ibata, R A, Gilmore, G, and Irwin, M J (1994). A Dwarf Satellite Galaxy in Sagittarius. *Nature* 370(6486) p 194
5. Most astronomy textbooks discuss rotation curves of galaxies and the implications for dark matter. See, e.g., Mihalas and Binney (1981). Updates on the status of dark matter searches frequently appear in popular science publications such as *Scientific American*
6. Bertin and Lin (1996) discuss the nature of spiral arms and theories of spiral structure in detail. The textbook *Voyages* (Fraknoi, Morrison, and Wolff, 1997) is a good source for a more popular-level explanation
7. Leitch and Vasisht (1997)
8. A fairly recent review article on the Orion molecular cloud is Genzel and Stutzki (1989). The press releases of the Space Telescope Science Institute <www.stsci.edu/> are a good source of updated information
9. On the possibility that Native Americans viewed and drew the supernova that led to the Crab nebula, see, e.g., Malville and Putnam (1989)
10. A recent review of the Milky Way bulge lore is in Gerhard (2000)

11. For a thorough discussion of the distance to the center of the Galaxy, see Reid (1993)
12. On the bar at the center of the Milky Way, see, e.g., Blitz and Spergel (1991)
13. The Australians' papers on Sagittarius A* are (1) Piddington and Minnett (1951) and (2) Bolton *et al* (1954)
14. Oort, J H (1968)
15. Lynden-Bell and Rees (1971)
16. A review article that describes the galactic center in some detail is Genzel, Eckart, and Krabbe (1995). See also the earlier but useful reviews: (1) Brown and Liszt (1984), (2) Genzel and Townes (1987)
17. For more detailed, but popular-level descriptions of black holes and their effects on objects that come close to the event horizon, see Fraknoi, Morrison and Wolff (1997) and Greenstein (1983)
18. Ghez *et al* (2000)
19. Ferrarese and Merritt (2000) p L12
20. See Irwin *et al* (1990), Lavery and Mighell (1992) (for the 1990 discovery), Ibata, Gilmore and Irwin (1994), Whiting, Irwin and Hau (1997), Karachentsev and Karachensteva (1999), Armandroff, Davies and Jacoby (1998)
21. Searle and Zinn (1978)
22. Ibata *et al* (2001)

REFERENCES

Allen, G 1793a. *Gentleman's Magazine* **63**, pp 9–12 and 126–127

Allen, G 1793b. *Gentleman's Magazine* **63**, pp 213–216

Armandroff, T E, Davies, J E and Jacoby, G H 1998. *Astronomical Journal* 116, pp 2287–2296

Armitage, A 1963. *William Herschel*. Garden City, New York: Doubleday

Baade, W 1963. *Evolution of Stars and Galaxies*. Reprint, 1975. Cambridge: MIT Press

Barnard, E E 1913. *Astrophysical Journal* **38**, pp 496–501

Batten, A 1988. *Resolute and Undertaking Characters: The Lives of Wilhelm and Otto Struve*. Dordrecht: D Reidel

Becker, B 1993. Eclecticism, opportunism, and the evolution of a new research agenda: William and Margaret Huggins and the origins of astrophysics. PhD Thesis, Johns Hopkins University

Berendzen, R and Hart, R 1973. *Journal for the History of Astronomy*, **iv** (1973), pp 46–56

Berendzen, R and Hoskin, M A 1971. *Astronomical Society of the Pacific* Leaflet No. 504

Berry, A 1961. *A Short History of Astronomy, from Earliest Times Through the Nineteenth Century*. New York: Dover Publications

Bertin, G and Lin, C C 1996. *Spiral Structure in Galaxies: A Density Wave Theory*. Cambridge: MIT Press

Bessel, F 1838. *Monthly Notices of the Royal Astronomical Society* **4**, pp 152–161

Blanpied, W A 1998. *Physics Today* **51**(2), pp 34–40

Blitz, L and Spergel, D 1991. *Astrophysical Journal* **379**, pp 631–638

Bok, B J 1965. *Publications of the Astronomical Society of the Pacific* **77**(459), pp 416–421

Bok, B J 1974. *Harlow Shapley and the Discovery of the Center of our Galaxy*. In Neyman, J (ed) (1974)

Bok, B J 1976. *Biographical Memoirs, National Academy of Sciences* **48**, pp 241–259

Bolton, J G, Westfold, K C, Stanley, G J and Slee, O B 1954. *Australian Journal* **7**, p 96

Brown, R L and Liszt, H S 1984. Sagittarius A and its Environment. *Annual Reviews of Astronomy and Astrophysics* **22**, pp 223–265

Bruck, M T 1991. *Irish Astronomical Journal* **20**(2), pp 70–77

Bruck, M T and I Elliott 1992. *Irish Astronomical Journal* **20**(3), pp 210–201

Brush, S G 1978. *Journal for the History of Astronomy* **9**, pp 1–41

Burbidge, E M 1994. *Annual Review of Astronomy and Astrophysics* **32**, pp 1–36

Burnham, S W 1900. *Publications of the Yerkes Observatory* **1**, pp i–xxvii

Christianson, G E 1995. *Edwin Hubble: Mariner of the Nebulae*. New York: Farrar, Straus and Giroux

Clerke, A M 1895. *The Herschels and Modern Astronomy*. New York: Macmillan

Condos, T 1997. *Star Myths of the Greeks and Romans: a Sourcebook Containing the Constellations of Pseudo-Eratosthenes and the Poetic Astronomy of Hyginus*. Grand Rapids, Mich.: Phanes Press

Crommelin, A 1917. *Scientia* **21**, pp 365–376

Curtis, H 1921. *Bulletin of the National Research Council* **2**(3), pp 194–217

Davis, H S 1898. *Popular Astronomy* **6**, pp 129–138, 211–220, 220–228

De La Rue, W 1861. *Chemical News* **4** pp 130–133

DeVorkin, D 2000. *Henry Norris Russell: Dean of American Astronomers*. Princeton: Princeton University Press

Dick, W R and Ruben, G 1988. *In Mapping the Sky: Past Heritage and Future Directions*. International Astronomical Union Symposium No. 133. Dordrecht: Kluwer Academic Publishers, pp 119–121

Dingle, H and Martin, G R (eds). 1964. *Chemistry and Beyond: A Selection from the Writings of the Late Professor F A Paneth*. New York: John Wiley and Sons

Drake, S 1957. *Discoveries and Opinions of Galileo. Translated with an Introduction and Notes by Stillman Drake*. Garden City, New York: Doubleday Anchor

Eddington, A S 1938. Forty years of astronomy. In Needham, J and Pagel, W (eds), *Background to Modern Science* Cambridge: Cambridge University Press, pp 167–178

Farber, E 1966. *Milestones of Modern Chemistry: Original Reports of the Discoveries*. New York: Basic Books

Felber, H (ed). 1994. *Briefwechsel zwischen Alexander von Humboldt und Friedrich Wilhelm Bessel*. Berlin: Akademie Verlag

Ferrarese, L and Merritt, D 2000. A fundamental relation between supermassive black holes and their host galaxies. *Astrophysical Journal* **539**, pp L9–L12

Fournier d'Albe, E E 1923. *The Life of Sir William Crookes*. London: Fisher Unwin

Fraknoi, A, Morrison, D and Wolff, S 1997. *Voyages Through the Universe*. Fort Worth: Saunders College Publishing

Genzel, R and Townes, C H 1987. *Annual Reviews of Astronomy and Astrophysics* **25**, pp 377–423

Genzel, R and J Stutzki. 1989. *Annual Reviews of Astronomy and Astrophysics* **27**, pp 41–85

Genzel, R, Eckart, A and Krabbe, A 1995. The galactic center. In *Seventeenth Texas Symposium on Relativistic Astrophysics and Cosmology* Böhringer, H Morfill, G E and Trümper, J E (eds), *Annals of the New York Academy of Sciences* **759**. New York: New York Academy of Sciences

Georgelin, Y M and Georgelin, Y P 1976. *Astronomy and Astrophysics* **49**, pp 57–79

Gerhard, O 2000. Structure and mass distribution of the Milky Way bulge and disk. In *Galaxy Disks and Disk Galaxies*. ASP Conference Series, Funes and Corsini (eds)

Gernsheim, H and Gernsheim, A. The photographic arts: photography. In *A History of Technology*, vol 4. Singer, C, Holmyard, E J, Hall, A R and Williams, T I (eds) Oxford: Clarendon

Ghez, A M, Morris, M, Becklin, E E, Tanner, A and Kremenek, T 2000. *Nature* **407**(6802), pp 349–351

Gill, D 1896. *Annals of the Cape Observatory* **3**, pp ix–lxviii

Gingerich, O 1970. "Harlow Shapley." In *Dictionary of Scientific Biography* pp. 345–352

Gingerich, O 1984. *Harvard Library Bulletin* **32**, pp 73–82

Gingerich, O 2000. Kapteyn, Shapley and their universes. In *The Legacy of J C Kapteyn: Studies on Kapteyn and the Development of Modern Astronomy*, Van Der Kruit, P C and Berkel, K Van (eds), Dordrecht: Kluwer Academic Publishers

Greenstein, G 1983. *Frozen Star*. New York: Freundlich Books

Halley, E 1718. *Philosophical Transactions of the Royal Society* **30**, pp 736–738

Haramundanis, K (ed). 1996. *Cecilia Payne-Gaposchkin: An Autobiography and Other Recollections*. Cambridge: Cambridge University Press

Hart, R and Berendzen, R 1971. *Journal for the History of Astronomy* **ii**, pp 109–119

Hastie, W [trans.] 1968. *Kant's Cosmogony*. New York: Greenwood

Heilbroner, R L 1953. *Worldly Philosophers*. New York: Simon and Schuster

Herschel, J 1841. *Monthly Notices of the Royal Astronomical Society* **5**, pp 89–98

Herschel, Mrs J 1876. *Memoir and Correspondence of Caroline Herschel*. New York: D Appleton and Co

Herschel, W 1784. *Philosophical Transactions of the Royal Society* **74**, pp 437–451

Herschel, W 1789. *Philosophical Transactions of the Royal Society* **79**, pp 212–255

Herschel, W 1791. *Philosophical Transactions of the Royal Society* **81**, pp 71–88

Herschel, W 1800. *Philosophical Transactions of the Royal Society* **90**, pp 284–292

Herschel, W 1817. *Philosophical Transactions of the Royal Society* **107**, pp 302–331

Henbest, N and Couper, H 1994. *Guide to the Galaxy*. Cambridge: Cambridge Univ. Press

Hoagland, H 1965. *Publications of the Astronomical Society of the Pacific* **77**(459), pp 422–430

Hogg, H S 1965. *Publications of the Astronomical Society of the Pacific* **77**(458), pp 336–446

Hoskin, M A 1976. *Journal for the History of Astronomy* 7, pp 169–182

Hoskin, M A 1963. *William Herschel and the Construction of the Heavens.* London: Oldbourne

Hubble, E 1922. *Astrophysical Journal* **56**, pp 162–199

Hubble, E 1923. *Publications of the Astronomical Society of the Pacific* **35**, pp 261–263

Hubble, E 1925. *Astrophysical Journal* **62**, pp 409–433

Hubble, E 1936. *Realm of the Nebulae*. (Reprint, 1982. New Haven: Yale University Press)

Hubble, E 1929. *Proceedings of the National Academy of Sciences* **15**, pp 168–173

Hubble, E 1954. *The Nature of Science and Other Lectures*. San Marino: The Huntington Library

Huggins, W 1856. *Monthly Notices of the Royal Astronomical Society* **16**, pp 175–777

Huggins, W and Miller, W A 1864a. *Philosophical Transactions of the Royal Society* **154**, pp 139–160

Huggins, W and Miller, W A 1864b. *Philosophical Transactions of the Royal Society* **154**, pp 413–435

Huggins, W 1864. *Philosophical Transactions of the Royal Society* **154**, pp 437–444

Huggins, W 1865. *Proceedings of the Royal Society* **14**, pp 39–42

Huggins, W 1868. *Philosophical Transactions of the Royal Society* **158**, pp 529–64

Huggins, W and M Huggins. 1889. Note on the photographic spectra of Uranus and Saturn. *Proceedings of the Royal Society* **46**, p 40

Huggins, W and Lady Huggins. 1909. *The Scientific Papers of Sir William Huggins*. Sir William Huggins and Lady Huggins (eds) volume 2 London: William Wesley and Son

Hughes, E 1951. *Annals of Science* **7**, No. 1, pp 1–21

Ibata, R, Irwin, M Lewis, G and Stolte, A 2001. *Astrophysical Journal* **547**, pp L133–L136

Ibata, R A, Gilmore, G and Irwin, M J 1994. *Nature* **370**(6486), p 194

Irwin, M J, Bunclark, P S Bridgeland, M T and McMahon, R G 1990. *Monthly Notices of the Royal Astronomical Society* **244**, p 16

Jaki, S 1990. *Cosmos in Transition: Studies in the History of Cosmology*. Tucson: Pachart

Jeans, J 1919. *Problems of Cosmogony and Stellar Dynamics*. Cambridge: Cambridge University Press

Jones, B Z and Boyd, L G 1971. The *Harvard College Observatory: The First Four Directorships, 1839–1919*. Cambridge: Harvard University Press

Joy, A H 1940. *Publications of the Astronomical Society of the Pacific* **52**(306), pp 69–79

Kapteyn, J C 1909. *Astrophysical Journal* **29**, p 46

Kapteyn, J C 1914. *Journal of the Royal Astronomical Society of Canada* **8**, pp 145–159. (Reprinted from *Scientia*, November 1913)

Karachentsev, I D and Karachentseva, V E 1999. *Astronomy and Astrophysics* **341**, p 355

Kellner, L 1963. *Alexander von Humboldt*. London: Oxford University Press

Kragh, H 1987. *Centaurus* **32**, pp 114–139

Krul, W E 2000. Kapteyn and Groningen: a portrait. In *The Legacy of J C Kapteyn: Studies on Kapteyn and the Development of Modern Astronomy* Van Der Kruit, P C and Van Berkel, K (eds) Dordrecht: Kluwer Academic Publishers

Kopal, Z 1981. *Astrophysics and Space Science* **79**, pp 261–264

Kopal, Z 1986. *Of Stars and Men: Reminiscences of an Astronomer*. Bristol: Adam Hilger

Laplace, P 1835. *Exposition du Système du Monde*. (Reprint, 1984. [Paris]: Fayard)

Lavery, R J and Mighell, K J 1992. *Astronomical Journal* **103**, pp 81–83

Leitch, E M and Vasisht, G 1997. *New Astronomy* **3**, pp 51–56

Lemaitre, G 1931. *Revue des Questions Scientifiques* **20**, pp 391–410

Lockyer, J N 1874. *Nature* **10**, pp 254–256

Lynden-Bell, D and Rees, M 1971. *Monthly Notices of the Royal Astronomical Society* 152, pp 461–475

MacPherson, H 1919. *Herschel*. New York: Macmillan

Malville, J M and Putnam, C 1989. *Prehistoric Astronomy in the Southwest*. Johnson Books: Boulder, Colorado

Mayall, N U 1970. *Biographical Memoirs of the National Academy of Science* 41, 175–214

Mihalas, D and Binney, J 1981. *Galactic Astronomy: Structure and Kinematics*. New York: W H Freeman and Company

Mills, C E and Brooke, C F 1936. *A Sketch of the Life of Sir William Huggins*. London: np

Moulton, F R 1924. *An Introduction to Astronomy*. New York: Macmillan

Newton, I 1729. *The Mathematical Principles of Natural Philosophy* by Isaac Newton, Translated into English by Andrew Motte, with an introduction by I Bernard Cohen. (Reprint, 1968. London: Dawsons)

Newton, I 1730. *Opticks*. (Reprint, with a foreword by Albert Einstein, an introduction by Sir Edmund Whittaker, an analytical table of contents by Duane H D Roller, and an introduction by I Bernard Cohen. 1979. New York: Dover)

Neyman, J (ed). 1974. *Heritage of Copernicus: Theories "Pleasing to the Mind."* Cambridge: The MIT Press

North, J 1965. *The Measure of the Universe: A History of Modern Cosmology*. Oxford: Clarendon Press

Oort, J H 1968. Radio astronomical studies of the galactic system. In *Galaxies and the Universe*, Woltjer, L (ed), New York: Columbia University Press

Oort, J H, Kerr, F J and Westerhout, G 1958. *Monthly Notices of the Royal Astronomical Society* **118**, pp 379–389

Panek, R 1998. *Seeing and Believing*. New York: Penguin Putnam

Paul, E R 1993a. *Life and Works of J C Kapteyn*, by Henrietta Hertzsprung-Kapteyn; an annotated translation with preface and introduction by E Robert Paul. Dordrecht: Kluwer Academic

Paul, E R 1993b. *The Milky Way Galaxy and Statistical Cosmology 1890–1924*. Cambridge, New York: Cambridge University Press

Piddington, J H and Minnett, H C 1951. *Australian Journal of Scientific Research* **A4**, p 495

Plotkin, H 1982. Henry Draper, Edward C Pickering, and the birth of American astrophysics. In *Symposium on the Orion Nebula to Honor Henry Draper*. Glassgold, A E Huggins, P J and Schucking, E L (eds), New York: New York Academy of Sciences

Raymo, C 1982. *365 Starry Nights*. New York: Simon and Schuster

Reid, M J 1993. *Annual Reviews of Astronomy and Astrophysics* **31**, pp 345–372

Sandage, A 1961. *The Hubble Atlas of Galaxies*. Washington, DC: Carnegie Institution of Washington

Sandage, A 1989. *Journal of the Royal Astronomical Society of Canada* **83**, pp 351–362

Seares, F 1922. *Astronomical Society of the Pacific Proceedings* **34**, pp 233–253

Searle, L and Zinn, R 1978. *Astrophysical Journal* **225**, p 357

Serres, M, Dagognet, F and Sinaceur, A (eds) 1975. *Auguste Comte: Cours de Philosophie Positive, Leçons 1 à 45*. Paris: Hermann

Shapley, H 1909. *Popular Astronomy* **17**(7), pp 397–401

Shapley, H 1915. *Contributions of the Mount Wilson Solar Observatory* **116**, pp 1–92

Shapley, H 1917. *Publications of the Astronomical Society of the Pacific* **29**, pp 213–217

Shapley, H 1969. *Through Rugged Ways to the Stars*. New York: Charles Scribner's Sons

Smith, R W 1982. *Expanding Universe: Astronomy's "Great Debate," 1900–1931*. Cambridge: Cambridge University Press

Struve, O 1957. *Sky and Telescope* **16**, pp 69–72

Struve, O [Wilhelm]. 1895. Wilhelm Struve. Zur Erinnerung an den Vater. Karlsruhe: G Braun'schen Hofbuchdruckerie

Struve, W 1847. *Etudes d'Astronomie Stellaire*. St. Petersburg: Imprimerie de l'Académie des Sciences

Trimble, V 1996. *Publications of the Astronomical Society of the Pacific* **108**, pp 1073–1082

van Berkel, K 2000. Dutch astronomers in the USA. In van der Kruit, P C and van Berkel, K (2000)

van Berkel, K and van der Kruit, P C 2000. Note on E R Paul's translation of H Hertzsprung-Kapteyn's biography of J C Kapteyn. In *The Legacy of J C Kapteyn: Studies on Kapteyn and the Development of Modern Astronomy* Van Der Kruit, P C and Van Berkel, K (eds), Dordrecht: Kluwer Academic Publishers

van der Kruit, P C and van Berkel, K 2000. *The Legacy of J C Kapteyn: Studies on Kapteyn and the Development of Modern Astronomy*. Dordrecht: Kluwer Academic

Weiss, L 1982. *Watch-making in England 1760–1820*. London: Robert Hale

Welther, B 1982. *Isis* **73**, p 94

Whiting, A B, Irwin, M J and Hau, G K T 1997. *Astronomical Journal* **114**, pp 996–1001

Wright, H 1966. *Explorer of the Universe: a Biography of George Ellery Hale*. (Reprint, 1994. Woodbury, New York: American Institute of Physics.)

Wright, T n.d. *Second or Singular Thoughts upon the Theory of the Universe*; edited from the manuscript by M A Hoskin. (Reprint, 1968. London: Dawsons)

Wright, T 1742. *Clavis coelestis: being the explication of a diagram entituled a "Synopsis of the universe or, The visible World epitomized;"* with a preface by M A Hoskin. (Reprint, 1967. London: Dawsons)

Wright, T 1750. *An Original theory or new hypothesis of the Universe, 1750: a facsimile reprint together with the first publication of A Theory of the Universe, 1734*, [by] Thomas Wright of Durham; introduction and transcription by Michael A Hoskin. (Reprint, 1971. London: Dawsons)

INDEX

393